The Cell in Health and Disease

Monographs in Clinical Cytology

Vol. 2
2nd, revised edition

Editor
George L. Wied, Chicago, Ill.

Co-Editors
Emmerich von Haam, Columbus, Ohio
Leopold G. Koss, New York, N.Y.
James W. Reagan, Cleveland, Ohio

Basel · München · Paris · London · New York · New Delhi · Singapore · Tokyo · Sydney

The Cell in Health and Disease

An Evaluation of Cellular Morphologic Expression of Biologic Behavior

John K. Frost

Professor of Pathology and Director, Division of Cytopathology,
Department of Pathology, The Johns Hopkins University School of Medicine,
and Pathologist, The Johns Hopkins Hospital, Baltimore, Md.

172 figures and 7 tables, 1986

Second, revised edition

Basel · München · Paris · London · New York · New Delhi · Singapore · Tokyo · Sydney

Monographs in Clinical Cytology

Vol. 2 (1st edition)
John K. Frost, Baltimore, Md.
The Cell in Health and Disease
XII + 148 p., 81 fig., 2 diagr., 15 charts, 1969.

National Library of Medicine, Cataloging in Publication
    Frost, John K. (John Kingsbury), 1922–
    The cell in health and disease: an evaluation of cellular morphologic expression of biologic
behavior
    John K. Frost. –– 2nd, rev. ed. –– Basel; New York: Karger, 1986.
    (Monographs in clinical cytology; vol. 2)
    Bibliography: p. Includes Index.
    1. Cytodiagnosis I. Title II. Series
    Wl MO567KF v.2 (QY 95 F939)
    ISBN 3–8055–4150–3

Drug Dosage
    The authors and the publisher have exerted every effort to ensure that drug selection and dosage
set forth in this text are in accord with current recommendations and practice at the time of
publication. However, in view of ongoing research, changes in government regulations, and the
constant flow of information relating to drug therapy and drug reactions, the reader is urged to
check the package insert for each drug for any change in indications and dosage and for added
warnings and precautions. This is particularly important when the recommended agent is a new
and/or infrequently employed drug.

Contents

Section B.
Functional Anatomy of the Interphase Nucleus

V. Functional Anatomy of the Interphase Nucleus

Contents

Contents   XIII

To *Moira*   with deepest devotion –
whose ever continuing patience,
dedication and untiring belief in why we are here,
have made this work possible;
she is its major part.

And to *Moira Anne, Rosanne Grace, Noreen Anne Regina,
Therese Anne Olivia, John Kingsbury, Sheila Anne Maureen,*
and *James Keane*
whose changes in *biologic behavior* throughout the years
have been so meaningfully and beautifully reflected in
morphologic alteration.

Preface to First Edition

It may appear naive to approach such an enormous subject, as that of the cell in health and disease, in a monograph of this size. There exist, however, distinct needs for an endeavor of such scope to help reevaluate the approach to *communication* in the morphologic and physiologic study of biologic processes, and to pull together diverse observations and experiences to a practical, discriminating, and *diagnostic* level.

Modern communication in biology is dependent upon language which, by its very nature, is an analogue process. In order to be specific, it must be complete; but it is all too frequently long, discursive, and laboriously replete. In many situations, when it is made concise it becomes ambiguous. In biology, especially in medicine, ambiguity of communication can be catastrophic. Without general acceptance and use of proper points of reference and appropriate nomenclature, communication is chaotic, thought and investigation are stymied, incorrect concepts are fostered and transmitted, and improper conclusions and diagnoses are reached.

Current knowledge regarding the relationships of morphology to biologic behavior, is *im*perfect; yet it must not be said that not enough is known about it to use it. On the contrary, our mandate is to use that which we know, in fullest realization of its limitations and capacities, and to extend its boundaries with careful investigation, thought, and experience. Accordingly, we must recognize islands of knowledge, communicate about them, enlarge them, build upon them, and mold them into usable wisdom for benefit of mankind in its fight against disease.

'What cannot totally be known, ought not to be totally neglected; for the knowledge of a part is better than the ignorance of the whole.'

Abu-1-Fida Ismail ibn Ali
(Preface to his geography of the Orient: Tagwim al-Buldan)
1321 A.D. [1].

To obtain an overall pattern for study and development of concepts in such a brief presentation, only certain key types of functional differentiation are presented in detail (i.e.: squamous, columnar). From this vantage point, important general processes are studied (e.g.: metaplasia, maturation) which can be applied in like detail to other cellular functions. To keep this presentation within the scope envisioned by its editors, light microscopy is used virtually exclusively with but few clarifying references to electron microscopic fine structure.

Fitting together findings and observations from clinical experience and from investigative studies, creates patterns for understanding of these processes of life. From such broad understanding do reasonable and valuable concepts of health and disease arise.

Acknowledgements: The patience and understanding of the editors and of the publisher, especially George L. Wied, MD, and Thomas Karger upon whose shoulders fall the brunt of making this a cohesive series, are thankfully acknowledged. The valuable efforts of Ivan L. Bennett, MD, George O. Gey, MD, David H. Hollander, MD, James W. Reagan, MD, Carl P. Swanson, PhD, and Owsei Temkin, MD, are greatly appreciated for critically reviewing various portions of the manuscript and offering advice regarding certain concepts which, in their form herein, must remain my responsibility. I am further grateful to many more colleagues and students, whose efforts and association have been most constructive and whose comprehension and inquiry is a constant source of stimulation and reexamination of concepts. I also express my appreciation to Miss Tome Tanisawa for her superb secretarial efforts early in the endeavor, to Mr. Frank LaCorte for his co-operation in enlarging my photomicrographs and, especially, Mrs. Beatrice Franza for her skilled typing and dedicated assistance in proofing of the manuscript.

J.K.F., 1969, Heatherfield, Brooklandville, Md.

Preface to Second Edition

The years elapsing since the First Edition have been marked by significant discoveries and momentous advances in our understanding of the cell – at all levels, from molecular biology through clinical application. This Second Edition has been extensively revised, enlarged, and updated to reflect those of greatest importance to the clinical microscopist; yet, it does not stop here, as there is sincere interest by many to increase their skills evaluating the biologic significance of the structure of cells. Therefore, this book is primarily designed for diagnostic microscopists – the *pathologists* and the *cytotechnologists,* practitioners or students and, in addition, it has been created for the *cell biologists, investigators, educators* and their students – those who are interested in the fine structural details of the cell and in the clinically relevant biologic behavior which morphologic changes in these structures reflect.

The *format* of this Second Edition is to develop key concepts of cell morphology in health and disease, and approaches to their proper morphologic interpretation. The book is small and condensed, and is designed to be read through first, from beginning to end, as each segment is built upon that which precedes; then, it is structured to be used in detail as a reference text.

An underlying theme of this book is that the cell has at its disposal a vast repertoire of strategies for survival. This repertoire is not inexhaustable, however, and similar structural change may occur from different causes, where the most minute detail of difference may be key to proper interpretation. It is the final complex presentation, encountered at that point in time by the researcher or diagnostician, which must be unfolded into its simplest building blocks, each to be recognized and understood in order to identify the basic biologic processes which are going on in the cell.

The entire gamut of our knowledge about the cell in health and disease, from the more clinical end of the spectrum (pathologists and cytotechnologists) *through* the more fundamental domains (cell morphologists and molecular biologists), has expanded so rapidly in the last two decades since the First Edition that it is virtually impossible for workers at one end of this continuum to keep up with those at the other. One purpose of this work is to bring to those in the field, regardless of their place in the spectrum, a better appreciation of current approaches to understanding biologic behavior through study of cellular morphology, so that each may develop a better grasp of the others' fields of work, needs, accomplishments, and nomenclature in order to realize more fully the significance of their own essential activities, to gain insights of the others', and to bring about better intercommunication.

Thus a major aim of this project has been to cut through the enormous mass of information which has accumulated recently, to discover and characterize general principles, and to describe the state-of-the-art in this exciting area. A minisabbatical at the Marine Biological Laboratory, Woods Hole, Massachusetts, has been of great value to me in placing into proper perspective this rapidly proliferating fundamental knowledge of molecular cell biology and of the submicroscopic structure of the cell. This has greatly helped to bring it into focus and proper perspective, as it clinically relates to light microscopy of the cell and tissue, and to the host in its environment.

To accomplish this, one completely new section has been added – 'Functional Anatomy of the Interphase Nucleus'. Of great concern to me has been the amount of detail which I should include, and yet not lose any of my clinical readers. Ultimately, I have decided that what is therein presented is essential for a reasonable understanding of the significance which the light microscopic appearance of a cell bears *to* its biologic behavior. Understanding of this section will greatly assist the reader in building concepts which will bridge the gap between that which is at the cutting edge of our knowledge in the areas of molecular cell biology *and* that which is needed for highest quality clinical application.

Rather than being encyclopedic of data, this work is intended to be a guide to thought, understanding, and building of reliable concepts. The still larger areas of ignorance which exist, with the many facts which cannot yet be explained, are merely unsolved problems providing excitement and purpose for continued work to unravel the mystery in this field of morphology bespeaking biologic behavior.

Thus in those areas which are poorly understood, rather than simply presenting disjointed facts, I have ventured hypotheses to be considered and to be constructively criticized – working hypotheses which have stood

me in good stead throughout these many years of diagnostic clinical cytopathology and of clinical investigation. As did the monumental works of
Alexander von Humboldt nearly a century and a half ago, in reflecting upon
his extensive experiences, this small presentation has attempted to support
generalization by details, and to explain details by generalization:

> 'Nature is a free domain, and the profound conceptions ...
> she awakens within us can only be vividly *delineated* by thought ...
> worthy of bearing witness to the majesty and greatness of the creation.'
> [Cosmos, 1844, (146)]

One score years ago the First Edition was written primarily for those
pathologists, medical students, students of cytotechnology and cell biologists who chose to study in our laboratories. The feedback received was
tremendously positive and I am deeply indebted to the thousands who have
constructively responded through these years, asking questions and probing
for reasons which were not apparent. In addition to updating my concepts,
as they have developed throughout the years and as new information has
come forth from both basic and clinical domains, I am hopeful that this
current work addresses the many points our students and collaborators
have raised.

From students of our first classes for pathologists and cytotechnologists
during the forties in San Francisco, from our formal schools of cytotechnology continuous since 1953 in San Francisco and in Baltimore, from our
medical students, from the annual evening classes and seminars for pathologists, medical students and basic scientists since 1956 in Baltimore, from
the thousands of pathologists who have come to intensively study in resi-
dence for two weeks at our annual Postgraduate Institutes for Pathologists
in Clinical Cytopathology at Johns Hopkins since 1959, from those who
have studied with us throughout the world with the World Health Organization, International Academy of Cytology, International Academy of
Pathologists, American Society of Cytology, American Society of Clinical
Pathologists and from the medical students, residents, fellows, and dedicated associates who have worked so closely with us in our laboratories on
our own patients – I have drawn heavily from their probing questions to
mold and remold concepts into those I hold today, and which I have tried to
make objective and logical, and to express as completely as is practical for
this work.

Again, *George L. Wied,* MD, and Dr. (med. h.c.) *Thomas Karger,* with
his expert staff, have been most helpful and patient with one who finds

deadlines most difficult to meet. I deeply thank my close associates, collaborators, and longstanding friends *Yener S. Erozan,* MD, *Prabodh K. Gupta,* MD, *Norman J. Pressman,* PhD, *Sue T. Shutt,* CT(ASCP), CT(IAC), *Arline K. Howdon,* CT(ASCP), CFIAC, *Gary W. Gill,* CT(ASCP), CFIAC, *Karen M. Plowden,* CT(ASCP), CT(IAC), and my wonderful and dedicated staff and coworkers, with whom it is such a pleasure to be working throughout these exciting years. I am additionally most appreciative to *A. Laurence Dee,* MD, to *Larry R. Gerace,* PhD, and to *Rosanne G. Frost,* MD, for carefully reviewing parts of, or the entirety of the manuscript. I alone bear responsibility for the final result, however, and for the advice I have accepted or that which I may have chosen unwisely to disregard. I extend my very deepest appreciation for the tireless efforts of my editorial secretary, Mrs. *Beatrice K. Franza,* whose skills, dedication, and devotion to our work together over the many years has made this work possible, to Mr. *Frank J. La Corte,* whose skills both as a photographer and as a cytotechnologist have made our many years of collaboration fruitful and enjoyable, and to my divisional administrator, Mrs. *Beatrice M. Clendaniel,* who has helped me work around many monumental tasks in a relentless schedule which does not yield to the desires to record one's thoughts. Finally, I am lovingly in profound debt to my family, who selflessly and tirelessly have indulged me far too much time separated from them, manuscripts taken on vacations, and other time-demolishing preoccupational acts in the long, arduous course of revising this little book.

J.K.F., Brooklandwood, Brooklandville, Md.

Section A
Cellular Morphology Bespeaks Biologic Behavior

I. Cellular Morphology Bespeaks Biologic Behavior

Life is not static. It occurs at a *level of activity* which *differs* from its *surroundings*, and is *maintained* by its own *biologic processes*.

In both health and in disease, biologic changes are constantly occurring during life. They continue for varying periods of time *after* death, until the status of the environment is attained.

Much of this biologic *behavior* and the level of its activity, stable or in transition, is dependably *reflected* in the structure of tissues and cells. The high degree of *recognition* of these morphologic features, the *significance* of their changes and their *predictability*, form the basis of histology and pathology – and, thus of clinical cytopathology.

Morphology *does* bespeak biologic behavior. The knowledgeable examiner can transcend mere descriptive morphology, by projecting it into the dynamic and living images of the normal processes in health, and of the abnormal processes in disease. In this manner greater *useful* knowledge is gained of the cells' biologic behavior, of their tissues' biologic behavior and, more globally, of the hosts' biologic behavior.

A. Separated and Isolated Cells

This essence of biologic behavior is reflected in the morphology of cells comprising *tissues* [289] and, with careful specimen clinical acquisition and cytopreparation, is retained after separation from their tissue mass. Not only does this information thus remain, but is often *accentuated* by exfoliation.

Greater details and *intra*cellular relationships of the *intact* isolated cell thus are available for careful inspection. While much of the *inter*cellular relationships and tissue pattern may be greatly diminished or lost by exfoliation, careful evaluation of that which is retained is most valuable.

Key information is displayed in the morphology of the intact, isolated cell additional to that which is available in the sliced fragments of cells tucked away within the surrounding elements of their tissues. This lack of obscuring from surrounding tissue, makes evaluation of the individual cell more complete.

The *intact* whole cell is available for examination rather than a thin slice of it. The unfixed cell usually thins and spreads out upon the surface of a glass slide, making its entire contents available to close and careful microscopic inspection (Figs. I-1, I-2). In this way, cellular samples can provide invaluable information for determining the health and disease of its host.

B. Patterns and Processes

There exists only a limited number of ways in which cells and tissues can react, morphologically, to the large number of causes for biologic change. They are thus insufficient in variety to match each of the myriad diseases or biologic processes.

A ratio of one process to one morphologic change, or of one disease to one simple pattern, is not common [119]. Complex patterns result from multiple processes. A number of the elements may appear similar in many respects, so that patterns superficially resemble each other; however, careful evaluation can detect subtle differences which are key to understanding the interacting biologic processes, to discriminating between them, and to establishing working diagnoses.

To this end, *complex patterns* are unweaved for recognition of simpler or more basic patterns, and to decipher the basic processes involved for better understanding and interpretation. Then, changes in these parts which make up the whole, are reweaved together again, and reevaluated globally in relationship to the whole.

Careful attention to *minute details* can serve to better separate processes, and even to identify causes and etiologic agents. Some critical features which are key to this evaluation are of the finest detail which require highest resolution of the light microscope (Fig. I-1), correlation with known electron microscopic fine structure (Figs. I-3, I-4), and keen observational experience to interrelate them with biologic behavior for full meaning.

The interplay of morphologic features with *time*, is another important factor to be considered when one can obtain sequential specimens for examination in otherwise inexplicable situations. Important new features, which are dependent upon time, are detectable in new specimens. These can lead to more thorough understanding of the whole, when it is reconstructed by weaving back the parts into the initial pattern.

Biologic processes which are quite different, therefore, can produce identical morphologic changes; conversely, certain morphologic changes overlap in their biologic specificity. When mimicked identity or when overlap of specificity does occur, however, observations by multiple, sequentially-timed specimens help to separate and identify the true biologic nature. This allows for recognition of the numerous subprocesses and of their complex interplay, as they enter into or exit from the biologic situations (see D, below).

All components of an overall morphologic picture, therefore, must be evaluated in complex, difficult situations. Each, both individually and together as a whole, offer value in identifying the biologic process(es) they bespeak.

C. Syndromes of Structure

Morphologic patterns are syndromes of morphologic changes. They sensitively indicate underlying biologic processes of health or disease, when recognized and placed in proper time interrelationship.

Figures I-1, I-2. Poorly differentiated adenocarcinoma of the lung, metastatic to the brain. Light photomicrographs.

Fig. I-1: Exfoliated cells in the cerebrospinal fluid. In their processing, they have flattened and thinned out upon the slide, making more fine morphologic *intra*cellular detail of the *entire* cell available in one plane of focus, and allowing for complete focusing up and down throughout the *entire* nucleus. This permits more complete and careful microscopic examination of the nuclear membrane, chromatinic rim, chromatinic net, parachromatin, and nucleoli in order to better determine biologic behavior. The large vacuoles (upper portion of cytoplasm) suggest secretion, but are not diagnostic thereof. These cells can be definitely recognized as constituting a diagnostic true tissue fragment (DTTF) (Table VI-1), but little more remains of their *inter*cellular relationships.

Fig. I-2: Tissue section. In their processing, the cells have rounded up and remain thick. Less is available for inspection on one plane of focus of each nucleus, than in a properly and carefully flattened cell, above; furthermore, as a result of the necessary sectioning, only a thin slice (i.e.: 5 μm) of each cell is available for *intra*cellular examination. Most of the *inter*cellular relationships remain intact, however, so that poorly formed glands are recognizable on the right and on the left, which have been lost in most of the exfoliated cells. The nuclear detail and information which is available for examination, however, is less in these smaller, rounded up, thinly sectioned nuclei than in the exfoliated cells.

I-1: Papanicolaou stain; I-2: Hematoxylin and eosin stain.

I-1, I-2: $\times$ 1,950.

(From Frost, in Rubinstein; 1972 [95].)

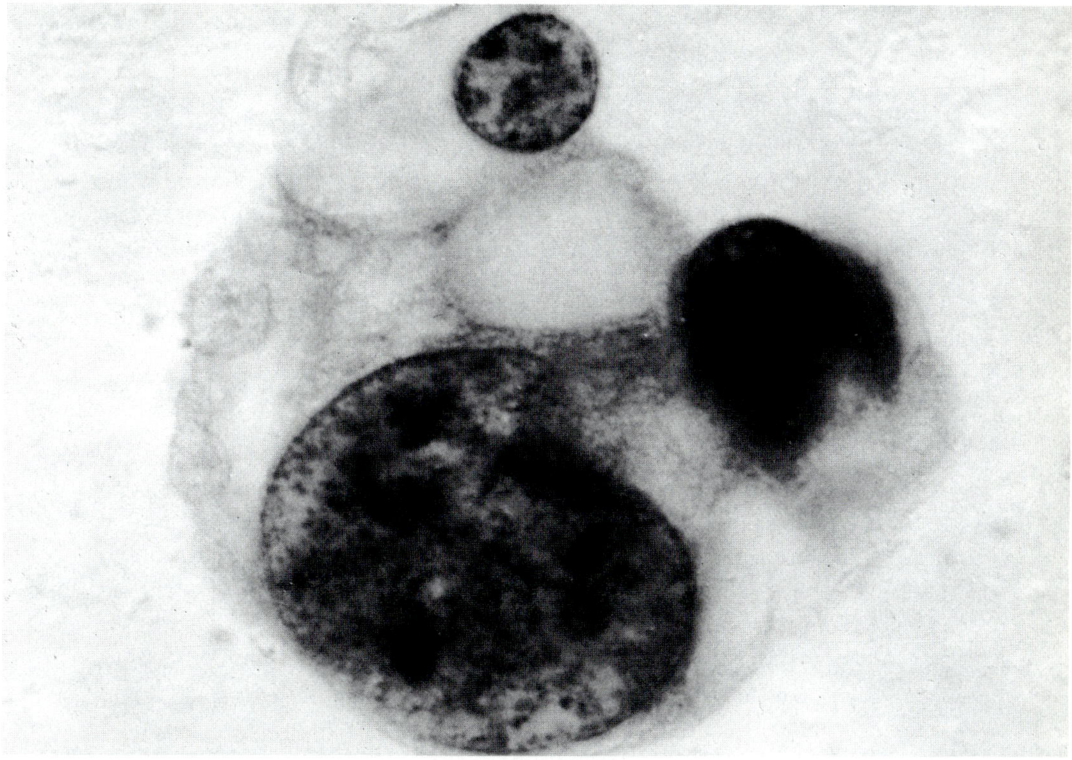

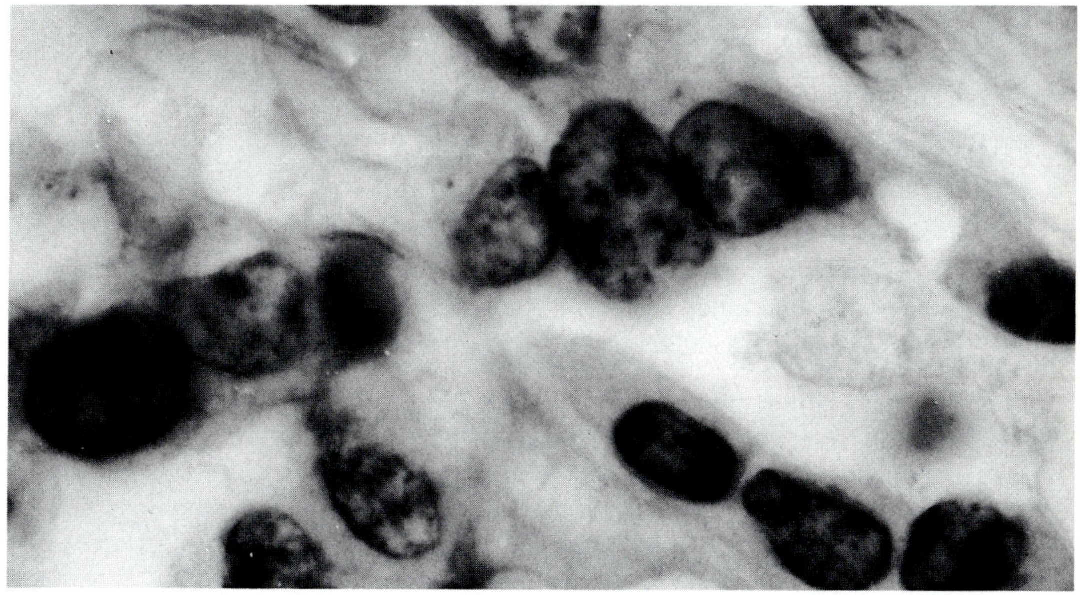

1

2

This is not unique to pathology, but occurs throughout the biologic spectrum. One recognizes a close friend by first observing certain objective details (i.e.: eye color, interpupillary distance, size of ear lobe) which may be virtually nonspecific, individually; but identification can be achieved when based upon the overall pattern which is created by these multiple nonspecific morphologic features. The more key features there are, the more definite can be the recognition, and the less chance there is for false association. Thus these key features, when occurring together in a syndrome, enable one to make the proper diagnosis, i.e.: the recognition of a friend. The recognition of disease in clinical medicine, in histopathology, and in cytopathology is brought about in like manner.

In the cellular study of health and disease, therefore, certain identical structural features or changes may be found in different biologic states. By themselves these structural features may not be morphologically separable, and are thus nonspecific; however, when they occur together in identifiable patterns, or with other morphologic changes, syndromes of structure emerge. These can be recognized by experience, and thus allow diagnoses to be made of the more specific biologic processes which the syndromes denote.

D. Time-Sequential Examination

As referred to above, successive phases of a process can be carefully investigated in greater detail by examining multiple cytopathologic samplings obtained in a time-oriented sequence. In many conditions (e.g.: healing pulmonary infarct) the changes of the process as it develops are detectable and identifiable, and are discriminable from other diseases (e.g.: cancer) whose morphology is virtually unchanging over weeks.

Tissue biopsies interrupt and alter the basic ongoing biologic processes, with morphologic patterns compounded by the added inflammation and healing. This frequently results in a confusing evaluation of the basic biologic significance of morphologic findings.

Cellular samples, in contrast, which have exfoliated from a surface and thus have left the host with*out* interrupting its integrity, can be examined with no artifacts whatsoever introduced. In this way, if time-sequential information is indicated, frequent multiple samples can be obtained without artifactually altering the processes of the lesion. Without the additional compounding and confusing processes, therefore, a clearer determination can be made of the basic biologic processes which are active.

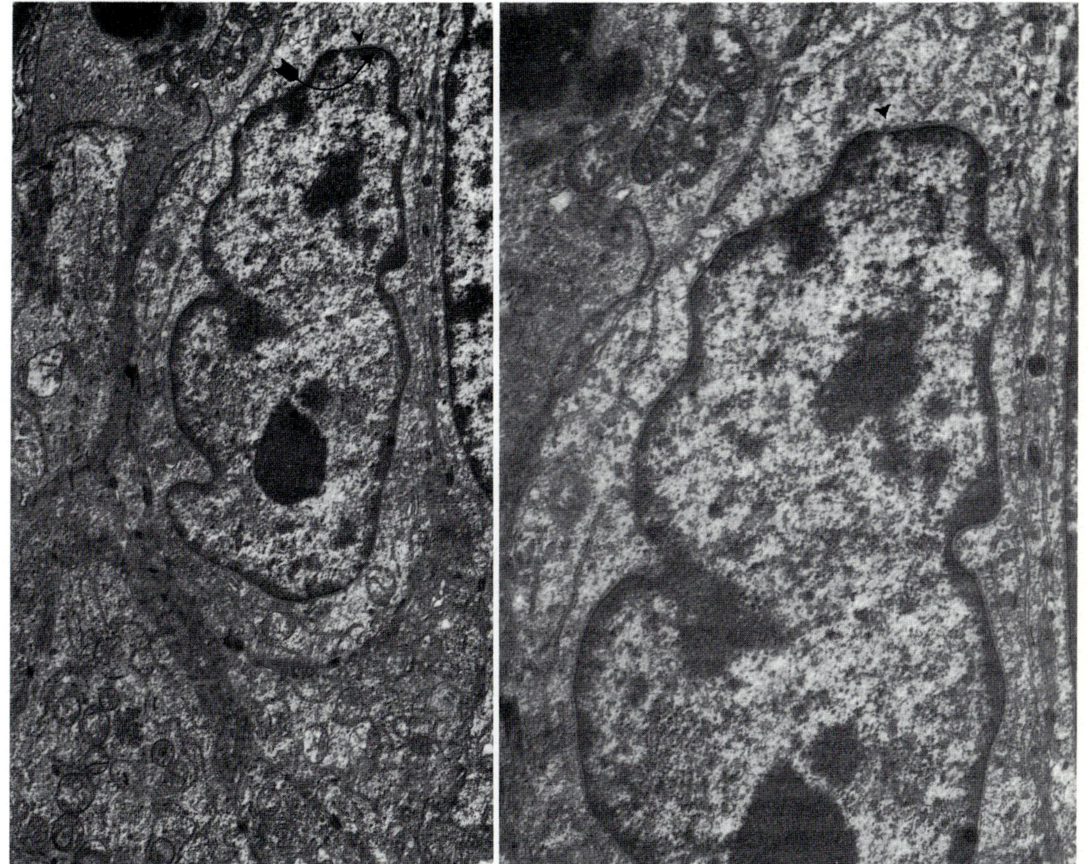

Figures I-3, I-4: Poorly differentiated primary carcinoma of the lung.

Fig. I-3: At this magnification the true nuclear envelope is just barely perceptible in only a few areas (e.g.: arrow head), but the chromatinic rim (historical 'nuclear membrane' of light microscopy) is prominent (e.g.: arrow). The latter, with its sharply defined *outer* profile clearly indicating the *shape* of the true inner nuclear *membrane*, has thickness varying with the amount of lamina and marginated heterochromatin which is condensed and adherent to the inner nuclear membrane to constitute the chromatinic rim.

Fig. I-4: A large nucleolus is at the bottom of the electron micrograph. The heterochromatin condensations of the chromatinic net are not uniform and are unevenly dispersed throughout the parachromatin. The interchromatinic areas between the condensed chromatin, are of *un*equal size and have varying amounts of finely divided chromatin threads and granules. This imparts varying degrees of clearing to greatly *un*equal areas of parachromatin of a well preserved cell viewed in the light microscope (see Fig. I-1). There are a number of nuclear pores which are barely discernible (e.g.: arrow head) even at these relatively low magnifications.

Osmium/methacrylate.

I-3: × 2,500; I-4: × 12,500.

(Frost, Erozan, Donovan; unpublished electron micrographs.)

Of the many other methods available for obtaining specimens, most lie somewhere between these two extremes of artifactual changes. Gentle surface aspiration or lavage removes only material which already has exfoliated and lies loosely on the surface. Gentle scraping or swabbing, in addition, loosens and removes superficial cellular layers without appreciably altering biologic processes.

Deeper cellular samples can be obtained, but with varying degrees of modifying biologic behavior. Brushing and moderate abrading, for instance, produce deeper cellular layers. In hard keratotic epithelium (e.g.: surface of the tongue) little artifactual alteration occurs; however, in extremely soft tissues (e.g.: floor of the mouth) the same sampling may markedly alter behavior, and thus morphology, by adding new processes (e.g.: hemorrhage, inflammation, healing). A more thorough abrasion through the epithelium, such as with the establishment of a Rebuck 'skin window' [233], removes cellular samples of deeper skin tissues, with increased trauma to compound the resulting picture.

Needle aspiration adds processes (e.g.: injury, hemorrhage, repair) which compound the interpretation of subsequent needlings. Fine needle aspiration does this to a virtually *insignificant* minor degree, however, as compared to that which results from large bore cutting needles or, even more so, from surgical exploration with biopsy.

E. Human-Automation Interaction

Based upon the pioneering and revolutionary works of Caspersson [53], cellular interpretation is moving ahead with valuable assistance from precise quantitation and automation. There are many *additional* features which engineering techniques are able to bring to the evaluative process and of which man is incapable of sensing or perceiving, which are greatly increasing human ability to interpret biologic behavior from cellular morphology.

Fig. I-5. Squamous cell carcinoma of the lung, sputum. These are not photomicrographs, but are full color digitized images composed of 0.25 μm diameter pixels, which are more easily seen in the enlargements *B* and *C*, each recorded in the complete visible spectrum: in red, green and blue. While this black and white reproduction of the full color

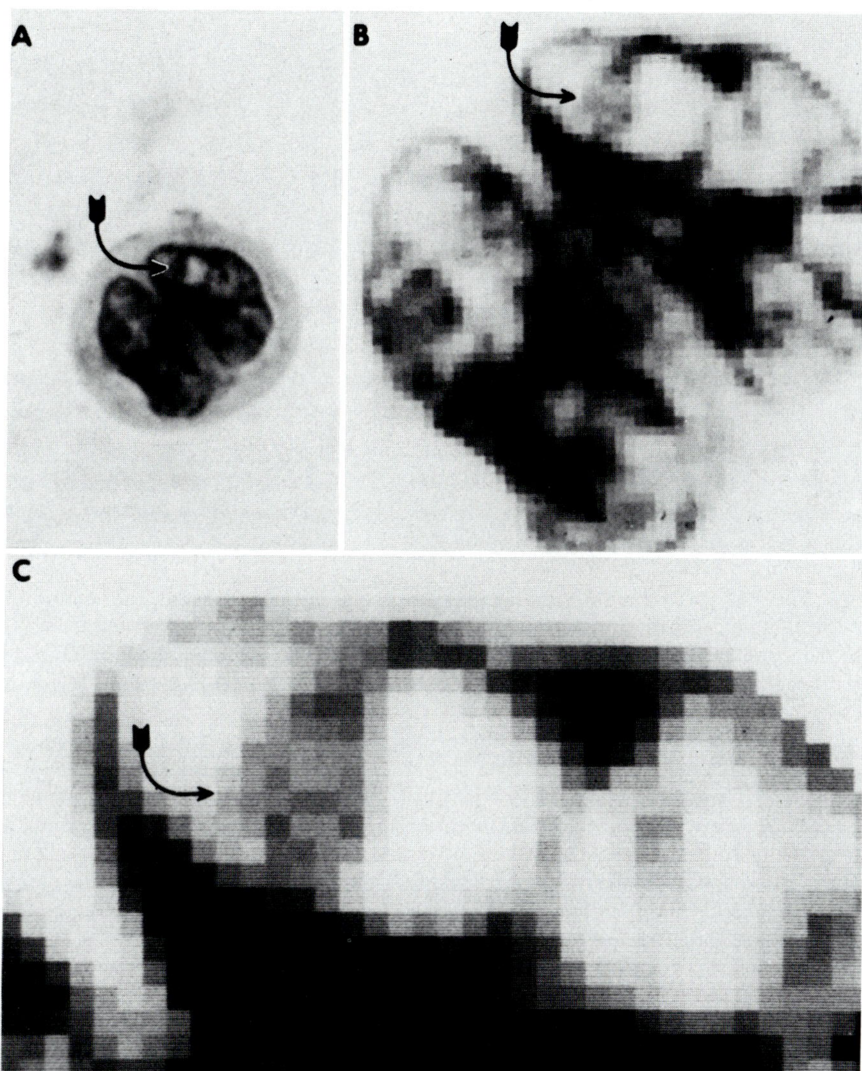

digitization does not illustrate the precise spectral information which has been captured (note nucleolus at arrow, and see Fig. I-6), it does give some perception of the accurate spatial relationships and photometric absorption data which reside in the computer memory for analysis and better understanding of biologic behavior.
Papanicolaou stain.
I-5*A*: × 1,600; I-5*B*: × 5,250; I-5*C*: × 13,000.
(From Frost, et al.; 1985 [102].)

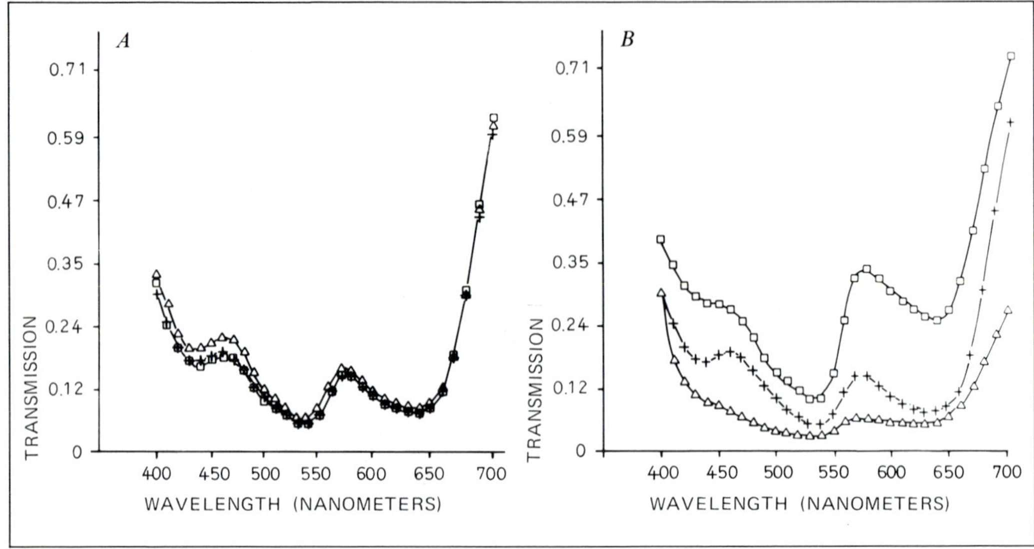

Fig. I-6. Spectral transmission characteristics of nucleoli. Spectral information of nucleoli of cancer cells over the entire range of the visible spectrum. Sensed and recorded by the Scanning Transmission Optical Microscope (STOM) of the Johns Hopkins Biomedical Image Data Analysis System (BIDAS) using a 0.25 µm diameter probe at the center of each nucleolus (see Fig. I-5).

A. Three independent readings of three different nucleoli, one in each of three nuclei of cancer cells exfoliated from *one* lung cancer. There is high sensitivity and reproducibility of the three independent readings.

B. One reading of one nucleolus of three cells, each from *three different* cancers (+ = large cell undifferentiated carcinoma from sputum; □ = transitional cell carcinoma from urine; △ = adenocarcinoma from peritoneal fluid). Even though they appear similar to the human eye, nucleoli of the three tumors have distinctly different spectral absorption and transmission characteristics. Analysis of such spectral signatures aid identification of structural composition, proteins, and biologic processes.

(From Frost, et al.; 1985 [102].)

The multidimensional features – of *spatial* orientation, relationships, size, shape; of *photometric* absorption, transmission, emission; and of *spectral* hue, signature, transition – are all of prime importance in evaluating a given scene on a microscopic slide (Figs. I-5, I-6). Human ability is being greatly augmented in those areas, where man is capable of arriving only at rough estimates, by more precise *quantitative* and *automated* techniques [31, 41, 70, 107, 110, 111, 128, 156, 180, 194, 217, 221, 305].

Enormous speed is attainable by a variety of automatic techniques, especially the so-called low resolution configuration and flow-through systems [64, 106, 110, 149, 282–284], the flying spot, video, and diode array detectors [128, 191, 221, 275, 304], and the ultrafast laser scanners under development [32]. The axial illumination, photodetector systems [107, 220] have highest resolution (0.25 μm), accuracy, and reproducibility. While not yet up to human capacity in many aspects, various systems surpass it in other respects [128]. Uniform and predictable staining is essential [20, 122, 124].

Of greatest value, therefore, is not the mere repetition of human capabilities, but to exploit and take full advantage of the many areas where automation is superior. In many of them it augments the senses and intellect of its human counterpart by detecting, quantitating, comparing and analyzing qualities of the cellular specimen to which the human is insensible and incapable of perceiving [102]. Development of its tremendous capacity for data gathering, cataloguing, storage and retrieval, data analysis, comparison and integration into the whole, and transmitting this information instantly over boundless distances of space [98, 215], is increasing at phenomenal speeds and becoming more and more clinically useful.

This rapidly is becoming a reality. Not only will it be serving the large research centers but, through telecommunication between independent work stations to remote data banks and consultative analysis stations, will assist the independent pathologist, diagnostician or researcher working in his own office and laboratory.

II. Biologic Behavior:
General Activity and Functional Differentiation

Tissue and cellular biologic behavior can be bisected, conceptually, into two areas: *general activity* and *functional differentiation.*

General activity is logically further divisable into four large areas: *euplasia, retroplasia, proplasia,* and *malignant neoplasia.* Each is further divisible innumerably, where indicated, to better understand the particular biologic process(es) under consideration. While their biologic significance is discussed in this chapter, their morphology will be presented in Section C.

Functional differentiation subdivisions are plentiful representing, on the one hand, the many recognized *cell types* (e.g.: columnar, muscle) and, on the other hand, the innumerable more specific functions (e.g.: secretion, movement) of which cells are capable [42, 135]. Their biologic significance will be discussed briefly at the end of this chapter, but their morphologic counterparts will be covered in greater detail in Section D.

A. General Activity

1. Euplasia (eu: Gr. = good, well, true, most typical;
plasis: N.L., Gr. = molding, formation)

This is the baseline 'normal' reference state of biologic activity, in which metabolism and growth are sufficient for continued existence in a working community of healthy, nonstressed cells and tissues (Fig. II-1).

The zone of activity of this reference state varies between major cell types. This difference can be dramatic between bone marrow cells, stratified squamous epithelial cells, and fat cells (Fig. II-2). Even within a given major cell type, however, there can be significant variance between sub-cell types. This can be especially true between strata of epithelium (e.g.: parabasal cell,

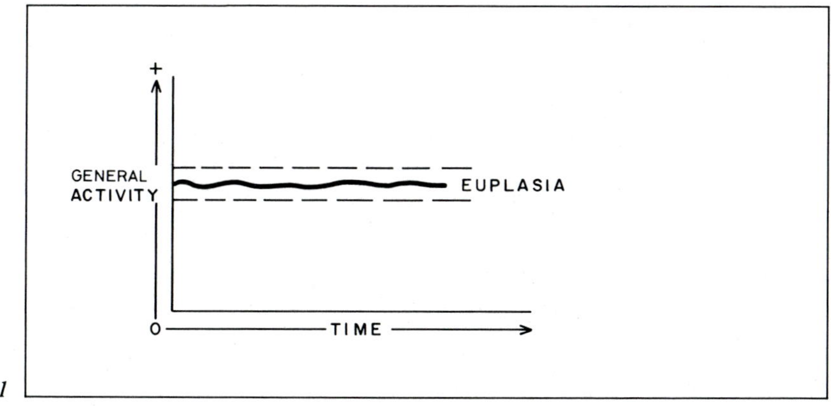

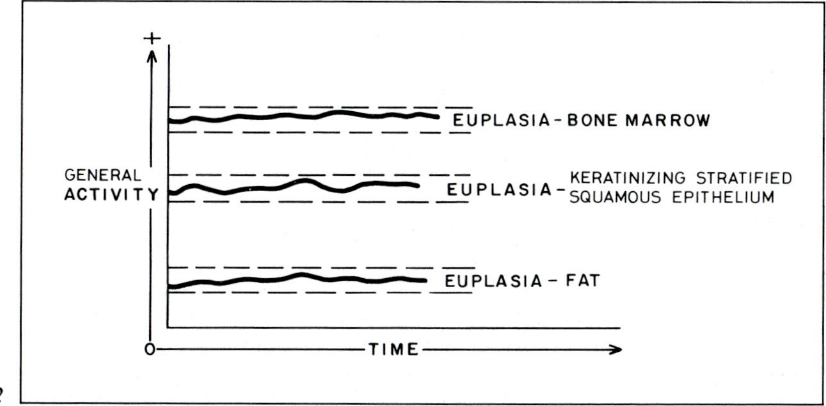

Figure II-1. The zone of *euplasia* diagrammatically represented for a given cell type which is at a level of general activity sufficient for existence in a working community of healthy, nonstressed cells.

Figure II-2. The zone of *euplasia* for each cell type varies between different major cell types. Represented here schematically is the Zone of Euplasia for an average euplastic cell from bone marrow which is at a higher level of general activity than for an average euplastic cell from stratified squamous epithelium. The average euplastic fat cell has even a lower level of general activity.

intermediate cell, superficial cell) and where there are different levels of maturation (Figs. II-3a, II-3b).

The prototypes of cells in classical histology [42, 135] are at this *euplasia* reference level of activity. Their morphology has thus become the accepted basic frame of reference for each cell type.

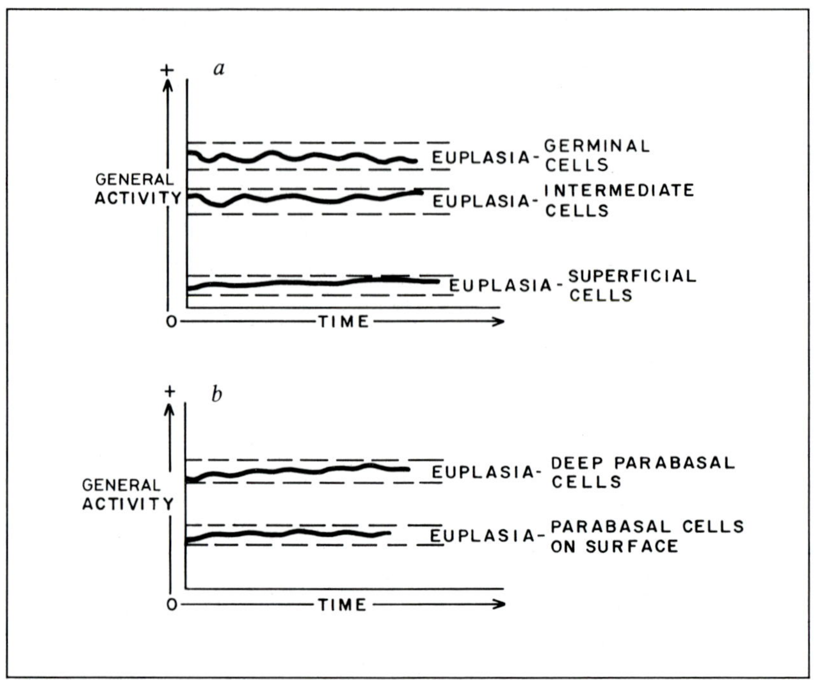

Figures II-3a, II-3b. The zone of *euplasia* varies for cells whose job within an organized tissue requires different levels of general activity within a major cell type. Here represented are average euplastic cells from each of three cell strata within an intact Euplastic Stratified Squamous Epithelium which has matured: a) to the level of the superficial cell, where the euplastic parabasal cells have an average level of general activity which is higher than that of the superficial cells lying on the epithelial surface; b) to the level of parabasal cells (parabasal atrophy) [99], where the parabasal cells on the surface are at a lower level of activity than those lying in the deeper strata.

2. Retroplasia (retro: Gr. = backward, back; plasis: N.L., Gr. = molding, formation)

Cellular general activity decreases below that of the normal baseline, euplasia, as a result of many processes, e.g.: aging, starvation, injury, degeneration, death, necrosis (Fig. II-4). Such biological retrogression [52, 112] may go on to more advanced stages of degeneration, death and necrosis (Fig. II-5).

The degree of morphologic change accompanying this retrogressive biological behavior, is dependent upon many factors which affect the cell

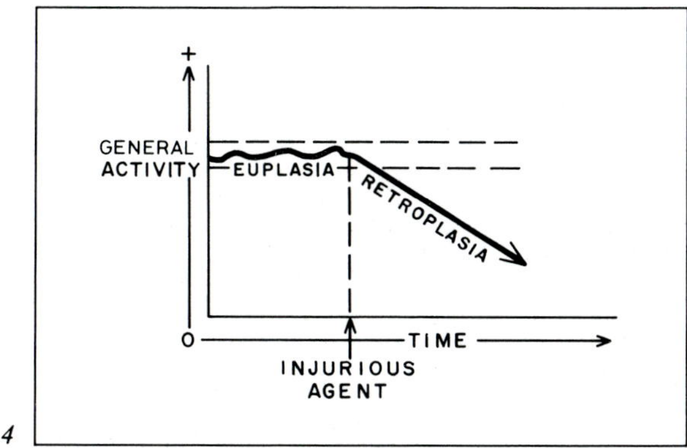

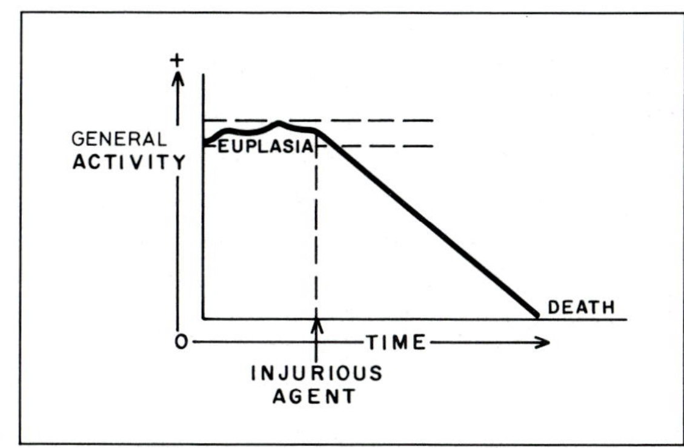

Figure II-4. Retroplasia. A cell in euplasia is injured by a harmful agent (e.g.: thermal, microorganism, radiation, chemical, mechanical). It decreases its general activity in the presence of many retrogressive changes. If the time elapsed and environmental conditions are sufficient, morphologic changes of retroplasia occur.

Figure II-5. A cell in *retroplasia* (Fig. II-4) continuing to decrease its general activity without recovery, until it dies. If the time elapsed is short, and the conditions for the cell during this time are preservative, only slight morphologic changes associated with degeneration and death may occur. If the conditions are sufficiently severe, however, retroplastic changes can be extreme.

during this period (e.g.: length of time, pH, enzymes, temperature). More morphologic changes are present in a condition of retroplasia which has existed over a long time, than one existing over a brief period. When death is immediate and is simultaneous with fixation (Fig. II-6) virtually *no* morphologic change occurs other than that produced by, and characteristic of, the fixative *per se.*

3. Proplasia (pro: Gr. = ahead, forward;
plasis: N.L., Gr. = molding, formation)

Cellular activity *increases* above that of euplastic reference activity in response to many factors (e.g.: nutrition, hormones, replication, repair, regeneration, stimulation, carcinogens). These result in many forms of increased cellular activity such as building of intracellular structures, formation of intercellular substance, secretion, and cellular replication (Fig. II-7).

Some types of progressive activity are regarded as normal, such as: physiologic replication (Fig. II-7), work hypertrophy, physiologic hyperplasia, increased secretion, regeneration (Fig. II-8). Others are considered abnormal (e.g.: atypical hyperplasia, atypia, dysplasia, anaplasia, intraepithelial neoplasia).

When the cycle of injury-degeneration-recovery-regeneration (Fig. II-8) is interrupted by another injury to the regeneration cell (Fig. II-9), *retro*gressive changes of the new injury are superimposed upon the *pro*gressive tissue and cellular changes (e.g.: severe inflammatory atypia). The reinjured cell repeats the cycle of injury-degeneration-recovery-regeneration. If the noxious agent is removed, the cell eventually resettles into a normal euplastic state (Fig. II–9).

Figure II-6. Death of a euplastic cell by *immediate* fixation. This allows virtually *no* time for retroplastic morphologic changes to occur.

Figure II-7. Proplasia of cell replication. Cell cycle of a schematic functioning cell with 48 hour generation time (Tg). M′ phase (mitosis) of approximately one hour. G_1 phase (first gap in DNA synthesis) of appoximately 34 hours. The cell is diploid during this time and creates its specialized functional differentiation with transcription (RNA synthesis), translation (protein synthesis, e.g.: microtubules, fibrils), cellular building (e.g.: organelles, polymerization of microtubular skeleton, fibrillary system, membranes), and function (e.g.: secretion, absorption, transport, covering, protection, motion). S phase (synthesis) when DNA is synthesized, increasing from diploid to tetraploid by replication. G_2 phase (second gap in DNA synthesis) of approximately four hours, where the now tetraploid cell ceases DNA synthesis and reorganizes itself for mitosis, M″, where each chromosome (now organized as a 2n pair of sister chromatids) will split longitudinally to generate two sister chromatids, one homologue will go to each daughter cell which will thus be diploid again.

Figure II-8. Proplasia of regeneration. An injurious agent causes general activity to decrease, with the resulting retroplasia of degeneration. Then, if the cell recovers, proplasia sets in as the regenerating cell convalesces, eventually becoming euplastic again.

+
GENERAL ACTIVITY
EUPLASIA
O
DEATH
TIME
IMMEDIATE FIXATION

6

M'
G₁
S
G₂M''
+
GENERAL ACTIVITY
(PREMITOTIC ACTIVITY)
PROPLASIA
EUPLASIA
O
TIME
INITIAL MITOSIS
FINAL MITOSIS

7

+
GENERAL ACTIVITY
(Recovery and Regeneration)
PROPLASIA
EUPLASIA
(Injury)
RETROPLASIA
EUPLASIA
O
TIME
INJURIOUS AGENT

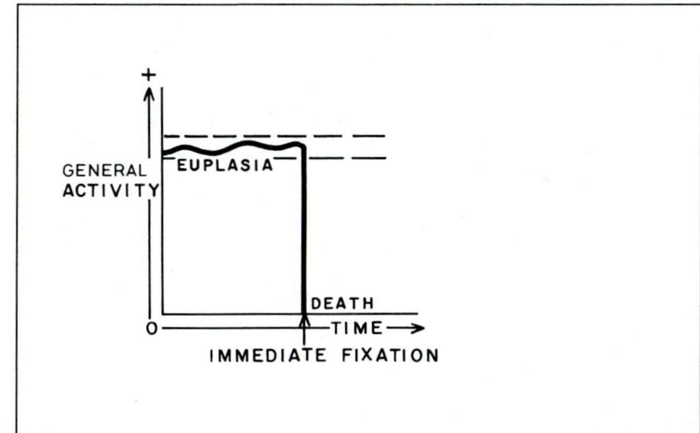

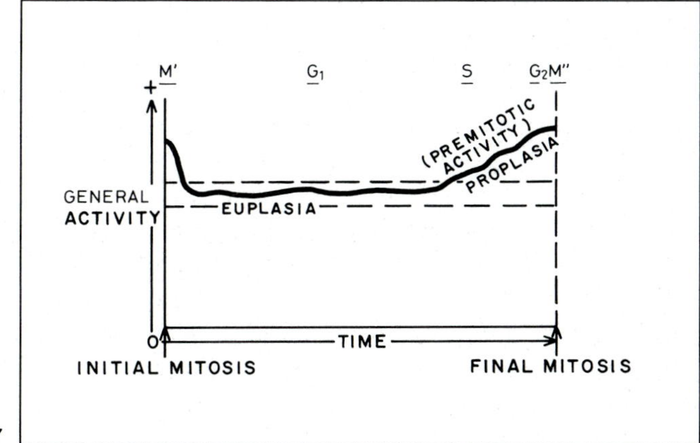

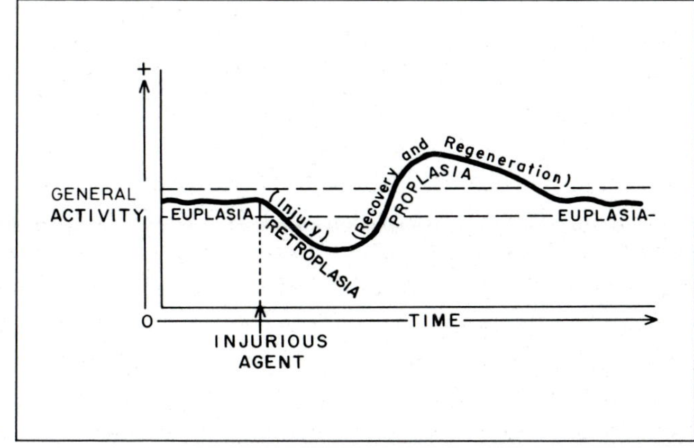

8

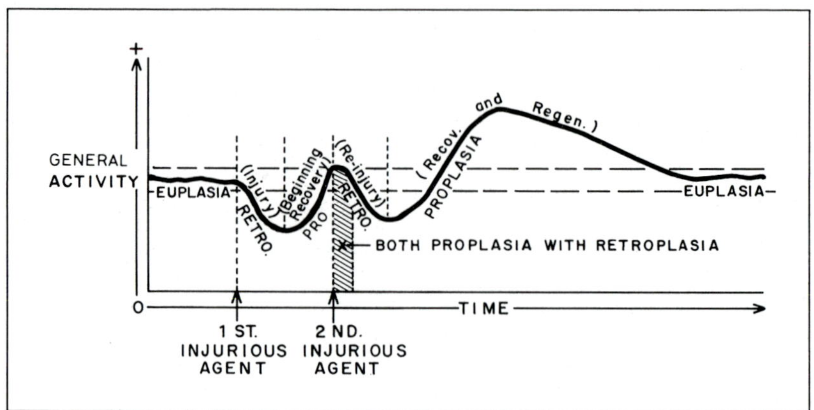

Figure II-9. The *proplastic* cell, recovering from the first injury, receives a second harmful stimulus. For a while there exists new *retro*plastic changes of re-injury superimposed upon the remaining *pro*plastic changes of recovery. For a second time the cell recovers and becomes *pro*plastic, frequently at a higher level of general activity from the double injury. This gradually decreases as the cell completes recovery and returns to *eu*plasia.

In *chronic irritation* the injury persists, or continues to occur intermittently (Fig. II-10). This results in a continuing cycle of degeneration-regeneration-degeneration-regeneration-degeneration, etc., with the cells and tissues prevented from achieving the normal euplastic state. The picture thus can become extremely complex morphologically, with the interplay of injury and recovery producing superimposition of retroplasia upon proplasia. If the cell is not overcome and does not die, but continues to survive the situation, the resulting effect can be that of gradual increase in level of general activity (Fig. II-10) so that a constant proplastic state of continuing stimulation is maintained. This situation is common in promoters, substances which facilitate the action of carcinogens in neoplastic development.

Some agents (e.g.: some viruses) produce an initial phase of stimulation which then results in, or is followed by, injury (Fig. II-11). When this stimulation-injury persists or continues to occur intermittently, an even higher level of general activity can be achieved (Fig. II-12). Such continuing reactions to incessant stimulation-injury, can produce abnormally active proplastic states. Morphologically, many of these are severe 'borderline' or 'shades-of-gray' lesions, and constitute some of the lesions designated as atypia, atypical hyperplasia, atypical metaplasia, and dysplasia.

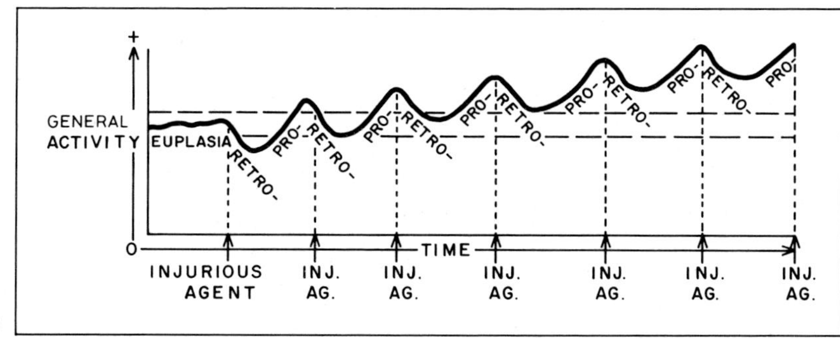

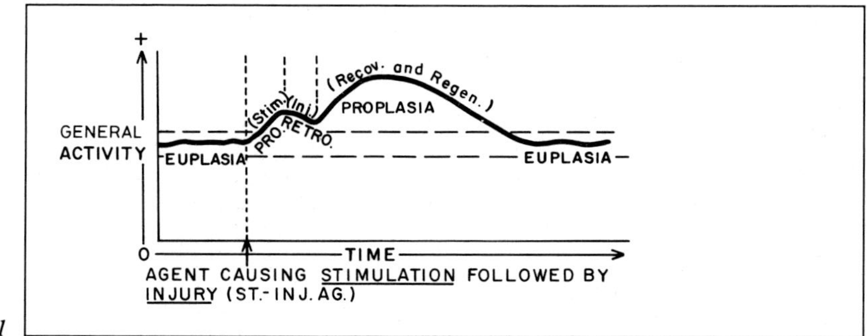

Figure II-10. Chronic irritation with continuing, recurring injury to a cell during its regeneration. This results in a cell which is maintained at a high level of general activity (proplasia) with periods of superimposed degeneration (retroplasia). This results in complex morphologic patterns of cells maintained under chronic irritation, in such a proplastic-retroplastic state of incessant injury and regenerative stimulation.

Figure II-11. An agent producing *stimulation* to the cell followed by injury, first causes general activity to increase (proplasia) during the phase of stimulation. Then, as cellular damage occurs, general activity decreases (retroplasia). If the cell recovers, general activity again increases during regeneration, with proplasia continuing until the recovered cell again settles into normalcy and becomes euplastic.

Carcinogens can affect tissues in many states of activity. There is evidence to indicate, however, that tissue maintained in such complex retrogressive/progressive states may be more susceptible to the effects of carcinogens [35, 158, 159, 244, 245, 248, 264]. Mitoses are frequent in these complex retrogressive/progressive states of chronic irritation, a state also found with virtually all promoters. Once initiation has occurred, it is irreversible and inherited by the daughter cells; thus, the lesion cannot reverse

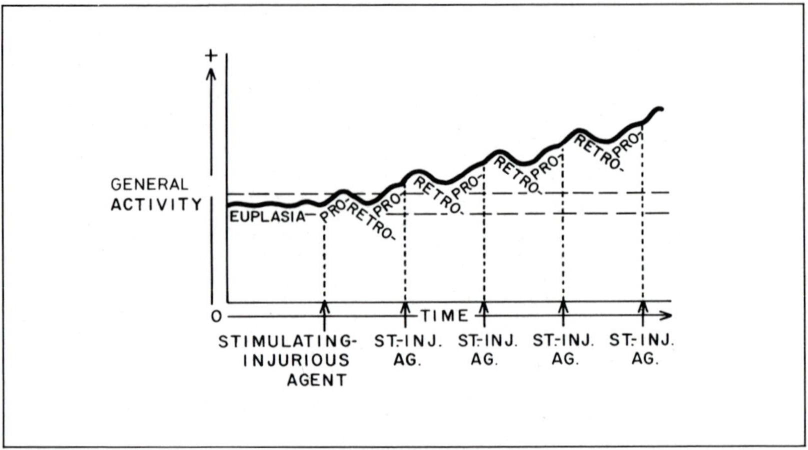

Figure II-12. Chronic irritation with a continuing, recurring stimulative-injurious agent (St.-Inj.Ag.). The cell is maintained at an extremely high level of activity, and the pattern can be extremely complex from this incessant stimulation to increase general activity to an even higher level.

to normal, so that it either dies, remains static, or proceeds on to a higher order of neoplasia (Figs. II-13, II-14).

Neoplastic progression takes varying lengths of time. It may, infrequently, occur rapidly or, usually, be a slow and relentless process which gradually changes through neoplastic mutations to new and more virulent clones which overgrow the less aggressive ones. This *heterogeneity* of cancers is reflected not only in their *morphologic pleomorphism,* but also in the *varied biologic activity* (e.g.: rapidity of *growth,* speed of *progression* through invasion and metastasis, propensity to *metastasize,* diverse *response* to therapy) which *different cells* of the *same tumor* exhibit.

Figure II-13. 'Co-carcinogenic set-xy': A situation which provides all of the necessary set of ingredients (i.e.: Factors x + y; e.g.: stimulative, injurious, genetic, irritative, chemical, radiation, microorganismic, age) for the production of 'cancer-xyz', except one (i.e.: Factor z). Factor z is added for a sufficient time (T_{ca}) to produce 'cancer-xyz'. The introduction of the missing carcinogen (Factor z) to the 'co-carcinogenic set-xy' (produced by Factors x + y) brings about 'cancer-xyz' (caused by Factors x, y, and z; no one of which is capable of causing 'cancer-xyz' by itself).

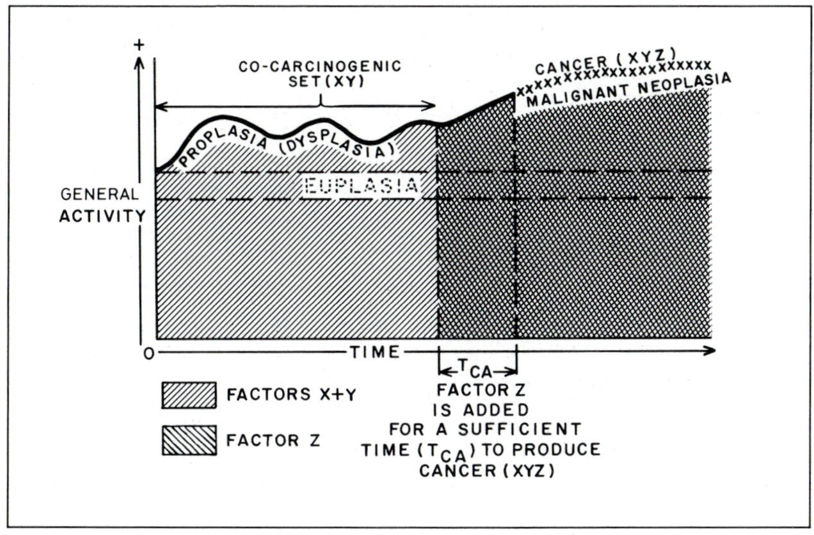

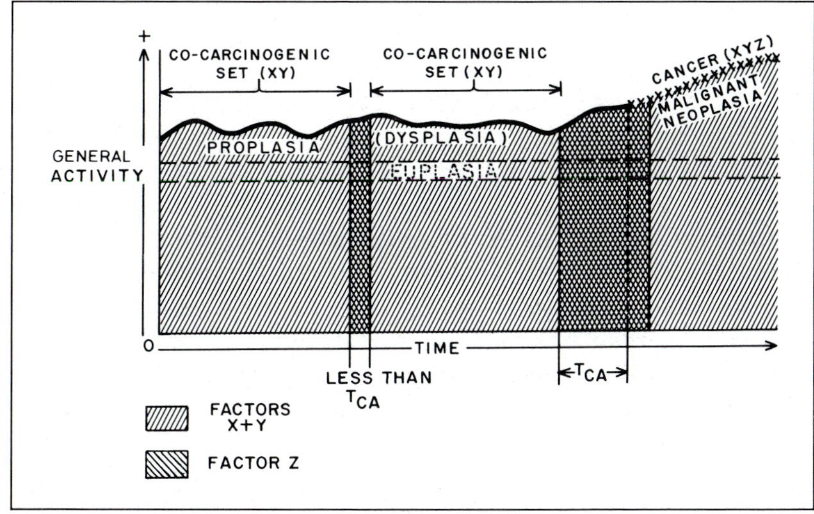

Figure II-14. Time of action. If it is necessary for Factor z to operate upon the 'co-carcinogenic set-xy' (produced by Factors x + y) for a minimum period of time (T_{ca}) to result in 'cancer-xyz': a) Cancer is not produced when Factor z is withdrawn after acting for a period of time less than T_{ca}. A period of altered general activity occurs, which is reversible to the previous state of 'co-carcinogenic set-xy'. b) When Factor z does act for the period of time (T_{ca}), initiation occurs (an irreversible mutation) and the early stage of 'cancer-xyz' results. Irreversible 'cancer-xyz' continues even after Factor z is withdrawn after it has acted for time T_{ca}.

In the early stages of such an irreversible cancer it can appear as a dysplasia, an *in situ* (e.g.: intraepithelial) neoplasia. This is a 'borderline' or a 'shades-of-gray' lesion which, in some cases, may not all be distinguishable morphologically from the earlier reversible lesion by our present techniques. They are, however, biologically quite different and there is great promise that quantitation and automation soon will be able to assist human assessment in making the necessary discriminations, including normalized hematoxylin DNA ploidy [41, 102]. Ewing [72] observed over seventy years ago that certain lesions may not be classifiable as either 'cancer or not cancer'. They may be neither one nor the other, but in the process of becoming cancer.

It is very important to morphologically recognize such co-carcinogenic or precancerous states which progress into malignant neoplasia to a significantly high degree [17, 82, 102, 113, 127, 186, 206, 207, 227]. It is additionally important, from the standpoint of both practical therapy and of basic biologic understanding, to differentiate those states which will progress to clinical cancer without additional carcinogenic assistance, from those which are reversible and will not progress to cancer at all, *and* from those which will not progress to cancer if the carcinogen(s) are removed (Fig. II-15). The significance of these distinctions is obviously great, not only from the viewpoint of basic biologic understanding but also from that of practical, clinical therapy and the management of the environment [102].

Many of these lesions have complex patterns of mixed activity and are difficult to differentiate with certainty, either histologically or cytologically, or impossible to so do at our present level of knowledge. Better elucidation of these 'borderline' or 'shades-of-gray' areas and of their biologic significance is obtained when both histology and cytology are utilized. When the patterns of the four basic levels of activity (i.e.: euplasia, retroplasia, proplasia, malignant neoplasia) are morphologically recognized and the relative importance of each is determined in the given situation, the process or processes as a whole are then reconstructed to the level of their presenting complexity (Chap. XVI).

Communication in the 'Shades-of-Gray' Area. Clear communication in this field (i.e.: between investigators, diagnosticians, therapists, patients) regarding the biologic significance of these processes and their complex morphologic patterns, is greatly hampered by a veritable babel of terminology [90, 93, 225]. Many terms in current use include: anaplasia, atypia, atypical hyperplasia, atypical metaplasia, atypical squamous metaplasia,

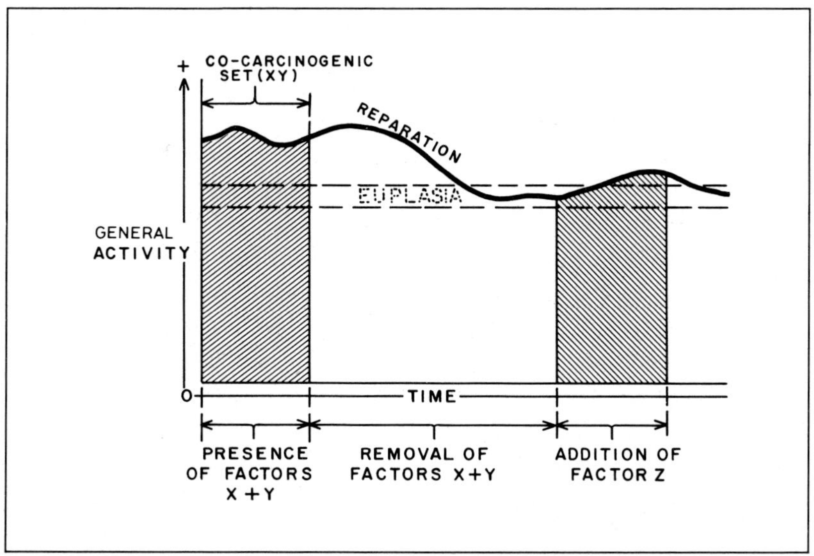

Figure II-15. If the 'co-carcinogenic set-xy' was not initiated, so that the lesion is reversible, or the lesion dies, removal of Factors x + y results in healing of the tissues with their return to euplasia. The introduction of the missing carcinogen (Factor z) after this, does not result in 'cancer-xyz'.

basal cell hyperactivity, dysplasia, hyperplasia, intraepithelial carcinoma, intraepithelial neoplasia, metaplasia, precancer, prosoplasia, reserve cell hyperplasia, spinal cell atypia, subcolumnar cell hyperplasia, subcylindrical atypia, subcylindrical cell hyperplasia, transformational zone, transitional zone – and countless further combinations of these and of other concocted and concatenated terms.

While many of these labels originally were applied with varying degrees of specific denotation, most have come to be used all too frequently as general connotative terms. These confuse *apparently* specific changes, which were given specific names in assumption of this specificity, with changes due to entirely different biologic processes than those which are connoted in the labels applied. In this way the same term often has come to connote, with implied exclusivity, a biologic process from a morphologic pattern which infrequently is associated with it. Consequently, nearly all are now employed differently by different workers in national as well as international usage, misused, and generally misunderstood.

When a specific process is discussed in this text (e.g.: metaplasia, hyperplasia) every attempt is made to use the appropriate term denotatively. When multiple processes are involved, the specific denotative term for each process is employed wherever possible.

The term *dysplasia* (*dys:* N.L., Gr. = ill, difficult, faulty, abnormal; *plasis:* N.L., Gr. = molding, formation) is a crucial case in point. In 1962, an international group of workers in the field suggested the use of this term to represent epithelial changes which are 'more' than normal but 'less' than cancer [303]. As the term had been and continues to be used by some to designate more specific epithelial patterns with neoplastic connotation [229], many have been hesitant to apply it in the former looser and more general fashion. When the actual biologic process(es) is incompletely known, others have used expressions which are more noncommittal (e.g.: 'shades-of-gray') [94], or denotative of more general properties (e.g.: atypia) [155]. Most recently in vogue in the gynecologic area is the use of the term cervical intraepithelial neoplasia (CIN) [237], which is still very connotative of presumably more specific (yet morphologically elusive) biological implication.

It must be understood clearly that the major difficulties are twofold: one with our *interpretation* of tissue and cells, and the other with our *communication* about these basic biologic processes between clinician, researcher, and pathologist diagnostician. This is *not* a problem of cytologic 'reporting by numbers', which was so acute around a score of years ago [93, 225]. Most pathologists today report cytopathology by disease and the process(es) involved in anatomic terms, and clinicians expect it and appreciate its added information about their patients.

Those who today employ the term *atypia, dysplasia,* or *intraepithelial neoplasia* synonymously, usually imply a particularly severe *proplasia,* which is not only *increased in activity* but is also *atypical* and has a *precancerous* or *preneoplastic* connotation – yet, one which has not thus far become biologically cancer (malignant neoplasia). Those who use *atypia* in this severe and neoplastic sense, as well as those who use dysplasia and CIN, usually are careful to separate it from inflammatory atypia without neoplastic connotation.

The greatest variable factor in their usage and understanding, is the degree of precancerous or neoplastic connotation which is believed to be present in the lesion before the term is employed. Some allow a very slight connotation, while others restrict it to where precancer or neoplasia is more substantially suggested.

Modifiers are employed with all three. *Atypia* and *dysplasia* usually are modified by slight, mild, fringe or minimal for the lowest grade of neoplastic concern; moderate for the middle grade; and marked, severe, grave or suspicious for the most serious. *CIN* is usually modified by assigning I for the minimal grade, II for the middle, and III for the most severe.

The two roughly approximate each other, except for individual variations in such a subjective matter as interpretation. There is, currently, *one* major difference: CIN III includes *in situ* carcinoma of the cervix in addition to the severe lesions designated as marked atypia or marked dysplasia.

In other sites (e.g.: bronchus, bladder, esophagus, larynx, mouth) *atypia, dysplasia,* and *intraepithelial neoplasia* also are among the most commonly used terms. Their use, and the employment of other terms, varies markedly throughout the world and, even, within given countries. Efforts by the World Health Organization and others are welcomed and badly needed, but should be based upon essential biologic information which is still being developed and gathered, and upon that coming forth from quantitative and automated systems.

4. Malignant Neoplasia (malignare: L.L. = to act maliciously;
neos: Gr. = new; plasis: N.L., Gr. = molding, formation)

Malignant neoplasia, or *invasive cancer,* is an unrestrained new growth of tissue or cells serving no physiologic purpose, with a tendency to spread and invade other tissues and to endanger the life of the host. It has different qualities than either proplasia or retroplasia, and is not merely increased or decreased general activity.

While it tends to have an average increased activity, *each cell* may have baseline activity, decreased activity, or increased activity and thus may show *either* euplasia, retroplasia, or proplasia. *Two* of them may be exhibited throughout a given tumor, or all *three* – in *addition* to malignant neoplasia.

Some cancers are rapidly growing, while others grow slowly. Any cancer can undergo degeneration. Thus cancers can also show euplasia, retroplasia, and proplasia *in addition to,* or even *in place of,* morphologically evidencing malignant neoplasia. Malignant neoplasia is, thus, a complex biologic dimension unto itself which is reflected in the morphologic structure of the malignant cell.

Invasion is a particularly deadly dimension, that of the ability for a malignant neoplasm to invade its host. Cancers exhibit varying degrees of this quality (e.g.: noninvasive or *in situ,* microinvasive, 'early' invasive, invasive, lymphatic invasion, lymph borne, blood vessel invasion, blood borne, metastatic) which are expressions of interplay between tumor activity and host response.

In great measure, this biologic activity of invasion in malignant neoplasia is reflected in morphology, not only histologic but also cytologic [70, 94, 99]. Morphologic borderlines between the stages of invasiveness exist, for practical purposes; frequently, however, they are not sharp either histologically or cytologically, or both – or, for that matter, biologically. The present incomplete state of our knowledge in biologic delineation and identification of the various stages of, and factors involved in, invasion is reflected in our imperfect and poorly agreed upon morphologic delineation and identification [194, 198, 258]. Great is our need *clinically* for utilizing information from all available parameters (e.g.: histopathology, cytopathology, tissue culture, electron microscopy, cytochemistry, immunodiagnosis, automated quantitative cytometry) to help in recognizing the major stages of the neoplastic progression in the development of malignant neoplasia.

Numerous parameters of biologic activity (e.g.: growth in tissue culture, cinekinetics, cytogenetics, ability to establish metastases *in vivo*), help to delineate one stage from another [23, 27, 84, 139, 249]. Furthermore, well-chosen cellular morphologic parameters (i.e.: criteria) parallel and characterize the biologic activity [43, 250] and help clinically delineate and identify the stage of development which a given tumor has attained [60, 62, 68, 69, 92, 94, 99, 111, 126, 150, 154, 155, 188, 193, 194, 198, 224, 226, 228, 246, 247, 273].

These morphologic findings associated with *biologic general* activity, which are more fully discussed in Section C, are found mainly in the *nuclear* structure.

B. Functional Differentiation

Conversely, those morphologic findings which are associated with *functional differentiation* activity, are found mainly in the cytoplasmic structures.

Cells have a multiplicity of potentials for functional differentiation. Theoretically virtually every normal diploid cell of the body is endowed with, and continues to bear, the capability to differentiate in all of the ways

Necrosis associated with *obstruction* and *inspissation* usually is not as uniform. In the firm central core of inspissated material, necrosis is most advanced. The cells in the surrounding material are usually of more recent exfoliation, so that their degenerative changes are less severe and they lie within a more amorphous and less dense proteinaceous substance.

Debris which has been engulfed by phagocytes should be carefully inspected. At times it can provide valuable additional information about the type of the necrotic process present, its extent, and the tissues which have been involved.

Change of the necrotic pattern in *sequential* samples, yields valuable evidence as to the type and progress of a biologic process. Thus, in addition to understanding a given necrotic pattern, it is of great value to study its development at multiple interval examinations.

Degenerative changes continue to occur after exfoliation. Under physiologic or proper preservative conditions, they will be minor if the time is short. If not, however, major structural alterations occur which are confusing with those occurring beforehand as a part of the biologic process. These background necrotizing changes can occur while *lying within the body* (e.g.: in body cavity fluid, mucus plug, enteric contents, urine in obstruction); or *later* as a part of *delayed* fixation (i.e.: storage, drying), or as the result of the usage of *improper* or less desirable fixative (e.g.: isopropyl alcohol, methanol, formalin, antifreeze).

While not always possible, attempt should be made to differentiate the necrotic changes and to determine when and where they occurred. When background necrosis occurs after exfoliation, all elements in the specimen which have been exposed to the same adverse effects show uniform degenerative and necrotic features.

Likewise, when cellular spreads have been allowed to thoroughly dry before wet fixation, the changes present are virtually uniform across the whole preparation; however, if the time elapsing between spreading and fixation is borderline, the more protected central and thicker areas may show less degenerative changes due to drying than the material about the edges of the spread. On the other hand, when degenerative changes are due to improper or slowly penetrating fixative or when the character of the material makes it extremely difficult for the usual proper fixative to penetrate, the center of the thicker areas shows more changes due to poor fixation (e.g.: acidophilia, loss of detail) than the thinly spread areas.

Degenerative cellular changes occur more rapidly and intensively in cells which have been sitting in *inflammation* and *hemorrhage*. Elements which are present in these conditions (e.g.: proteolytic enzymes, degraded blood products, nonisotonicity, extremes of pH) appear to *presensitize* cells so that subsequent conditions (e.g.: slight drying, questionable fixative, storage time), which could be otherwise innocuous or produce only border-line effects, now produce marked degenerative charges.

The resulting increased morphologic changes due to degeneration, can be severe and catastrophic. A brief delay in fixation of this type of material shows more extensive drying effects than in noninflammatory and nonhemorrhagic material.

This is frequently the case in material taken from a chronically infected cervix, while material from a normal cervix will show no effect from the same amount of air-drying. Likewise, this plays a major role in the state of poor preservation encountered in material from the first tap of a body cavity, when fluid had been sitting in the cavity for many days to months.

E. Obstruction

Lumina and orifices obstruct under many conditions (e.g.: inflammation, infection, hemorrhage, foreign bodies, stones, neoplasms) and at many sites (e.g.: tubules, glands, bronchi, ureters, diverticuli). The appearance of

---

Figures IV-8 through IV-11: Tuberculosis, lung, fine needle aspiration.

Fig. IV-8: Caseous necrosis extends along the right, with erythrocytes and lymphocytes throughout the field and epithelioid cells in the lower left.

Fig. IV-9: A higher power of the caseous necrosis. Note the granular, 'cottage cheese' appearing proteinaceous material with*out* evident bacteria, but containing nuclei of histiocytes and neutrophils in various stages of degeneration.

Fig. IV-10: Epithelioid cells with abundant, highly textured, chromatophilic cytoplasm and histiocytic-type nuclei with prominent nucleoli. While the nuclei tend to be fairly central, they are definitely off-center and even indented. The cytoplasm has little or no debris, and does not demonstrate the usual phagocytic activity of the pulmonary macrophage. This is a quasi tissue fragment, with very little nuclear molding.

Fig. IV-11: A tubercle gains access to the bronchus. A multinucleated giant cell (Langhans' giant cell) is separating off into the sputum, followed by a mass of epithelioid cells, followed by another Langhans' giant cell at the leading edge of the darker amorphous caseous portion of the tubercle (just barely included along the right border of the photomicrograph).

IV-8 through IV-10: Papanicolaou stain; IV-11: Hematoxylin and eosin stain.

IV-8, IV-11: × 190; IV-9, IV-10: × 1,300.

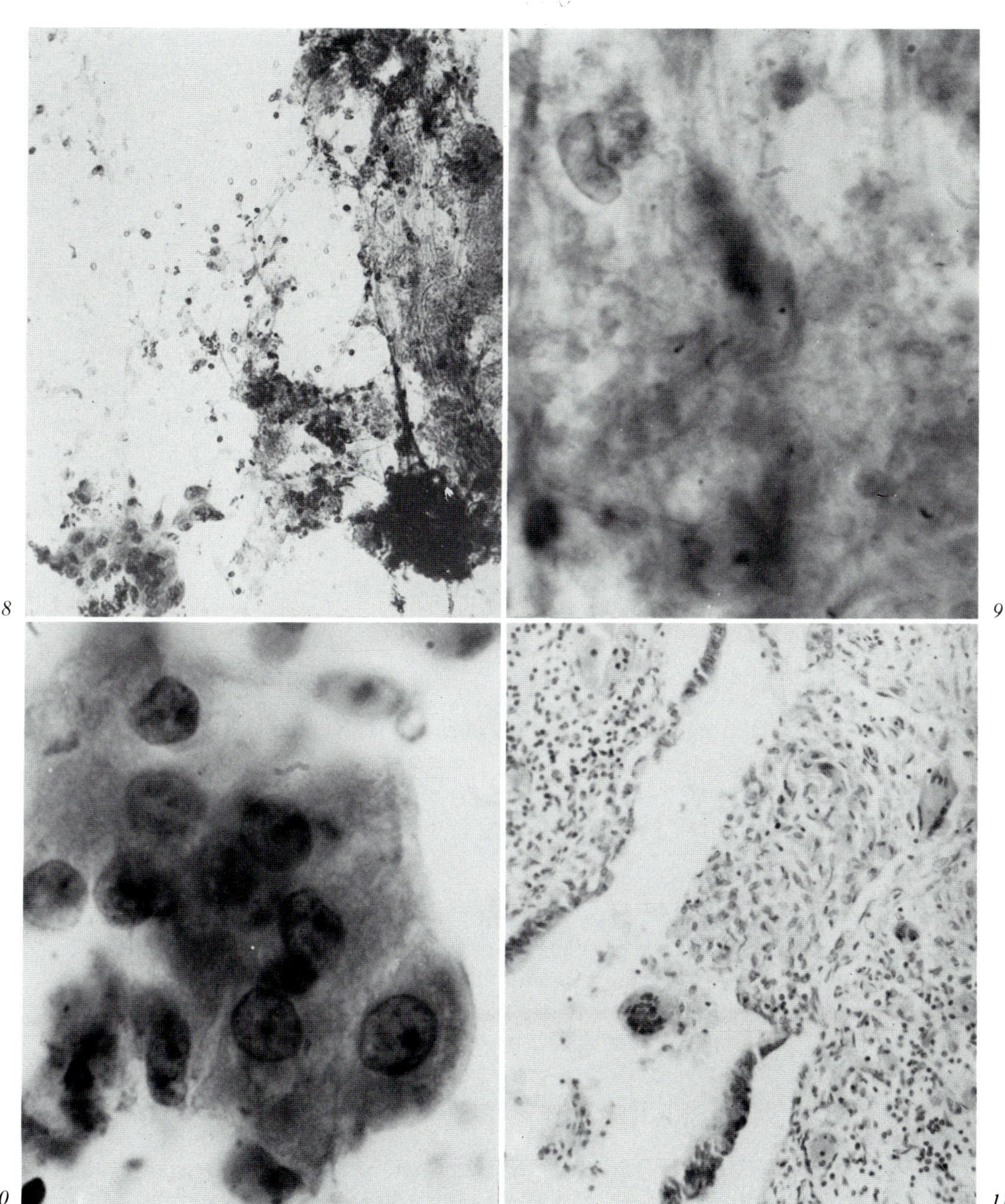

8 9

10 11

the cellular specimen is changed from the obstruction and, with time, it frequently becomes marked and characteristic.

When a lumen is obstructed, material builds up behind this barrier. It usually becomes *infected* in the presence of impaired drainage and circulation, due to the presence of microorganisms in a buildup of nutritive proteinaceous debris within the contents. At times it is *hemorrhagic* but, nearly always, necrosis is a major component.

When the wall of the plugged structure absorbs fluid, the material becomes dehydrated and *inspissated.* When the wall pours fluid into the lumen or when proteolytic enzymes actively denature and destroy the protein of the material, it becomes *liquified.* It can become *caseous* under proper variations in these and other conditions.

Various other factors which influence its appearance include characteristics of the *obstruction* itself (e.g.: complete or partial; continuous or intermittent; ball-valve) and type of *wall* (e.g.: transportive – ureter, mammary duct; secretory – bronchial, gallbladder, cervical gland; absorptive – alveolus). Also the nature of the *inflammation* (e.g.: acute, chronic, necrotizing), the type of *infectious* agent, and the *duration* of the process(es) all serve to help determine its character.

1. Inspissative Obstruction

Inspissation with obstruction causes *strands* of desiccated mucus to appear in the specimen background. The strands contain varying amounts of degenerated cells and debris throughout their substance. The most advanced degree of cellular *degeneration,* however, is present in those elements at the *center* of the strands, while the better preserved ones are toward the margins. This biphasic degeneration signals probable *obstructive* necrosis.

2. Extreme Inspissation

This produces hard or rubbery material which can form a cast retaining the shape of the structure in which it was formed (e.g.: bronchiole, renal tubule, cervical gland). It is firmly packed, frequently appears hyaline, and tends to remain separate and distinct from the rest of the background. Material from this densely inspissated type of obstruction, occurs only intermittently in specimens, due to regeneration time after discharge of the cast.

The *Curschmann's spiral* of the respiratory tract [63, 104] is a classical example of this inspissated cast formed in a bronchiole and coughed up into sputum. It consists of a hard, central core of desiccated, compressed, proteinaceous material which usually appears hyaline (glassy) and golden. It may appear black and pigmented, however, due to numerous small bubbles of entrapped air.

There is a peripheral shroud about a Curschmann's spiral which consists of a filmy veil made up of strands of less inspissated mucus which appear to be glutinous. As more mucus is secreted and the plug is slowly forced out of the bronchiole (as toothpaste from a toothpaste tube) it curls into a spiral from bronchiolar muscular and ciliary action. Its sticky peripheral shroud or filmy veil entraps material, cells and debris, which exfoliate and exude from the walls of the bronchioles and bronchi through which the spiral passes.

Characteristics of this *shroud* and its *contents,* therefore, can divulge much about the biologic process(es) going on in the more distal regions of the tracheobronchial tree. In active bronchial *asthma,* for instance, Curschmann's spirals characteristically contain a thick and heavy mantle of neutrophils, with or without eosinophils. In chronic bronchitis (e.g.: in cigarette smokers), squamous metaplasia and bronchogenic carcinoma, the spirals usually have a clear veil or one indicating hemorrhage (e.g.: hemosiderin-laden).

At times epithelial cells, which may be diagnostic of the condition of the mucous membrane over which the spiral traveled (e.g.: normal, metaplastic, neoplastic), will be entrapped in the shroud. A thorough inspection of the entrapped shroud contents may be rewarding of information about the biologic process(es) going on in the deepest respiratory passages.

3. Less Inspissation

Material may escape around the obstruction and release bits of its necrotic matter periodically into the specimen. When material behind the obstruction is more thoroughly liquified, the necrotic cellular debris is distributed more uniformly throughout the cellular spread and may give an amorphous liquid appearance to virtually the entire background.

4. Release of Obstruction

Reduction or *removal* of the cause of the obstruction, releases the backed-up material and allows the inflammatory process, in the wall of this previously obstructed structure, to diminish. These changes are reflected in the cellular preparation: The degree of necrosis diminishes, obscuring of other processes is less, the serous background of edema decreases, inflammation changes from acute to chronic and, eventually, the processes of repair and healing predominate. As healing continues, necrosis and inflammation diminish and, eventually, disappear.

5. Intermittent Obstructions

Usually from a ball-valve phenomenon or from a recurring process, intermittent obstructions produce periodic patterns of obstruction and release. The appearance and degrees of these two phases, may vary cyclically. The cytologic patterns of obstruction (i.e.: necrosis, inflammation, inspissation or liquification) yield to clearing (i.e.: evidence of regeneration and repair) which then yields again to the obstructive pattern, etc.

6. Physiologic Obstruction

A *physiologic deterrent* to the removal of material without an actual mechanical blockage, produces various phases of these obstructive patterns as the process(es) progresses. Thus, in *bronchiectasis* there is pooling of contents, stagnation, inflammation and necrosis. In *cavitating tumors,* occurring frequently in the highly keratinized squamous cell carcinomas or in Hodgkin's disease, many of these features of a physiologic obstruction are found. The large ulcerating lesions in areas which can become physio-

---

Fig. IV-12: A detached ciliated tuft (DCT) (center of photomicrograph) from a ciliated columnar cell. Two neutrophils (below) provide a size gauge. In the presence of heavy inflammation, this pinched-off ciliary tuft is evidence of a viral infection (i.e.: ciliocytophthoria). In the absence of infection, DCT gives evidence that ciliated cells have shed their cilia, as happens cyclically in the cervix and tubes. From the nasal mucous membranes, these structures have been observed and reported [50] to continue their beating for hours on a warmed stage. As the lumenal surface is angled, the terminal plate is not visualized.
Papanicolaou stain.
IV-12: × 1,900.

Fig. IV-13: Pollen. Pollen appears in many preparations, especially in the respiratory tract (both upper and lower); but also in the female genital tract from douches, in the gastrointestinal tract after ingestion, and in mounting media from the air of the laboratory. Pollens differ greatly in morphology, depending upon the plants from which they come; but most have a characteristically *irregular, thick, glassy wall* over the entire outer circumference, with *pale, hematoxylinophilic nucleoid* material either diffusely scattered or in indistinct clusters within the pollen body.
Papanicolaou stain.
IV-13: × 950.

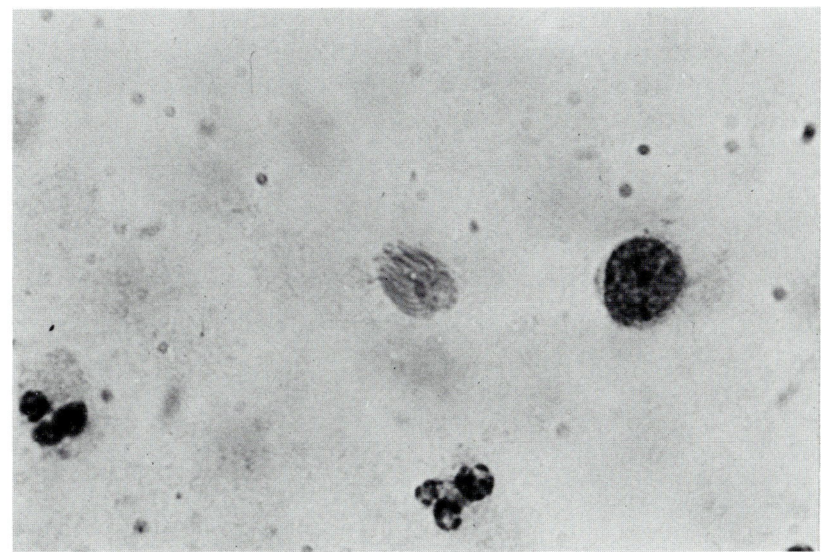

12

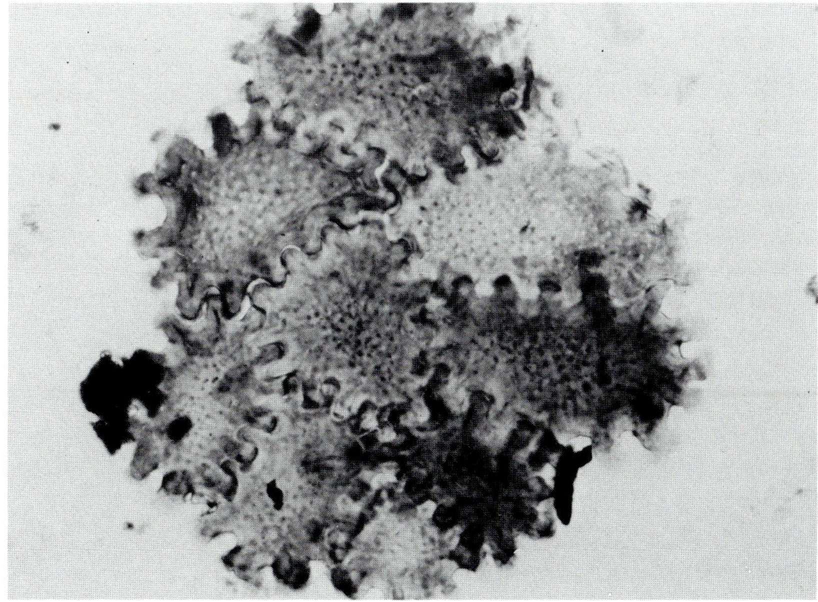

13

logically stagnant (e.g.: stomach, urinary bladder, cervix uteri, colon) may yield evidence of the necrotic liquifying contents which are retained in the crater of the ulcer, etc., plus a striking outpouring of serum and leukocytes.

F. Specific Elements

A careful search of the specimen background frequently reveals more specific elements which can be identified, and their pathogenic role evaluated. Thus identification of bacteria, fungi, protozoa, or helminths may lead to recognition of an etiologic or supporting role which they are playing in the disease process(es). Cellular changes indicating a possible virus as an etiologic agent, may appear in the background as debris (e.g.: ciliocytophthoria [50, 203]) (Fig. IV-12). Foreign bodies (e.g.: sutures, minerals, asbestos, lipid, pollens) (Fig. IV-13) may serve to indicate environmental conditions to which the patient has been subjected, and which may play a role in better understanding the biologic process(es) encountered [104, 131, 286].

These *background* changes, therefore, are thoroughly evaluated *along with* the *cellular* changes covered in the following pages. In this way all evidence, which might assist in better understanding the significant biological process(es) of health and disease which are involved, becomes available for the most complete assessment possible.

Section B
Functional Anatomy of the Interphase Nucleus

V. Functional Anatomy of the Interphase Nucleus

A. Purpose

The cell is a fascinating, beautifully structured, incomprehensively labyrinthine, and intricately and effectively operative unit of life – in health. In disease, it can be strikingly bizarre and increasingly complex.

Much can be learned by simply observing properly fixed and prepared material with light microscopy. Yet, there is significantly *more* which can be understood about biologic behavior by *meticulous* examination of this *same* light morphology, by basing it upon what has been revealed through electron microscopy (Figs. I-1 through I-4), and by placing it in proper biologic perspective through numerous methodologies including immunochemistry, molecular biology, automated microscopy, and extensive long-term clinical follow-up studies.

The purpose of this chapter is to bring to the microscopist, using the finest resolution of the wave lengths of *light,* an accurate appreciation of the finer structure of the cell and of its biologic significance as revealed by these other techniques. In that way, one can imply and deduce more accurately, from light microscopic observations, the cellular and molecular biologic activities taking place.

This is for the ultimate use by the *light* microscopist to understand the greatest amount possible about human health and disease, to render accurate diagnoses and prognoses, to help interpret and direct research, and to assist in skillfully guiding proper therapy and assessing its progress. In this regard and with *research,* human *diagnosis* and *care* as the ultimate goal, this brief review of the cell's structure and function is presented.

The *nucleus* is key to our understanding of the general biologic activity of the cell. Thus, this chapter is devoted mainly to the *nucleus,* with *cytoplasm* where it specifically is indicated in the current task. The task at hand is to *integrate current understanding* of the major nuclear functions and the ultra structural organization for its physiologic activities (e.g.: formation of mitotic chromosomes, interphase chromatin packing and unfolding, repli-

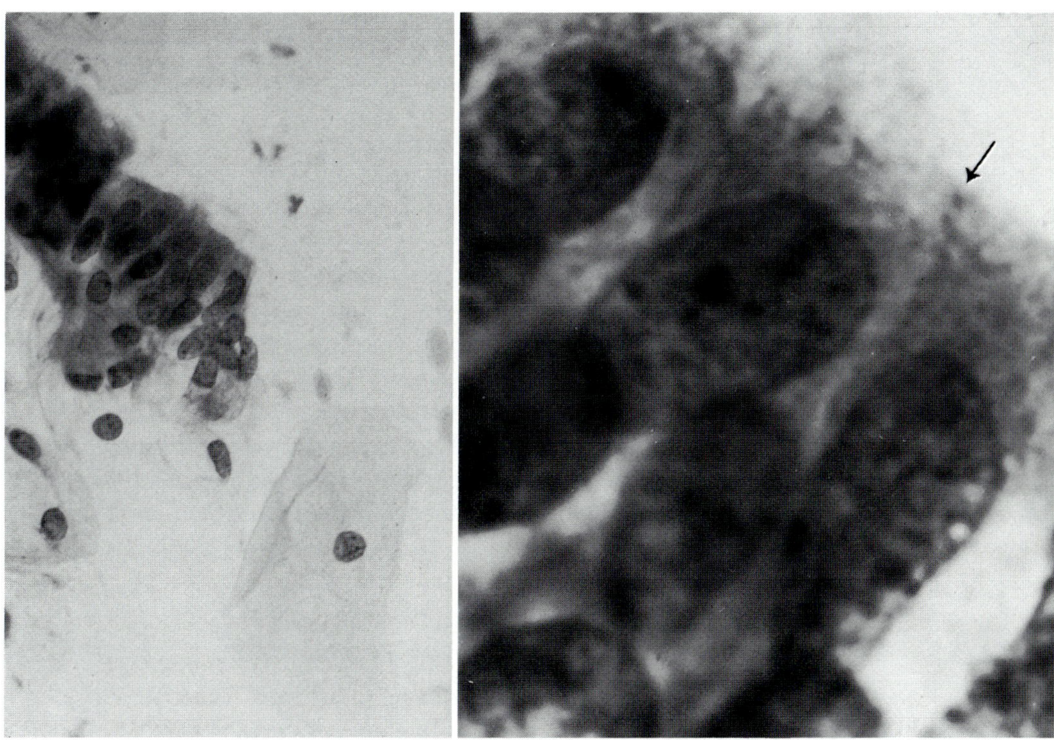

Figures V-1, V-2: Columnar cells. Fast (vagino-pancervical) smear.
Fig. V-1: Most of the cells in this sheet of endocervical epithelium are oriented with their epithelial luminal border along the upper right of the tissue fragment. An intermediate squamous cell lies in the lower right corner.
Fig. V-2: The tight junction between the first and second cell from the right (arrow), appears as a terminal bar at the luminal border. Note the prominent chromatinic rim (historical 'nuclear membrane'), one or two prominent nucleoli per nucleus, and the pattern of the chromatinic net uniformly dispersed throughout the parachromatin. Enlargement has been extended beyond the sharp limits of true resolution by light microscopy, in order to approximate the magnification of the electron micrograph, Figure V-3.
Papanicolaou stain.
V-1: × 470; V-2: × 3,250.

cation of DNA, DNA transcription synthesizing RNA, transport of RNA and other molecules, protein production, control of cellular activity) *into the framework* of human diagnostic *light* microscopy.

Advances in our knowledge of the cytoplasm have been equally significant. This involves functional differentiation, however, and is less critical for the purposes of this book than accurately assessing general biologic behavior. To maintain the size desired for this text, therefore, the reader is referred to Section C and to the references for the fine structures and physiology of the cytoplasm.

B. The Historical 'Nuclear Membrane' of Light Microscopy

The deeply hematoxylinophilic structure bounding the nucleus in light microscopy, has been referred to historically as the 'nuclear membrane' (Figs. V-1, V-2). It is now clear, however, that the nuclear membranes of the nuclear envelope are too thin to be visible to the light microscopist, and that this classical 'nuclear membrane' of light microscopy is *not* the true *nuclear membrane* (Fig. V-3), but is a complex structure comprised of the *inner nuclear membrane* of the nuclear envelope, the *nuclear lamina,* and the *peripherally condensed heterochromatin* – the *chromatinic rim* (Fig. V-4).

C. The Nuclear Envelope

The nuclear envelope is essentially a huge, dynamic, membrane cisterna which surrounds the entire nucleus during interphase. The cisterna contains enzymes and other materials in a position where they are optimally available to the chromosomes which are attached, in an orderly fashion, to its nucleoplasmic surface.

The nuclear envelope disappears, as such, throughout mitosis. At the time of anaphase it is extensively disassembled. During late anaphase and telophase, it is reassembled about the periphery of the condensed chromosomes of the daughter cells.

This membrane cisterna is a complex structure composed of many subunits, which are usually grouped conveniently into the *outer nuclear*

Fig. V-3: Columnar cells, bronchial epithelial biopsy. A low power, survey electron micrograph approximately matching the magnification of the light photomicrograph in Figure V-2, for comparison. At this magnification the lamina and inner nuclear membrane are not separately resolvable from the much thicker chromatinic rim which is visible in the light microscope as the historical 'nuclear membrane' (see Figs. V-1, V-2). In a few areas, the outer nuclear membrane and perinuclear cisterna can just barely be discerned. A prominent nucleolus (n) is present in most of the nuclei and the chromatinic net is uni-

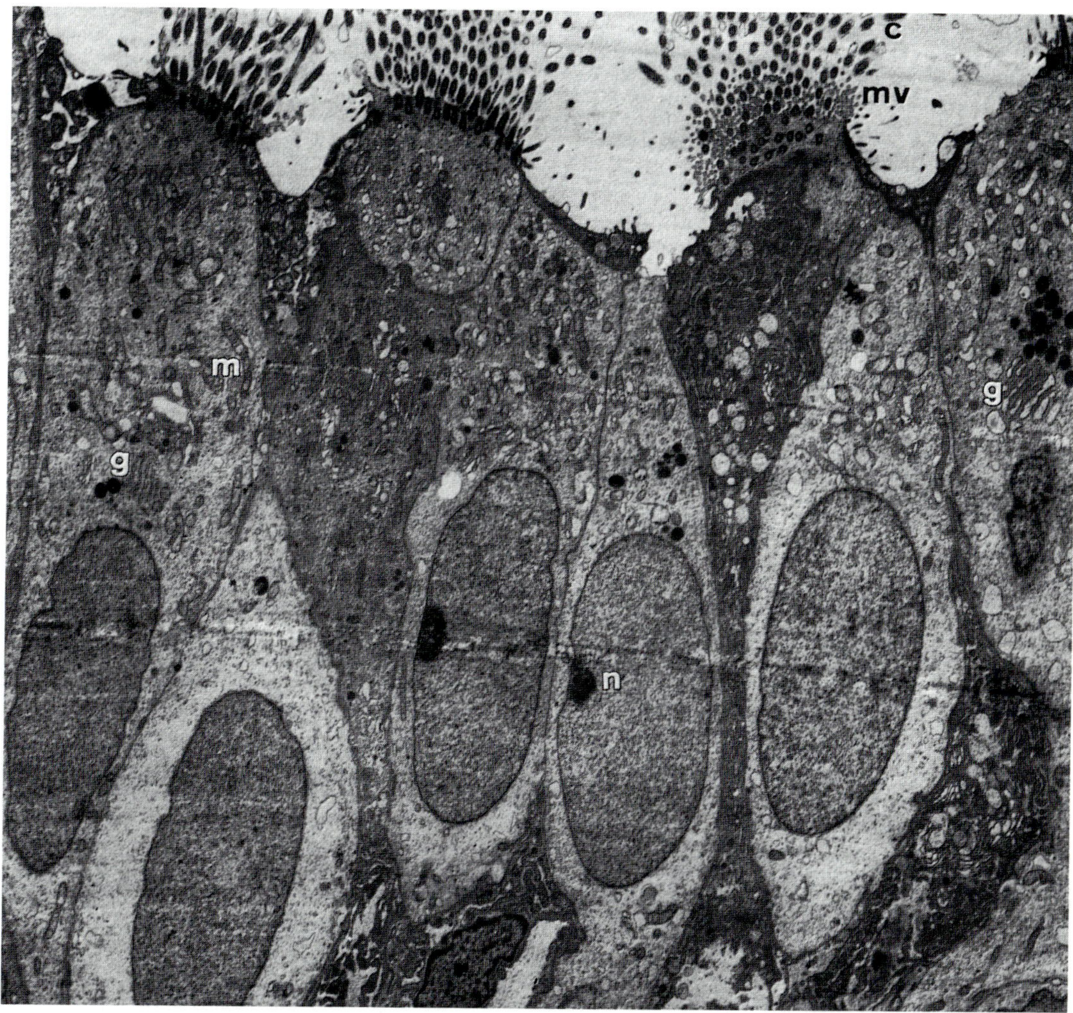

formly dispersed throughout the parachromatin. In the cytoplasm, most of the mitochondria (m) are elongated with prominent cristae forming transverse septa. Tubules and vacuoles of endoplasmic reticulum with its prominently clear cisternae, run throughout the cytoplasm. A Golgi complex (g) is recognizable above two of the nuclei as stacks of parallel, membranebounded cisternae. On the luminal surface are secretory vacuoles and secretory products, numerous branching microvilli (mv), and cilia (c) with their two central fibrils, nine outer doublet fibrils, and prominent rootlets.

Osmium/methacrylate.

V-3: × 4,000.

(Frost, Erozan, Donovan; unpublished electron micrograph.)

membrane, the *inner nuclear membrane,* numerous *nuclear pores,* the *perinuclear cisterna,* and the *nuclear lamina* (Fig. V-4):

– The *outer nuclear membrane,* on the cytoplasmic surface, is continuous with the membranes of the endoplasmic reticulum. It frequently contains ribosomes, as does the *rough* endoplasmic reticulum.

– The *inner nuclear membrane,* on the nucleoplasmic side, is continuous with the outer nuclear membrane at every pore. The nuclear lamina is firmly attached along its inner surface. In contradistinction to the outer nuclear membrane, the inner nuclear membrane contains *no* ribosomes.

– Numerous *nuclear pores,* usually roughly regularly spaced about the nucleus, are delineated completely by continuous reflections of the inner and outer nuclear envelope. The orifice of each pore is sealed by a complex protein structure which carefully monitors and controls transit of particles between the nucleus and the cytoplasm.

– The *perinuclear cisterna,* or perinuclear space, lies between these three structures and is continuous with the cisternae of the endoplasmic reticulum.

– The *nuclear lamina* is a protein coating on the inner (nucleoplasmic) side of the inner nuclear membrane and the margins of the pores, and to which are attached the peripherally marginated heterochromatic portions of the chromosomes. It is difficult to optically resolve the lamina from the inner nuclear membrane and from the outer regions of the peripherally condensed heterochromatin, unless the lamina is excessively thick.

1. Outer Nuclear Membrane

This electron dense, thin-walled lipoprotein membrane measures 5.0–7.5 nm (0.005–0.0075 µm) in thickness. It is too thin to be resolved by the wavelengths of light and, thus, is invisible to the light microscopist.

Fig. V-4: Columnar cell nucleus, bronchial epithelial biopsy. In this higher magnification, one can clearly resolve the finely irregular outer nuclear membrane (om); the perinuclear cisterna (nc); the inner nuclear membrane, lamina, and maginated condensed heterochromatic chromatinic rim (cr); the nuclear pores (np); the condensed heterochromatic chromatinic net (cn); the pale parachromatin with its finely divided euchromatin and heterochromatin barely resolvable as thin threads and granules (pc); and the nucleolus (n).

Osmium/methacrylate.

V-4: × 25,000.

(Frost, Erozan, Donovan; unpublished electron micrograph.)

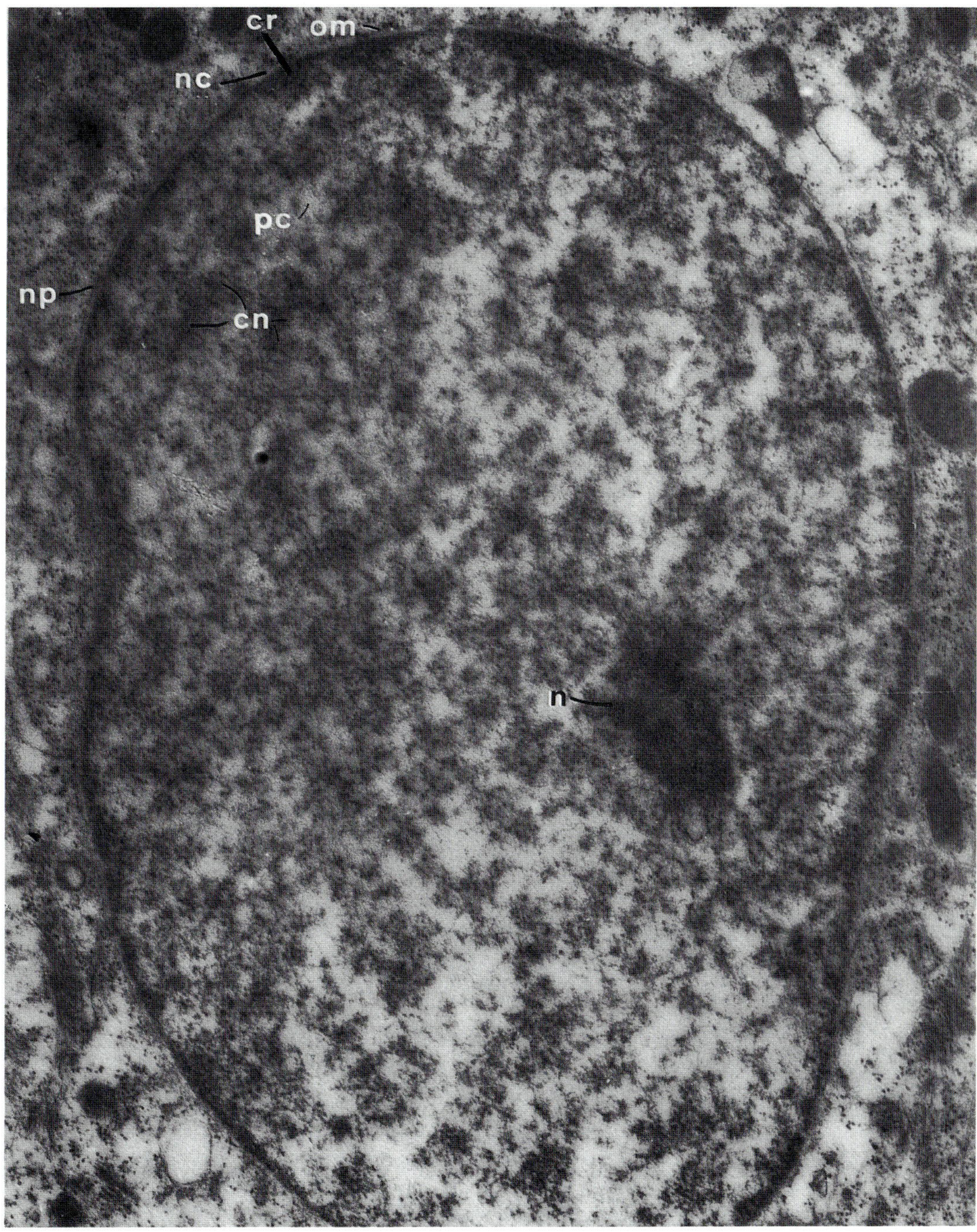
cr
om
nc
pc
np
cn
n

Figures V-5 through V-8: Nuclear envelope region, bronchial epithelium (see Fig. IX-9).

Fig. V-5: Euplasia, columnar cell. The outer nuclear membrane (om) is thin and erratically vacillating with occasional attached ribosomes (r). The inner nuclear membrane and lamina (im-l) are much straighter and more rigid, having very few irregularities or perturbations as does the outer nuclear membrane. There are some fine strands of protein in the perinuclear cisterna (pc) (possibly for support; possibly enzymes, etc.). The nuclear pore complex (np) on the right is a symmetrical protein structure monitoring transfer of substances between nucleus and cytoplasm. There appears to be at least one beaded fibrous bandlike structure extending from the right side of the pore annulus into the nucleus, as if guiding and assisting proribosomes to gain access to the pore [87]. The chromatinic rim (cr) consists of tightly condensed chromatin in a beadlike configuration attached to the inner nuclear membrane by the moderately thickened lamina.

Fig. V-6: Early retroplasia, columnar cell. The nuclear envelope is wrinkled, both inner and outer membranes. The chromatinic rim is extremely hyperchromatic and markedly condensed. There is a great loss of the normal granular structure of its condensed chromatin from compaction, and from *smudging* of the individual granule outlines and of the inner edge of the chromatinic rim (chromatin/parachromatin interface). The lamina is strikingly dark and prominent, as are the nuclear pores (np). There is increased protein in the perinuclear cisterna, with bleb formation at the left and on the right. The ribosomes of the outer nuclear membrane are smudged and indistinct in outline. There is marked clearing of most of the chromatin out of the parachromatin, apparently into the chromatinic rim and chromatinic net, with rounding up and agglutination of the few chromatin threads which remain in the parachromatin.

Fig. V-7: Proplasia, squamous metaplasia. The chromatinic rim is much thicker than in euplasia (Fig. V-5), yet the chromatin granules are distinct and not hyperosmophilic, as in retroplasia (Fig. V-6); nor is the inner border of the chromatinic rim either blurry or rounded as it is in retroplasia (Fig. V-6). The parachromatin is cleared, but the chromatin threads which remain therein are delicate, and appear to be well preserved and not agglutinated as in Fig. V-6. There is a prominent lamina. Both inner and outer nuclear membranes are thrown into undulating waves, increasing the surface area of the nuclear envelope. There is an accumulation of ribosomes on the outer nuclear membrane within the small courtyard (hof) protected by the nuclear folds, and with modest increase in the amount of proteinaceous material in the nuclear cisterna. There is a prominent nuclear pore (np) in midfield with a central granule and two fibrillary structures running from each side of the annulus into the nucleus.

Fig. V-8: Malignant neoplasia, invasive carcinoma of the lung. There are four nuclear pores (np) in this short expanse of envelope, indicating high metabolic activity. The thickness of the chromatinic rim varies markedly, and it is very osmophilic; but the granules constituting the rim are discrete, and are not blurred together as they are in Figure V-6.

Osmium/methacrylate.

V-5, V-8: × 63,000; V-6, V-7: × 65,000.

(Frost, Erozan, Donovan; unpublished electron micrographs.)

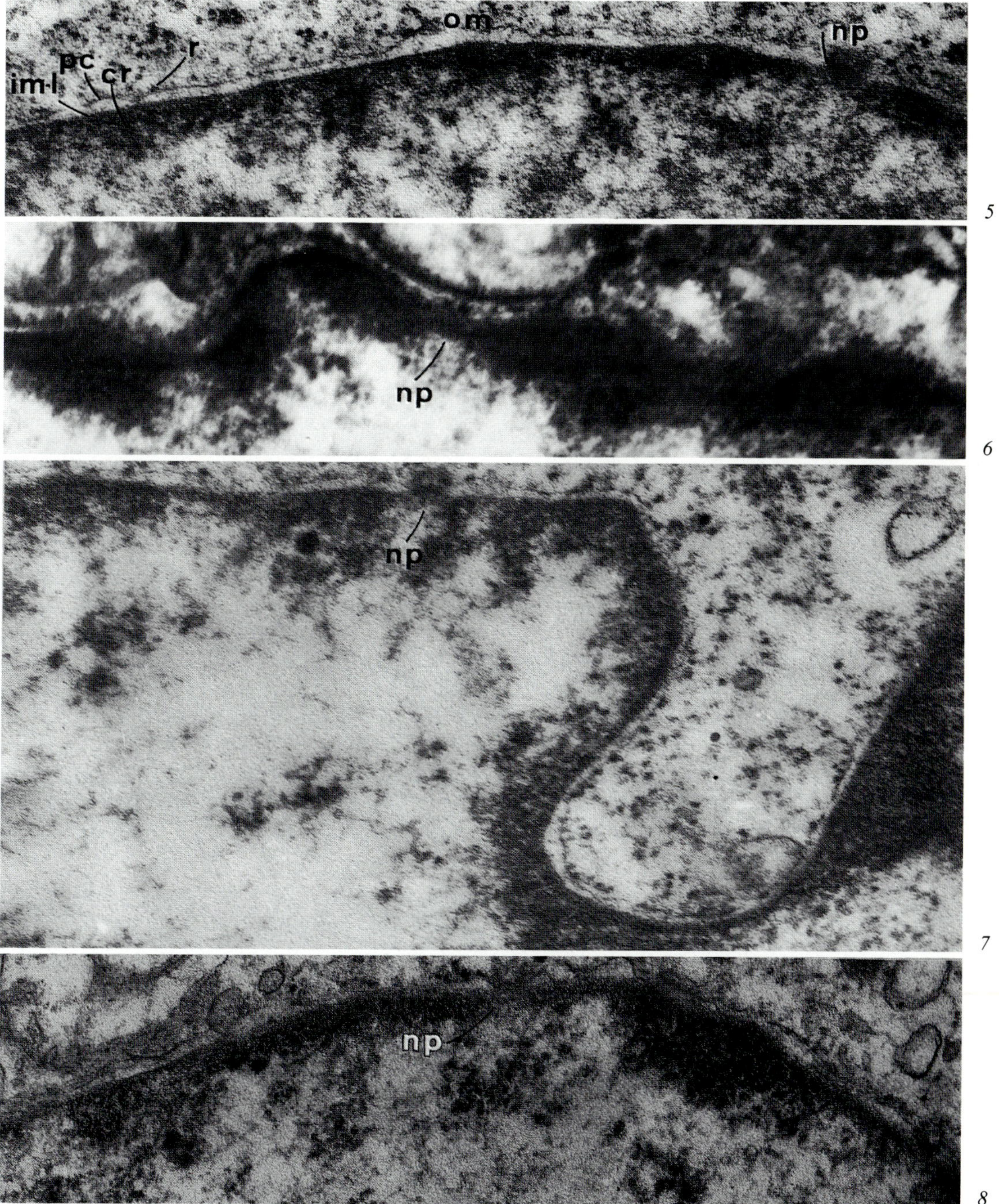
om
np
pc
r
im-l
cr
np
np
np

It is composed mainly of lipids and proteins. Its lipid content is somewhat lower than that of most other cellular membranes, but resembles the lipid pattern of the membranes of the endoplasmic reticulum. It thus stains with osmium, lead and other heavy metals.

Ribosomes frequently occur along this outer nuclear membrane, appearing similar to the rough endoplasmic reticulum (Fig. V-7). Occasional outbuddings of this outer nuclear membrane connect with the endoplasmic reticulum, giving the morphologic impression of a continuous system. Evidence from studies in cellular physiology bear out aspects of this concept of continuity, but also indicate significant differences in function.

The outer nuclear membrane, like the endoplasmic reticulum, is wavy; at times, it is extremely so (Fig. V-8). It can form blebs or secretory-type vesicules out into the cytoplasm, resembling the functions of the endoplasmic reticulum. It is much more prone to significant perturbations in shape than is the inner nuclear membrane which, usually, is held more rigid by the subjacent lamina and the peripherally condensed heterochromatin.

2. Inner Nuclear Membrane

Like the outer nuclear membrane, this is an electron dense lipoprotein membrane, of essentially the same thickness and composition. Also being too thin to be resolved by the wavelengths of light, it is invisible to light microscopy. It folds outward at the nuclear pores, meeting and continuing without interruption, as the outer nuclear membrane (Figs. V-5, V-7, V-8).

The exact constitution and properties of the inner nuclear membrane, are difficult to assess at all times with certainty. This is principally because the closely approximated subjacent lamina and the peripherally condensed heterochromatin, are difficult at times to separate for detailed analysis either by electron microscopy or by molecular biochemical techniques.

The *shape* of the inner nuclear membrane is more *uniform* than that of the outer nuclear membrane, apparently due principally to the rigidity afforded it by the subjacently attached lamina and the peripherally condensed heterochromatin. There are *no* ribosomes on the inner nuclear membrane. It is the inner nuclear membrane which gives the shape to the profile of the outer surface of the chromatinic rim in the light microscope, supplying information about the shape of the nuclear membrane. It is also the structure which keeps the outer surface of the chromatinic rim crisply defined, while the inner surface of the chromatinic rim (chromatin/parachromatin interface) blurs with degeneration (Fig. V-6) (Chap. VII).

3. Perinuclear Cisterna

This space about the nucleus is an electron lucent area between the outer and the inner nuclear membranes. Delicate fibers course between the two membranes (Fig. V-5), possibly affording structural support in maintaining its width and shape.

Variations in the width and shape of the space do occur, however. This is due more to the frequent undulations and irregularities of the outer nuclear membrane than to the inner nuclear membrane, which is held more uniform by the rigid subjacent lamina and the peripheral condensed heterochromatin of the chromatinic rim.

The perinuclear cisterna appears to be a channel in continuity with the cisternae of the endoplasmic reticulum, but with its own specialized contents rich in enzymes. It is situated in a key position to have great influence upon the genetic activity of the chromosomes which have attachments to the inner surface of the inner nuclear membrane, and upon the transfer of molecules between the two compartments, nuclear and cytoplasmic.

Rather than merely serving to restrain the nuclear contents, there may be a greater role for the two membranes and their cisterna: to contain a rich store of enzymes and other metabolites which are readily available to directly facilitate the active genetic and transport mechanisms going on within their environment.

4. Nuclear Pores

The inner and outer nuclear membranes of the nuclear envelope, are permeable to passive or active transfer of certain ions and molecules across the membrane. All particles, however, apparently are directed to the nuclear pores for nucleo-cytoplasmic transport.

At various locations fairly symmetrically distributed over the nucleus, numerous pores or passages in the nuclear membrane-cisterna-membrane complex are present for nucleo-cytoplasmic transport of particulate material. The nuclear membranes are molded intact around these areas, being reflected at the pore annulus toward each other, and being continuous with each other without interruption. The single large perinuclear cisterna thus is intact, continuous with the cisternae of the endoplasm, retained at uniform height by fine fibers, and interrupted only by a 'forest' of many short columns, which are the nuclear membranes reflecting about the annulus at the pore orifices.

a. Pore Orifice. A pore orifice consists of the inner nuclear membrane reflecting outward where, without interruption, it becomes the outer nuclear membrane. The inside diameter of this orifice (membrane to membrane) has a range from 60 to 120 nm, depending upon cell type or level of biologic activity; but tends to be constant in a given nucleus or within a population of one cell type. To carefully regulate the nucleo-cytoplasmic traffic, each orifice is completely plugged by a beautifully constructed proteinaceous pore complex [236].

b. Pore Complex. A pore complex is firmly attached to the nuclear membranes and the lamina, and completely fills its pore orifice. It includes:

1) An inner and an outer annulus, or ring, each composed of eight granules measuring 10–25 nm in diameter, symmetrically and evenly dispersed about the periphery of the orifice.

2) A central granule, rod, or point (Fig. V-7) from which radiate eight structures.

3) Eight cone-shaped structures, appearing to be groups of aggregated fibrils, extending from the central structure to the two annuli of eight granules and the nuclear membrane of the pore orifice.

4) Bundles of 4–8 nm wide filaments, which are attached to the eight granules of the inner annulus and which extend deeply into the nucleus, perhaps serve as a guide and an aid to transit of particles toward the pores (Figs. V-5, V-7), much as skeletal fibers do in the cytoplasm [6].

The number of pores per nucleus varies markedly, according to cell type and to biologic activity. While some huge ova have over 50 million pores and some quiescent cells have less than 100, most cells appear to average about 2,000–5,000 per nucleus. In the most quiescent cells, such as the late post mitotic cells of the erythroid series and of the more mature cells of keratinizing stratified squamous epithelium, they are extremely few.

The nucleo-cytoplasmic transit of particles across the nuclear envelope is poorly understood. The nuclear envelope appears to be passively permeable to the diffusion of many *ions* and *small molecules.* The passage of *particles* across it, however, is by *non*passive processes, and the pore complexes appear to be the exclusive, or at least the highly preferred, sites of transit.

Evidence strongly suggests that the whole diameter of the nuclear pore is not used for such transport events, but may be confined to a carefully monitored channel in the center of the pore complex with a diameter, when patent, of only 10–20 nm. Profound changes may occur in the shape and chemistry of structures during transit through the nuclear pore. For example, when the *acidophilic* proribosomes from the nucleolus have passed through the nuclear pores into the cytoplasm, they have become *basophilic* ribosomes. In times of increased biologic activity [120], material under active transport can be seen by electron microscopy to become congested and build up on one or both sides of the nuclear pore.

5. Nuclear Lamina

This electron dense proteinaceous structure is firmly adherent to the nucleoplasmic inner surface of the inner nuclear membrane. It markedly varies in thickness from below 5 nm to over 30 nm, depending upon cell type and level of biologic behavior. It is interrupted at the nuclear pores, but is firmly attached to the inner annulus of the pore complex.

It is difficult to separate and discriminate the lamina from the inner nuclear membrane, both biochemically and electron microscopically (Fig. V-7). Likewise, the lamina cannot be completely separated from the pore complex.

The lamina of numerous mammalian cell types consists essentially of a layer of strongly bound-together proteins, of which the predominant three have been recognized and named [117, 118, 257]. Lamin A (70,000 daltons) and lamin C (60,000 daltons) are structurally similar and bind strongly to *chromatin.* Lamin B (67,000 daltons) binds to the *inner nuclear membrane.* Recent evidence suggests that there additionally may be distinct (but related) lamins expressed in certain cell types (e.g.: germ cells). During mitosis, the nuclear membrane is entirely disassembled into small vesicular fragments which become widely scattered throughout the cytoplasm [117]. Lamin B remains membrane-associated to these fragments. Lamins A and C become monomeric and 'dissolved' throughout the cytoplasm.

Toward the end of mitosis, during late anaphase and telophase, the membrane vesicles with lamin B approach the vicinity of the periphery of the condensed chromosomes. Here the vesicles, gathering about the telophasic spherical mass of chromosomes, fuse with each other and with the nuclear pores, to reassemble into the spherical nuclear membrane.

In the meantime, also toward the end of mitosis, lamins A and C bind to sites on the chromatin and bring them to the nucleoplasmic surface of the inner nuclear membrane. In this way the terminal ends (telomeres) and the centromeric regions (with the kinetochores) of each chromosome [239], are firmly attached to the inner surface of the inner nuclear membrane by protein-protein interactions.

The chromatin which will then become the rest of the peripheral condensed heterochromatin, is then also bound to the inner surface of the inner nuclear membrane. It is less firmly attached and/or it lies more superficially (i.e.: toward the center of the nucleus), as it is able to be separated by less stringent investigative techniques. The binding appears to be at certain sites on the chromosomes to certain loci on the inner nuclear membrane. In this way, specific chromosomal areas are brought in at a time and portions of the same chromosome appear to be secured in the same vicinity of the nuclear membrane. Attachment sites probably occur periodically along the chromatin fibers, at about every 200–300 Kilobase pairs of DNA. The loops of each chromosome are confined to their topological domains within the interior of nucleus, not wound throughout the nucleoplasm at random [181].

Thus the nuclear lamina proteins appear to play a major role in the reassembly of the nuclear membrane during late anaphase and telophase, in the seeking and sorting out of specific areas of each chromosome at the end of mitosis, in the organized reattachment of the chromosomes to the inner nuclear membrane at the end of mitosis, and in the binding to or release from the nuclear membrane of elements of the chromosomes as dictated for biologic activity during interphase.

D. The Nuclear Matrix

There exists a skeletal scaffolding structure throughout the nucleus, referred to as the nuclear matrix, which remains *after removal* of all of the DNA, of the nuclear membrane and of the histone proteins [36, 37, 66, 115, 137, 151, 165, 171, 261, 293]. It is made primarily of nonhistone proteins and consists of a delicate fibrillary network.

The nucleoli are embedded in this scaffold. In some preparations, the nuclear pores are included; in others, only round rings suggesting annulae remain, with thin fibers extending inward therefrom.

In some preparations, the lamina remains as a part of the scaffold; in others, it is mainly removed. The lamin proteins appear, immunocytochemically, to be located exclusively in the region of the peripheral nuclear lamina, and not out into the more central scaffold.

How much of these structures is artifact of preparation and how much is the essential functional protein scaffold, is not clear at the present time. It appears to represent structural protein which gives additional rigidity and shape to the biologically *in*active heterochromatic chromatin of the chromatinic net and chromatinic rim. *In addition,* however, there is evidence that it provides *functional* definition and restriction to the biologically *active* euchromatic elements of the parachromatin, assisting enzymatically

and dynamically in breaking and rejoining the DNA molecule, in positioning its strands, and in imparting energy to it during transcription or replication [65, 66, 170, 171, 293].

The matrix appears to define, and to hold together, the base of each
loop of chromatin fiber which extends out into the nucleoplasm of the
parachromatin. In this way, each thus-defined domain of a chromatin fiber
may function in its biologic activities much as does supercoiled circular
DNA. While there is some evidence that DNA transcription of RNA from
DNA gene loci may occur at the periphery of these skeletal structures, most
work indicates that DNA transcription and replication both occur *away*
from the heterochromatin of the skeleton, in these chromatin loops out in
the *nucleoplasm* of the *parachromatin* [138].

E. The Chromatin

Historically, the term chromatin first was used to denote those nuclear
structures which stained heavily, (i.e.: are chromatophilic). With use it
came to connote basophilic chromatin (i.e.: basichromatin), those nuclear
structures which stain with basic dyes (e.g.: hematoxylin). Currently, *chromatin* designates the *DNA-protein complex* of the chromosomal material of
the *interphase nucleus.* In the *mitotic* and *meiotic cell,* this material is called
chromosomes. The DNA of the chromatin is the genetic material of the
nucleus. It contains, in genetic code, essentially all of the information
inherited from cell to cell. A relatively small amount of additional genetic
information for mitochondria is in the relatively minute amounts of mito-
chondrial DNA within those organelles.

A *gene* is a segment of DNA which is responsible for the eventual
synthesis in the cytoplasm of a given protein (i.e.: polypeptide chain). A
gene is considered to include the entire sequence which is transcribed and
represented in the messenger RNA (mRNA) for that protein, including all
coding sequences (exons) and intervening sequences (introns) as well as the
regions preceding (leader) and following (trailer) the coding regions. A
genomene is one haploid set of chromosomes with the genes they contain.

The *DNA* of this DNA-protein complex, stains specifically with certain
reactions or dyes (e.g.: Feulgen, propidium iodide, acridine orange) [64,
180, 211]. On the other hand, hematoxylin stains the *protein* of the complex
[19, 211]. Hematoxylin is not a simple dye. It is a complicated mordanted

dye with many compounds which stain, but whose most active staining component is one of its oxidation or 'ripening' products, hematein.

At the *acid* pH employed during staining or destaining, hematoxylin is *soluble* and appears *red-purple.* Upon 'blueing' at a *basic* pH, hematoxylin becomes *blue-purple* (so-called 'black'), in which state the stain becomes *insoluble* and permanently 'locked' into the structures.

Chromatin, therefore, has a strong affinity for hematoxylin, or is *hematoxylinophilic* (red-purple to purple to blue-purple, depending upon its pH). This is in striking and valuable contrast to the weaker affinity for hematoxylin exhibited by the *nucleolus,* which diagnostically can be made to stain *acidophilically* (i.e.: red to orange to yellow; gold to brown) if consistent staining techniques are utilized [100, 124].

However, nucleoli can *also* be stained by hematoxylin when there is overstaining by hematoxylin or improper staining with weak or old counterstains, or they are so small or weakly acidophilic that they are obscured by massive nucleolus associated chromatin. This biologically important discrimination between *chromatin* and *nucleolus* is based *entirely* upon spectral parameters (i.e.: *hematoxylinophilia vs. acidophilia,* respectively) under proper staining conditions, and does *not* involve spacial or photometric parameters (e.g.: size, shape, number, position, degree of chromatophilia). When definite recognition of these structures is important, therefore, proper and reproducible staining [7, 8, 100, 123, 124] is mandatory.

1. DNA Molecule and Associated Proteins

Purified DNA (deoxyribonucleic acid) is a long, thin, thread-like molecule measuring about 2.0 nm (20 Å; 0.002 μm) in diameter [3, 294]. It is formed by the helical coding of two polynucleotide chains (or strands) of DNA, each consisting of a backbone of 5′-3′ sugar phosphate links, from which the purine-pyrimidine base pairs (guanine + cytosine) (adenine + thymine) project inward to connect the chains.

The strands intertwine by gently twisting about each other in a double helix, making a complete turn approximately every 3.4 nm along the molecule. The distance between adjacent nucleotides is 0.34 nm, so that there are about ten nucleotide base pairs (bp) per turn (9.3–11.0; aver. 10.4 bp).

The diameter of the molecule is a constant 2.0 nm. Thus, the nucleotides across the two sugar-phosphate backbones face inward and are always paired, the larger purine (i.e.: guanine or adenine) paired with the smaller pyrimidine (i.e.: cytosine or thymine). The proportion of the amounts of guanine (G) and cytosine (C) to adenine (A) and thymine (T) are always the same. Thus any DNA can be expressed by its ratio of (G + C) to (A + T), ranging from 26 percent to 74 percent, but characteristic for each species.

The strands intertwine into a right-handed double helix. The coils turn clockwise looking along the helical axis. There are 10.0 to 10.7 base pairs per coil of the helix, depending upon factors of physical configuration and chemical condition (i.e.: torsion, hydration). In this way the coiled DNA is configured so that on the surface of the molecule there is a *wide* and *deep* helical groove measuring 2.2 nm across, alternating with a *narrow* and *shallow* helical groove measuring 1.2 Å wide.

This *helical coil* configuration is critical to the subsequent higher order coiling which occurs with packing into condensed heterochromatin (e.g.: chromocenters, granules, chromatin clumps of light microscopy) of the chromatinic rim and chromatinic net. Furthermore, in these grooves the DNA-complexed proteins can associate closely or intercalate with this huge DNA molecule, reacting or nestling deeply into its interstices. In the same way, certain dyes (e.g.: acridine orange) [64, 149, 180] can intercalate into this DNA-protein complex.

The *proteins* most closely and frequently complexed with the DNA double helix, are the *histones* of the nucleosomes. Some of these relatively simple water soluble proteins with high proportions of basic amino acids, are highly conserved throughout transcription, as well as throughout evolution of the eukaryotes. This implies that their role may be central and unchanging. Other proteins are also closely associated with chromosomal DNA. These include *enzymes, repressors, activators* (e.g.: transcription factors), and proteins of the *matrix* and *lamina.*

2. DNA Condensation and Packing

Chromosomal DNA with its associated proteins, exists in the interphase nucleus in configurations ranging from extremely *extended* forms (e.g.: those in *active* transcription or replication) to highly *condensed* structures of *inactive* heterochromatin. In mitosis and meiosis it reaches very high levels of condensation, with packing of around 7,000 times (Fig. V-9), into the traditionally recognized chromosomes of the species (46 in man: 22 pairs of autosomes and two sex chromosomes) [18, 21, 22].

With this 7,000:1 packing ratio, 1.0 μm of chromosome length represents 7,000 μm (7.0 mm) of DNA length. There are 1.8 meters (1,800,000 μm) of human DNA (6×10^9 base pairs) which must fit into a nucleus whose diameter may be less than 6.0 μm.

A *critical concept* regarding the ways available to the cell for accomplishing this, has been suggested with increasing clarity by many workers using multiple and diverse experimental approaches [18, 21, 22, 44, 45, 56, 57, 65, 78–80, 85–87, 117, 138, 143, 171, 177, 192, 200, 212, 234, 259, 277–279, 290, 293] and is currently reaching acceptance:

Chromosomal DNA in the chromatin of the interphase nucleus and in the chromosome of the mitotic or meiotic cell, at least in the lower order packing, is confined to precisely-ordered confirmations which depend upon biologic activity and environmental conditions. There is recent evidence suggesting that higher order chromosome organization may not be so precisely defined [181]. Rather it may have probabilistically 'favored' configurations, and the higher order structure may depend upon the state of differentiation of the cell.

It is obvious from all of this that DNA condensation and packing is *not* random, but is highly organized. Because of the different configurations of chromatin clumping and chromatin pattern with different biologic behavior and even though only parts of the story are known today, what is known will be here summarized and ordered from the smallest unit, the nucleosome, up to the largest or highest order packing and clumping which is visible by the light microscope as orderly structure in health and macabre formations in disease.

3. Nucleosome

The nucleosome is the basic subunit of chromatin. Its existence was suspected for some time; but it was initially described biochemically in 1973, and morphologically with the electron microscope in 1974.

Fig. V-9: Human chromosome at metaphase. This chromosome pair is of Group G (i.e.: 21–22), two of the five pairs of chromosomes with satellites from their short arms, which contain the nucleolus organizer regions. These are acrocentric chromosomes, as their centromeres are extremely close to one end, producing very short arms. The chromosomal DNA and associated proteins are condensed extremely tightly, at a packing ratio of 7,000 to 1. The thinnest fibers in this figure (0.5–1 mm in width) are the so-called 10 nm chromatin fibers. Many under tension are seen to unravel as 'beads-on-a-string' due to the bulging of the nucleosomes (halfway toward 2-o'clock of the figure) (b). The somewhat larger fibers (1.5–2.5 mm width) are the so-called 30 nm chromatin fibers, which are the basic fibers of the interphase chromatin as well as of the mitotic chromosomes.

V-9: × 100,000.

(Electron micrograph; courtesy of Bahr, G. [18].)

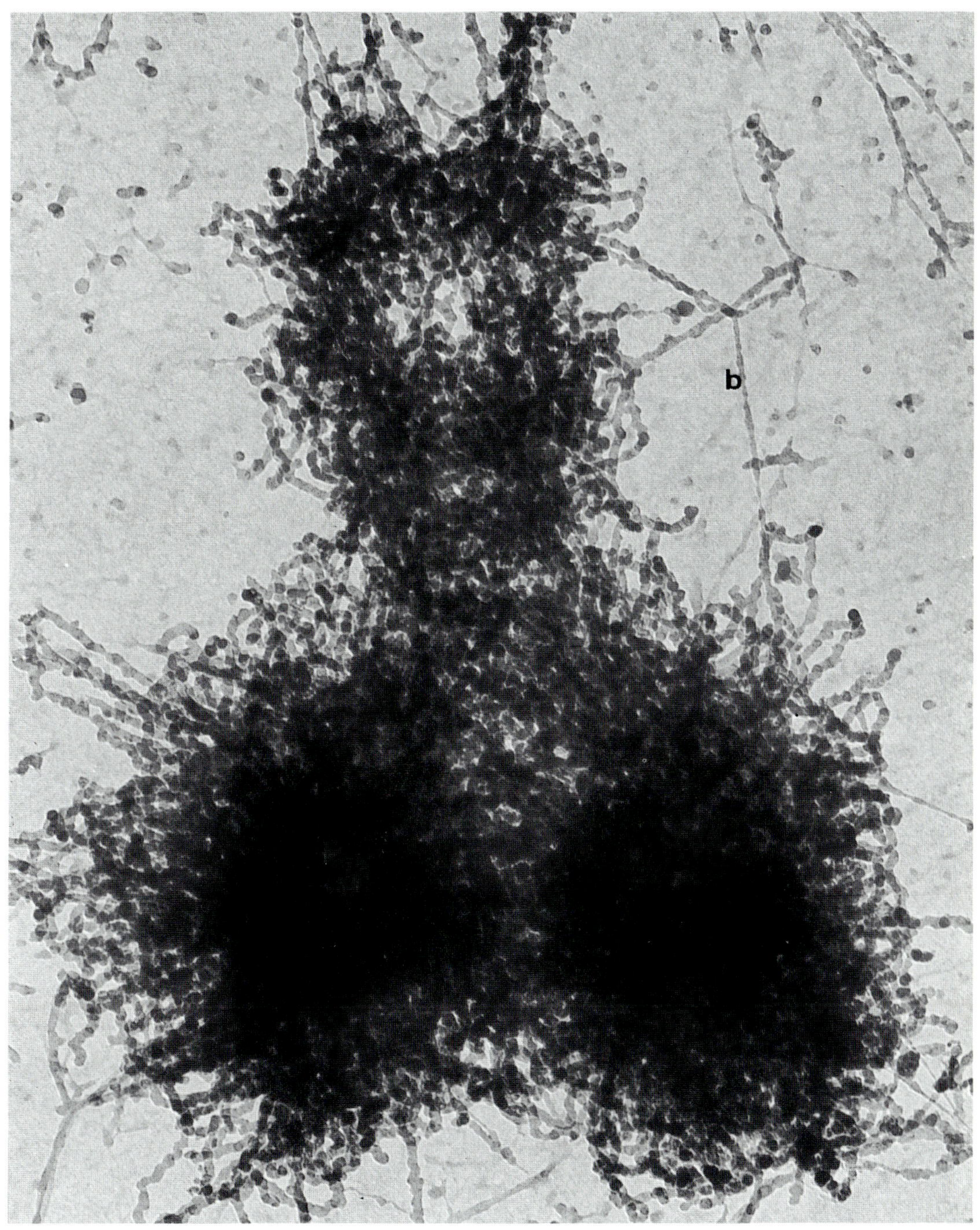

It is not a static entity, but undergoes changes associated with activity (e.g.: transcription, replication, repair), as well as in the formation of higher order condensation structures. In the later formations, or heterochromatin, the chromatin is biologically inactive.

The nucleosome is the first order of coiling (above the helically coiled duplex DNA molecule) in the hierarchy of confirmations into which chromosomal DNA is packed. Biologically, it is an extremely important unit in the processes of transcription and replication, as well as in chromatin condensation.

The nucleosome measures approximately $6 \times 11 \times 11$ nm. It is essentially in the shape of an oblate cylinder, having a diameter of 11 nm and measuring 6 nm along its flattened axis. To appreciate its size, one nanometer (nm) is one billionth of a meter. It is thus one millionth of a mm, one thousandth of a micron (μm), or 10 angstroms (Å). An average sized Barr body (sex chromatin) measures about 1 micron (μm) in diameter.

The core of the nucleosome consists of eight molecules of histones, two each of H2A, H2B, H3, and H4. Around this octomer core, the DNA molecule (core DNA) coils in approximately two turns involving about 146 base pairs (bp) of DNA, in a packing ratio of about 6. The strand of DNA between the coils, the linker DNA, varies in length from 8 to 114 bp per nucleosome. With a structure periodicity of the core DNA around 10 bp per coil, this indicates approximately 14 coils per nucleosome, or about 7 DNA coils per one turn about the nucleosome core.

Histone H1 is not one of the octomer core histone molecules. On the other hand, it plays a major role in the assembly and disassembly of the nucleosome, as well as in the higher order coiling encountered in the packing above this level. The DNA coils are locked into place about the core by H1, apparently sealing its central globular region (residues 40–116) to histone H2A of the core, and binding its two terminal regions to the DNA molecule at its points of entry and exit about the core. These two linker DNA segments, going into and coming out of the nucleosome, extend between adjacent nucleosomes.

4. 10 nm Chromatin Fiber – The Nucleosomal Fiber

In this fashion, strings of nucleosomes are bound tightly together in the nucleosome fiber of approximately 10–12 nm diameter. At times, due to physical factors (e.g.: stretching [22]) or chemical factors (e.g.: low ionic strength, low H1 concentration), the linker segments between the nucleosomes relax and the nucleosomes bulge out along the nucleosomal fiber as 'beads-on-a-string' (Fig. V-9). In other situations (e.g.: relaxation, increased ionic strength, adequate H1) linker segments are brought up, the nucleosomes close in together, and a uniform 10–12 nm diameter fiber is formed.

This so-called 10 nm fiber, or nucleosomal fiber, is the basic chromatin fiber. Thus, while histone H1 is not considered to be a nucleosomal histone (i.e.: H2A, H2B, H3, and H4), it does appear to be the key protein to the formation of the nucleosomal fiber and to many of the higher order configurations.

5. 30 nm Chromatin Fiber

The next higher order of coiling in chromatin condensation, is a family of fibers measuring 20–40 nm in diameter (Fig. V-9). As the actual measurement depends upon many factors and conditions (e.g.: relaxed, stretched, on film, in solution, ionic strength) under which it is visualized and measured (e.g.: electron microscope, low-angle x-ray scattering), these are all felt to be one 'family' in the next higher order of chromatin packing, and are usually referred to as the 30 nm chromatin fiber. This fiber is the basic constituent of interphase chromatin and mitotic chromosomes.

To form this 30 nm fiber, the currently favored model is for the 10 nm nucleosomal fiber to be further coiled into a tight helix, with a hydrated outside diameter of about 34 nm and inside diameter of around 10 nm across a long central hole running down the core of this solenoid. Proteins probably fill the hole. There is about a $40\times$ compaction along the axis of the solenoid core. Thus the packing ratio of the 30 nm fiber is about 40, with about 6 nucleosomes per turn of the coil.

There are approximately six nucleosomes per turn of the nucleosome fiber, but their orientation is not known for certain. Some evidence suggests that the flat faces of the oblate cylinder of the nucleosome are oriented approximately parallel to the axis of the solenoid, making them perpendicular to the long axis of each nucleosomal fiber. Other work indicates that they lie with their flat faces approximately perpendicular to the solenoid axis and, thus, parallel to the axis of the nucleosomal fibers. The latter configuration would appear to facilitate packing, with the linker segments drawn tightly across the periphery of the interior core, and would not necessitate a shift of 90° in orientation of the nucleosomes in their fiber in going from the 10 nm to the 30 nm configuration.

Histone H1 plays a key role in the formation and stabilization of this structure. They, and perhaps other proteins to be identified, appear to fill the hole running down the solenoid and spaces around the coiled nucleosome fiber, holding the 30 nm fiber firmly together even under stress. With reduction of H1 concentration and manipulating ionic strength, the 30 nm fiber disperses into the 10 nm nucleosomal fiber [22, 192, 279].

6. Higher Order Packing of Chromatin

Very little is known about the mechanism and fine structural details of the higher order packing of chromatin. This is unfortunate for the light microscopist, as it is at this level of packing that the heterochromatin becomes *visible* in the light microscope as chromatinic rim, chromatinic net, and nucleolus associated chromatin in threads and chromocenters (e.g.: granules, clumps). One assumes that in health it is as orderly as is the lower order packing forms (above), but there is little to indicate mechanisms responsible for the grotesque and macabre forms encountered in malignant neoplasia (Chap. IX).

Histone H1 appears to play a role also in forming and stabilizing these higher order conformations in addition to its central responsibilities for structure of the lower order packing. When H1 concentration is reduced, the structures break down to lower order configurations. Ionic strength manipulations also affect higher order structure formation reaching, curiously, a point at 80 mM NaCl where the H1 of the higher order configurations becomes masked to anti-H1 antibody [222, 274]. The contact sites between H1 and the core histones, other than the primary H2A contact at the level of the nucleosome, are more prominent with chromatin in the higher order structures [44].

Different *subfractions* of H1 differ greatly in their ability to condense nucleosomes [166]. Changes in their relative amounts have been noted in nondividing cells [213], during embryonic development [255], and in different organs [256]. These could well be reflected in fine differences of the higher order structure observable in light microscopy.

Much recent work on the topoisomerases appears to indicate a major role for these proteins, not only as a structural component of the mitotic chromosome scaffolds, and the nuclear matrix, but also in the packing of chromatin in the interphase nucleus [65, 66, 170, 171, 293]. Topoisomerases function to catalyze the breaking and rejoining of the DNA molecule during biologic activity (i.e.: replication, transcription). They are also key in altering DNA morphology, and in establishing and maintaining the topology of the chromatin structure.

If one could extrapolate from the configuration of what is known in the first three levels of chromatin condensation, one might anticipate when an order might be reached which is large enough to be visualized by the light microscope. The *nucleoprotein fiber,* averaging 4 nm in diameter, goes to the *10 nm fiber,* with an average increase in width of 2.5×. Folding to the *30 nm fiber,* is an increase of 3×. Hypothesizing an increase of 3.3×, would bring the *fourth order* fiber to 100 nm width, just barely at the theoretical lower limits of light microscopy resolution. The next order, with an increase of 3.5×, brings the hypothetical *fifth order* fiber to 350 nm (i.e.: ⅓ μm) which is well within the visible range of a good light microscope. For size references, each square in Figure I-5 is ¼ μm. The Barr X-chromatin body of a normal intermediate cell is approximately 1 μm in diameter [99].

7. Mitosis and Chromosomal Packing

Increasing the ionic strength of the fluid in which the chromatin fiber (nucleofilament) is bathed, results in its progressive folding to thicker structures. Evidence suggests that this progresses from the 10 nm nucleofilament consisting of a row of nucleosomes touching each other [78, 80], through loose filaments about 25 nm wide consisting of nucleosomes in an open zigzag arrangement [277] and 'super beads' of 20–30 nm apparently formed by folding of the 10 nm nucleofilament into a solenoid with about 6 nucleosomes per helical turn [78], into still higher-order helical structures with a fairly constant pitch but with increasing numbers of nucleosomes per turn (Fig. V-9) [279], and the 50 nm fiber consisting of radially arranged loops of 25 nm fibres [391].

The mechanisms of this chromatin compaction and the reasons for the structure and for the morphology of the resulting chromatin aggregations observed by light microscopy, thus are poorly understood. This is especially true of the orders above the 30 nm fiber, where virtually nothing is known of the mechanisms for this highest order.

Structural configuration of these higher orders of condensation most probably is dependent upon many other factors, about which very little is known. These include the nuclear matrix substances in which the chromosomal material is embedded or suspended, or to which it is adsorbed, and by the types and methods by which it is fixed [192].

8. Topological Organization of the Chromatin

As the nuclear membrane reforms in late anaphase and telophase (see C. 5., 'Nuclear Lamina'), organization of the chromosomes of mitosis into the chromatin of interphase, begins.

The monomeric proteins, lamin A and lamin C, move from 'solution' in the cytoplasm to the region of the condensed chromosomes. Here they find the centromere and the two telomeres (terminal ends) of each chromosome, bringing them to the inner surface of the newly forming inner nuclear membrane. They attach them in an orderly fashion to the nucleoplasmic surface of the inner nuclear membrane and to lamin B, which has remained associated with the membrane vesicles out in the cytoplasm during mitosis. Thus commences reconstitution of the nuclear lamina.

The portions of the chromosome which will become part of the peripheral condensed heterochromatin of the chromatinic rim, then move toward the inner nuclear membrane and become attached by the lamina proteins. This includes satellite DNA, which may come up to the membrane with its closely associated centromeric regions.

In man, the nucleolus organizer sites are located on the 5 satellite chromosomes, apparently in the satellite regions (Fig. V-9). There is some evidence from serial section reconstruction [45] to suggest that a part of each nucleolus may always be in contact with the nuclear envelope.

There is evidence that the peripheral heterochromatin of a given chromosome, marginates and attaches to the inner nuclear membrane in the vicinity of the attached telomeres and centromeres of the same chromosome. Thus there is suggestive evidence that in the normal human nucleus there may be, hypothetically, 46 areas of variable size on its inner nuclear membrane which are each taken up by the marginated heterochromatic portions of one chromosome attached to the inner nuclear membrane by the proteins of the lamina, constituting the chromatinic rim. The organization of loops of chromatin is not clear at the present time, but the 'radial loop model' on the structure of the mitotic chromosome is quite strongly favored [179]. This loop organization of the mitotic chromosome (Fig. V-9) must reflect that of the interphase chromatin in some fashion.

Loops of chromatin extend out from the protein of the skeletal structure into the nucleoplasm of the *parachromatin*. They vary markedly in length. Those which have been carefully measured [138] are 10–180 Kilobase pairs (Kbp: one thousand purine-pyrimidine base pairs; see E. 1.: 'DNA molecule and Associated Proteins') (3–60 µm) long. This corresponds closely with lengths of biologically key segments, e.g.: the length of replicons, which are units of a genome in which DNA is replicated, 10–50 Kpb; DNA regions which bear families of adjacent coordinatingly transcribed genes (e.g.: mouse β-globingene) 30–40 Kbp; mean lengths of DNA loops in the metaphase chromosome of HeLa cells (42 Kbp) [160]. Thus a single DNA loop coming from the proteinaceous matrix, may well represent a family of genes of coordinated expression, or domain, with the state of folding of such a chromatin loop determining the expression of the genes it contains. It is probable that this interphase chromatin loop organization reflects the organization of the chromatin loops of the mitotic chromosome (Fig. V-9).

There are secondary areas of chromatinic condensation in the large loops, each embedded in protein from the skeletal framework. There are undoubtedly longer loops than those measured by current techniques, whose secondary areas of heterochromatic condensation form the basis for the threads, granules and clumps of the chromatinic net.

Thus, loops of chromatin extend *from* their protein attachment sites on the nuclear skeleton *into* the surrounding nucleoplasm of the parachromatin. These loops extend both from the peripheral heterochromatin, attached by the lamina proteins to the inner nuclear membrane (i.e.: from the chromatinic rim), and from the secondary heterochromatic islands in proteins of the matrix (i.e.: from the chromatinic net).

Both DNA transcription of RNA and DNA replication, occur out in these loops of chromatin bathed in nucleoplasm *within* the *parachromatin.* Nascent RNA chains with associated protein are found within the parachromatin, hanging off of the chromatin loops [138, 161], as they are formed, and away from the skeletal structure. DNA replication in these constrained loops, which are firmly bound as they emerge from and return to the matrix, may be comparable in a topological sense to that of the circular DNA of a virus [268] in its replication.

Fixation of the ends of an active segment of DNA into an effective circle, or loop, would then allow supercoiling to occur [260, 262]. This apparently would add energy to the DNA molecule, open up areas for increased accessibility to the molecule for biologic activity, and separate strands to facilitate transcription and replication.

F. The Nucleolus

The *nucleolus* is involved with the production of *protein.* Its discovery has been attributed to Fontana (1781), an Italian biologist. It is frequently the most conspicuous structure in the nucleus (Figs. V-1 through V-4) and, as a result, has provoked much interest. It is only in the last decade or so, however, that its true function and nature have been elucidated.

The nucleolus is actually the site for the production and processing of the ribosomes up to their penultimate forms (proribosomes). They then pass through the nuclear membrane, assemble into their final form, and undertake the production of protein either as free ribosomes, mainly for intracellular proteins, or *attached* onto membranes (i.e.: rough endoplasmic reticulum, outer nuclear membrane) for the formation of protein [242] for extracellular use (e.g.: secretion) or use in the membrane system (e.g.: nuclear enzymes, histones).

In the Papanicolaou stain, the nucleolus is *acidophilic* (red-orange-yellow; golden-brown). This is in diagnostic contradistinction to the *hematoxylinophilia* of the chromatin.

When the ribosome then enters the cytoplasm, it is *basophilic* (blue; green) in contrast to its acidophilia in the nucleus. Thus a basophilic cytoplasm is a hallmark of a cell actively making protein, as is a prominent acidophilic nucleolus. This paradox is apparently due to change in configuration of the proteins associated with the proribosomal and ribosomal RNA, as it transverses the nuclear pore, with acidophilia changing to basophilia, or in the RNA/protein ratios in the two compartments.

1. Formation of Nucleoli

Nucleolar organizers are definite sites upon specific chromosomes from which nucleoli are formed. They are particular chromosomal regions, corresponding to a cluster of ribosomal RNA genes, which are repeated tandemly, one after another, along the DNA molecule. Thus, in a diploid nucleus there are twice as many nucleolar organizers as there are in the haploid genome. The repeating unit in mammals is about 43,000 bp (43 Kbp) of DNA. Only 13 Kbp of it is transcribed, with a nontranscribed spacer of 30 Kbp.

In man, nucleolar organizers occur on the five pairs of acrocentric satellite chromosomes (nos. 13, 14, 15, 21 and 22) (Fig. V-9), so that there exists the possibility for a normal diploid cell to have a *maximum* of *ten* nucleoli. In actuality, this high number is rarely attained because of repression or fusion of nucleoli, except in a few well-chosen situations.

In extreme growth, such as in embryonal tissue, the number of nucleoli per cell can be seen to increase up to ten in early mitosis, during *prophase* [15]. At this time in the cycle the nucleoli dissociate, dissipate, and then disappear. The chromosomes separate and condense into the individual chromosomes of mitosis.

Nucleoli again reappear in late mitosis. Ten first appear in *telophase,* and then fuse together or are repressed, to reduce the number down to that which is characteristic for that cell type at that level of biologic activity which is required during interphase.

During normal mitosis, nucleolar material is not recognizable in the light microscope. It dissipates throughout the cell as large RNA and protein macromolecules, which retain their critical associations and do not degrade throughout cell division. They then regroup at the sites of the nucleolar organizer, reassociate into their order of maturation, and reorganize back again into the complex machinery of the nucleolus.

During mitosis there appears to be a close association between the acrocentric satellited chromosomes bearing nucleolus organizing regions, a phenomenon known as satellite association. There is evidence for interconnecting fibers between these regions [71]. There is a strong component of the fibrillary nuclear matrix which is associated with nucleoli, and could conceivably have a role in their fusion and in reducing their number.

2. Morphology of the Nucleolus

The nucleolus is a highly organized factory – from pattern to product. The high concentration of the tandemly repeated rRNA genes, their very intensive transcription, and the complicated and lengthy processing of the products, creates the characteristic morpholgy of the nucleolus.

Examined by the *electron microscope,* a nucleolus consists of three regions, or components [73]. In the *medulla* is the more central *pale staining component* (Fig. V-10), and a surrounding *dense fibrillary component*; more peripherally located is the *granular component* of the *cortex*. The latter at times is in a reticular form, as a branching, anastomosing, coarse threadlike structure called the nucleolonema. At other times, depending upon the state of the cell's protein metabolism, it is compacted as a diffuse granular body about the outside of the nucleolus (Fig. V-10).

In the *light microscope* it appears as a diffusely acidophilic body (Figs. V-1, V-2), with both the Papanicolaou stain and the hematoxylin and eosin stain, except for hematoxylinophilic nucleolus associated chromatin. The anastomosing branching *nucleolonema* are recognized with silver impregnation (corresponding to the cortical *granular component* of electron microscopy). In the meshes of the latter are entrapped clearer, rounded areas called *pars amorpha* or *fibrillary centers* (the medullary *pale staining* and *fibrillary components* of electron microscopy, Fig. V-10).

3. Biosynthetic Function of the Nucleolus

The ribosomal RNA (rRNA) is transcribed from DNA templates in the nucleolar organizer region of each of the ten acrocentric chromosomes with satellites (i.e.: number 13, 14, 15, 21, 22 pairs) (Fig. V-9). In each human diploid cell there are approximately 400 rRNA gene copies spread out in small clusters on the ten different chromosomes. The templates are repeated numerous times at each site and occur in tandem fashion along the DNA strands of the nucleolar organizing regions of the satellite areas.

These areas loop out into the nucleus; the base of the loop is fixed in the nuclear matrix of the nucleolus so that the active euchromatic portion of the loop is held in the *pale staining component.* Here the rRNA is transcribed, promptly packaged with its associated proteins (which are synthesized in the cytoplasm and then move into the nucleus) and appears as the fine 5 nm thin ribonucleoprotein fibers of the *fibrillary component.* The fibrillary transcription products with their associated proteins, are then processed in the *granular cortex* as 15 nm granules.

The *processing* in the *nucleolus* of these fibers is very intricate. The very long rRNA transcription product (45S rRNA precursor) from the ten potential nucleolar organizing regions, is trimmed of excess segments (which are discarded and broken down so that their building blocks can be reused) and the remainder is cleaved into three fractions (i.e.: 18S rRNA; 5.8S rRNA; 28S rRNA). The *18S rRNA* fraction is processed and becomes associated with most of the 40S subunit proteins while it is in the nucleolus. It then leaves the nucleus via a nuclear pore withing 30 minutes of initial transcriptional creation of the 45S rRNA. This, along with its associated proteins, becomes the *40S small subunit* of the ribosome.

Processing in the *nucleolus continues* for the other two units of the 45S rRNA cleavage (i.e.: the *5.8S rRNA* and the *28S rRNA*) which are joined by

---

Fig. V-10: Nucleolus, carcinoma of the lung. The *pale staining component* (p), composed principally of fine fibers, contains DNA from the nucleolus organizer regions of the satellite chromosomes (Fig. V-9). The *dense fibrillar component* (f) is composed of the RNA transcripts with their associated proteins, the many fine 5 nm ribonucleoprotein fibers. The *granular component* (g) contains the 15 nm diameter particles which represent the most mature of the ribosomal precursor particles. There are moderate amounts of nucleolus associated chromatin (nac) which, in this cell, is a part of the chromatinic net; in other cells, it is a part of the chromatinic rim (Fig. V-3).

Osmium/methacrylate.

V-10: × 63,000.

(Frost, Erozan, Donovan; unpublished electron micrograph.)

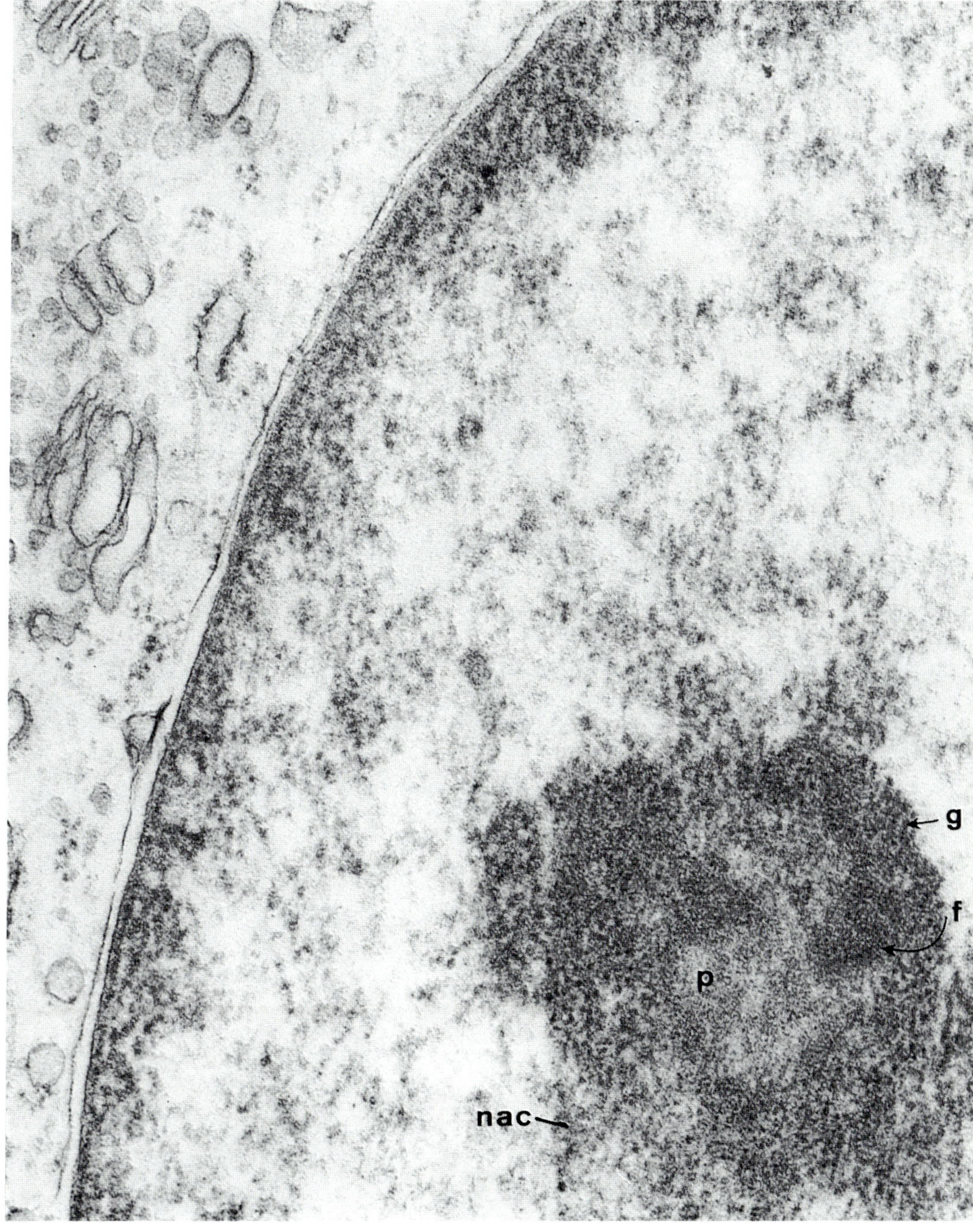
g
f
p
nac

a small *5S rRNA* molecule. This small 5S rRNA is made in the nucleus but *outside* of the *nucleolus*, from its DNA template on the long arm of chromosome 1. Together, these three rRNA molecules (i.e.: 5S rRNA, 5.8S rRNA; 28S rRNA), along with their associated proteins, create the precursor of the *60S large subunit* of the ribosome. This large subunit leaves the nucleus via a nuclear pore about 60 minutes after initial transcriptional creation of the 45S rRNA. The two subunits (*40S small* and *60S large*) attain their final functional forms as they are transported through the nuclear pores.

Once in the cytoplasm the *40S small subunit* finds a molecule of *messenger RNA* (mRNA) with a code to be translated into a protein, associates with it and joins a *60S large subunit* to form a functional ribosome. They then proceed to 'read' the mRNA molecule, from the 5′ end to the 3′ end, codon by codon. Specific transfer RNA (tRNA) molecules link to specific amino acids, bring them to the ribosome, and place them in the proper order dictated by the mRNA. Amino acid after amino acid is bound into polypeptide chains by peptide bonds, as protein is synthesized for the structure and work of the cell – its differentiation to perform function (see Section D).

This protein thread then forms near the surface between the two units of the ribosome, penetrates the large unit and exits from its other side [39]. If the nascent protein emerging from the large unit is to be membrane bound, it finds its membrane (i.e.: outer nuclear membrane, endoplasmic reticulum), passes through it and enters into its cisterna. In this way, the ribosome remains attached to the membrane as it is producing the protein (e.g.: rough endoplasmic reticulum, outer nuclear membrane).

4. Transport of RNA

Regardless of the type of RNA (e.g.: ribosomal RNA, rRNA; transfer RNA, tRNA; messenger RNA, mRNA; heterogeneous nuclear RNA, hnRNA), the intranuclear vectorial transport of particles of RNA with their associated proteins, from their sites of transcription throughout the nucleus and toward the nuclear periphery, is virtually unexplored [137]. Analogy with parafibrillary transport of particles in the cytoplasm [6] is inviting, but purely speculative and hypothetical at this time.

At the membrane, there is some circumstantial evidence suggesting that RNA may be transported along a fibrillary net to the pores (e.g.: fibers radiating from the annulus of the pore) and ultrastructural evidence is compatible with this outward passage of RNA through the pore complexes

(Figs. V-5, V-7, V-8; VIII-4), but total proof is not complete (Fig. V-5) [87]. The number of pores in the envelope increases with increased biological activity, as does the surface area of the nuclear envelope by its becoming wavy or undulated in configuration (Figs. V-5, V-7, V-8) [295].

There is weak evidence, gained from viral RNA transcripts, that there may be some RNA processing steps occurring at the nuclear periphery which involve elements of the nuclear envelope [287]. There are studies suggesting that enzymes (e.g.: nucleoside triphosphatase) located at or near the nuclear membrane, may be implicated in RNA transport through the nuclear envelope. This enzyme's activity increases with increased RNA transport out of the nucleus. It may modulate RNA transport by changing the degree of openness of the pores [141, 142]. There is also evidence that there may be cytoplasmic factors affecting release of RNA from nuclei [307].

There is a change in the RNA associated proteins as the RNA exits from the nucleus. This suggests that a substantial rearrangement of the RNA molecule occurs during transit, perhaps guided with or without its key structural proteins through the protein meshwork of the pore complex, leaving behind an old, picking up a new, or altering its own protein complement – or a combination thereof. This protein shift is reflected in the shift from 'acidophilia' of the proribosomes of the nucleolus to 'basophilia' of the ribosomes in the cytoplasm, as they go from the nuclear to the cytoplasmic compartments, even though both are pyroninophilic.

5. Nucleolus Associated Chromatin

There are varying amounts of chromatin associated closely with the nucleolus, mainly about its periphery. This consists simply of varying amounts of heterochromatin, and is a part of either the chromatinic rim (Fig. V-3) or chromatinic net (Fig. V-10), depending upon its location in the cell. As such, nucleolus associated chromatin reveals no more in light microscopy about the biologic activity of the cell than does the chromatinic rim or the chromatinic net both, or either of which it is a part.

G. Summary of Nuclear Structures by Light Microscopy

1. Chromatophilia

There are two major structures of the normal nucleus which are chromatophilically sharply discriminable: Hematoxylinophilic structures and acidophilic structures (Figs. V-2, V-3).

a. Hematoxylinophilic Structures (red-purple to blue-purple). These are the structures of *chromatin,* the DNA and its associated proteins of the material of the chromosomes. The three subregions of chromatin, which it is of great value to recognize by light microscopy, are:
–   the *chromatinic rim,*
–   the *chromatinic net,* (these two are *darkly* hematoxylinophilic and consist of condensed biologically *in*active heterochromatin with associated protein),
–   the *parachromatin* (very *palely* hematoxylinophilic, consisting of nucleoplasm with *very* small amounts of *finely* divided biologically active euchromatin *and* inactive heterochromatin, the delicate balance of the ratio between the two depending upon cellular activity).

b. Acidophilic Structures (red-orange-yellow; gold-brown). These structures are the *nucleoli,* having to do with the production of ribosomes from ribosomal RNA (rRNA) and associated proteins. There is usually hematoxylinophilic nucleolus associated chromatin about them, which is part of the chromatinic rim or chromatinic net.

2. True Nuclear Membrane Shape

The *shape* of the *true nuclear membrane* is the *shape* of the *profile* of the *outer* surface of the *chromatinic rim* (Figs. V-1 through V-4). The shape of this profile is exactly determined by the *shape* of the true inner nuclear membrane. The thickness of the nuclear membrane is so thin that it can*not* be resolved by the light microscope.

3. Chromatin

a. Chromatinic Rim. This is a complex of structures consisting of the *inner nuclear membrane,* the *nuclear lamina,* and the *peripherally condensed biologically inactive heterochromatin* (Fig. V-4). It is the peripheral hematoxylinophilic structure which has been incorrectly referred to historically as the 'nuclear membrane' (Figs. V-1, V-2).

The *thickness* of the *chromatinic rim* is directly proportionate to the amount of marginated condensed heterochromatin, and inversely proportionate to the degree of its compactness and condensation (Figs. V-5 through V-8) [138]. The *shape* of its inner surface is a reflection of the composition and character of the marginated heterochromatin. The *condition* of its inner surface (i.e.: clean-cut *vs.* blurred interface of the chromatin/parachromatin borders) reflects the state of preservation of the cell's DNA-associated protein (Figs. V-5, V-6).

b. Chromatinic Net. This is a network of condensed heterochromatin and proteins of the nucleus, extending from the chromatinic rim throughout the nucleoplasm (Fig. V-4). Like the chromatinic rim, it is heavily hematoxylinophilic.

It consists of thread-like and granular structures, whose shape and size are helpful in determining biologic activity, even though their nomenclature is inexact. The thread-like structures range from fine fibrils, through fibers, to agglutinated coarse strands (Figs. V-2, V-3). The granular particles grade from the barely visible finest *dusting* (ground glass) through *granules* (i.e.: *fine, moderate, coarse*) to *aggregates* (moderate to massive clumps).

The term *chromocenter* has more historical and anatomical meaning than useful biological denotation. It refers to prominent granular particles on up to massive clumps, of both chromatinic rim and net.

Ground glass dusting of barely visible fine particulate particles in a *well-*preserved cell, is to be carefully *discriminated from* the *gelatinous* appearance of gelled or laked chromatin in a *degenerated* cell (e.g.: herpes).

All borders of the chromatinic net, as well as the above-mentioned inner borders of the chromatinic rim, are chromatin/parachromatin interfaces which sensitively reflect the state of cell preservation. When DNA associated protein is well preserved, the chromatin/parachromatin interfaces appear clean-cut and well delineated (Fig. V-5). When protein degenerates, the chromatin/parachromatin interfaces become blurred and lack sharp definition (Fig. V-6). Thus the *inner* border of the chromatinic rim is in sharp contradistinction to the *outer* border of the chromatinic rim, the latter representing the profile of the inner nuclear membrane and remaining sharply defined throughout cell degeneration.

c. Nucleolus Associated Chromatin. This is a part of the heterochromatin. It is continuous with the chromatinic rim, chromatinic net or both, depending upon the location of the nucleolus in the nucleus and the orientation of the cell for viewing. Its size and shape, and the state of preservation of its chromatin/parachromatin interfaces, have the same biologic connotations as do the chromatinic rim and the chromatinic net.

d. Parachromatin. This is the very palely hematoxylinophilic domain between the chromatinic net, chromatinic rim and nucleolus in light microscopy. In electron microscopy it is referred to as interchromatin areas. Historically this pale hematoxylinophilia virtually has not been recognized, nor have the important functions and significance of the region. Many terms have been used to designate it (e.g.: nuclear sap, nucleoplasm, achromatin, interchromatin, euchromatin, parachromatin). Most of these terms convey seriously incorrect concepts. The latter term, however, which has been used by hematologists for many scores of years to signify this area [169], is adequately and purely descriptive of this light microscopic domain which lies very closely around (*para-*: Gr. = beside) the highly chromatophilic (*chroma:* Gr. = color) structures of the chromatinic rim, chromatinic net, and nucleoli.

Virtually all of the biologically active euchromatin which is not in the nucleolus, and a small amount of the inactive heterochromatin, exists entirely in the parachromatin. Here it lies in the nucleoplasm between the aggregations of heterochromatin of the rim and the net, in the form of nucleosomal and nucleoprotein fibers, constituted as loops of chromatin extending out into the nucleoplasm from the structural proteins of the nuclear matrix.

Here in the *parachromatin* is where transcription (DNA to RNA) and replication (DNA to DNA) take place. The amount of hematoxylinophilia of the pale parachromatin is proportionate to the amount of finely divided chromatin present. With increased biologic activity, the biologically inactive heterochromatin apparently moves out of the parachromatin into either the chromatinic rim or the chromatinic net, leaving behind less total chromatin *but* a *higher* percentage of biologically active *euchromatin* (Chap. VIII). Thus even though it is a pale hematoxylinophilia with less total chromatin, this 'cleared' parachromatin is richer in the biologically active euchromatin which is bathed in more nucleoplasm to provide a more adequate substrate for its biologic activities.

4. Nucleolus

This acidophilic structure is the factory of the ribosomes. Here, in cells which are in various stages of *protein production,* either intrinsic (e.g.: cellular structure) or extrinsic (e.g.: secretion), the ribosomal RNA (rRNA), with its associated acidophilic protein, matures in preparation for its transit through the nuclear pores into the cytoplasm.

Large, prominent and/or multiple nucleoli signify protein production. They can *only* be differentiated from chromocenters by the former's *acidophilia* and the *hematoxylinophilia* of the latter. Position, shape, quality or any other features do *not* discriminate between the two, so that proper staining is *essential.* Overstaining with hematoxylin and/or understaining with weak counterstains, will make their identification impossible. Except in lymphomas and leukemias, the mere presence of nucleoli is *not* a good malignant criterion, but merely bespeaks protein production.

Section C
General Activity: Its Morphologic Characteristics

VI. Morphologic Characteristics of Euplasia: Normal Baseline General Activity

A. General Considerations

Euplasia is the form of that reference state of biologic activity which is sufficient for continued existence in a working community of healthy non-stressed cells and tissues (see Chap. II). Euplasia is characterized by three major trends of regularity in the morphologic pattern: - *roundness, uniformity,* and *predictability.*

1. Roundness

Euplastic cells have a profound tendency for structural roundness in the absence of normal forces to the contrary (Figs. VI-1, VI-2). This expresses a natural tendency for moldable matter to reduce surface area to a minimum, by assuming a spherical shape (e.g.: a drop of oil rising in water). Nuclei, nucleoli, and chromocenters (i.e.: aggregations of chromatin) of euplastic cells thus tend to be *round* or (if irregular) their corners are *rounded* and are not sharply angled or spiculated.

Exceptions to this, focus attention upon more specific cellular activity and upon intracellular or extracellular forces responsible for these deviations from the expected norm. Nuclear irregularities found in the region of the cytocentrum (or hof) from pericentriolar kinesis (Fig. IV-1) or molding of a nucleus around vacuoles of secretion (Figs. V-1; VI-1), bespeak internal cellular activities and pressures. Likewise, the polygonal shape of superficial squamous cells and the honeycombing of 'en face' columnar cells (Fig. VI-1), evidence external forces from the surrounding tissues.

Yet even when forced out of round, euplastic irregular structures usually continue to exhibit this tendency of roundness by rounding off the corners of their otherwise distorted shapes. Thus one usually does not find the sharp angles between concavities of nuclear membrane wrinkling,

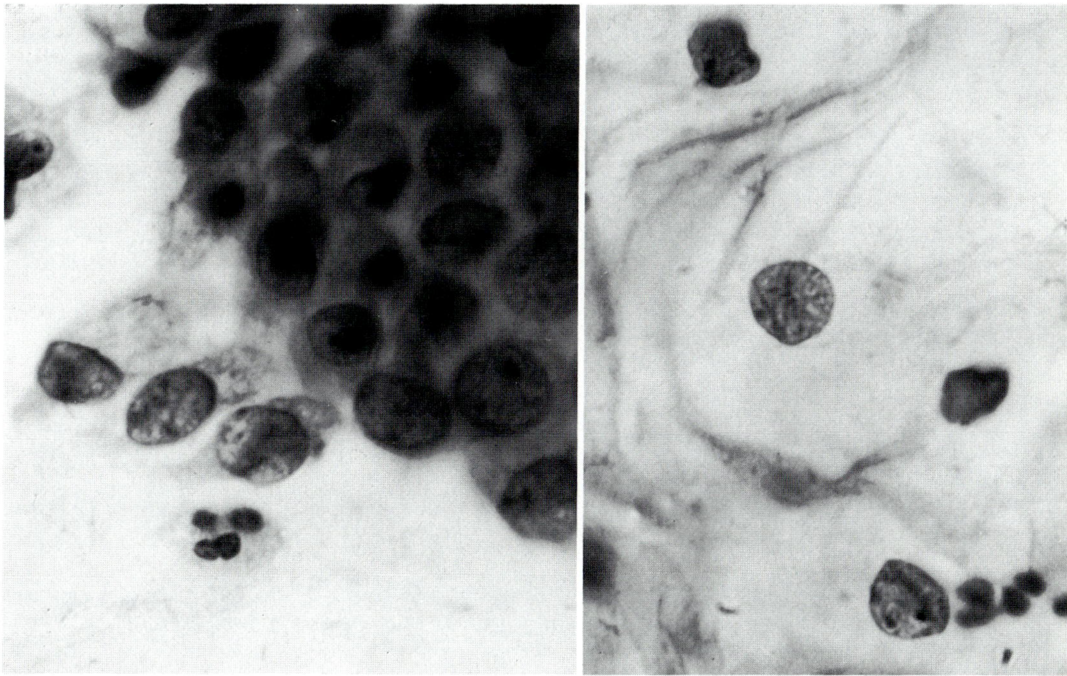

Fig. VI-1: Euplastic (normal) columnar cells exfoliated in a diagnostic true tissue fragment (DTTF) as a simple (single cell) sheet. There is *roundness* of the chromocenters and of the nuclear membrane (as adjudged by the shape of the outer surface of the chromatinic rim), except where there are obvious forces (e.g.: secretion; left-most cell). There is *uniformity* in the thickness of the chromatinic rim and in the dispersion of the chromatinic net throughout the parachromatin. There is *predictability* from one nucleus to another of this diagnostic true tissue fragment (DTTF) in: thickness of the chromatinic rim, chromatin pattern, parachromatin pattern, numbers of nuclear structures (e.g.: chromocenters, nucleoli), degree of chromasia. Note the faint, delicate, hematoxylinophilia of the parachromatin, which is so characteristic of euplasia. The nuclei are eccentrically located near the tail end of the three lower columnar cells, most of the others being *en face* and appearing as a 'honeycomb'. For *size,* note that the entire neutrophil (center) is about the size of a euplastic nucleus of the columnar cells.

Fig. VI-2: Euplastic (normal) intermediate squamous cells. Note that the chromatin of the chromatinic net tends to be more threadlike, while the chromatin of the columnar cell nuclei (Fig. VI-1) tends to be more granular. Other features of the columnar cell nuclei mentioned above, also apply to these squamous cells.

Papanicolaou stain.

VI-1, VI-2: × 1,200.

which can characterize retroplasia (Chap. VII), nor the sharp spicules and unpredictable sharply angulated irregularities of malignant neoplasia (Chap. IX).

2. Uniformity

Most structures of euplastic cells occur in a characteristic uniformity. Nuclear *chromatin,* regardless of its shape (e.g.: threads, granules), is dispersed in a uniform pattern between the parachromatin (Figs. VI-1, VI-2). The chromatinic rim is of uniform configuration and thickness. The shape of the *nuclear membrane* (profile of the outer edge of the chromatinic rim) is uniformly round or oval, unless indented from obvious forces, i.e., for functional reasons (e.g.: secretion) or molded from outside forces (e.g.: by neighboring cells growing in tissue) (Fig. VI-1).

This uniformity is characteristic of euplasia in tissue sections, in exfoliated tissue fragments, and in the individual cells. While the degree and type of uniformity may vary from cell type to cell type, the trend for uniformity is characteristic. When nonuniformity does appear in euplasia, the physiological reasons for it are usually morphologically recognizable (e.g.: products of secretion, cytocentrum activity, adjacent structures).

3. Predictability

Morphologic features characteristic of a given cell type can be predicted in euplasia. One can anticipate certain features with great reliability (e.g.: number and shape of nucleoli and of chromocenters, chromatinic pattern, nuclear shape, cellular shape, nucleo-cytoplasmic ratio), predicting them from one cell to the next cell in the same tissue or tissue fragment (Fig. VI-1). Comparison, thus, is between neighboring cells of identical type which lived under the same influences and had reached the same level of maturation at the time of exfoliation. They look predictably alike in key features.

B. Nuclear Structure

1. Nuclear Size and Shape

The *size* of a nucleus in a cell of euplasia, is about the *size* of an entire well preserved *neutrophil* (i.e.: nucleus and cytoplasm) (Figs. VI-1, VI-2). This valuable reference cell is nearly ubiquitous in histologic and cytologic material. Very few cells are an exception to this rule (e.g.: megakaryocytes).

Nuclei of intermediate cells (Fig. VI-2), columnar cells (Fig. VI-1), muscle cells, mesothelial cells, connective tissue cells, macrophages, etc., all fall within this norm when they are not stressed, the size of an *entire* well preserved neutrophil.

The *shape* of the nucleus is *round* in euplasia. This follows the general rule of *roundness* found with other plastic, moldable structures in euplasia.

2. Hematoxylinophilic Chromasia

The degree of *hematoxylinophilic chromasia* of a euplastic cell's nucleus, serves as a reference level for that cell type. If the cell is well preserved, the hematoxylinophilic chromasia is directly proportionate to the amount of DNA-associated protein which is present. This chromasia is imparted mainly by the chromatinic net and chromatinic rim; however, the hematoxylinophilic chromasia of the parachromatin (Figs. VI-1, VI-2) also is biologically significant, albeit faint.

3. Chromatin

Chromatin fibers of the interphase (i.e.: intermitotic, metabolic) euplastic cell, are finely divided and dispersed in the *parachromatin, imparting to it a faint hematoxylinophila.* In the *chromatinic net* and the *chromatinic rim,* they are condensed and aggregated, imparting to those structures a dark hematoxylinophilia. The latter two structures are organized in the light microscope as threads or rounded granules which are uniformly dispersed between the areas of parachromatin (Figs. VI-1, VI-2). This pattern of uniformity of the *chromatinic net dispersion within the parachromatin, the chromatin/parachromatin pattern,* is highly significant biologically. It varies according to cell type, at times significantly (i.e.: Chaps. XI, XIV); but within a cell type, it remains a good indicator of euplastic general activity.

4. Parachromatin

The *parachromatin* of euplasia contains extremely finely divided and uniformly dispersed elements of the chromosomes (Fig. VI-3). Much of it is euchromatic, in its active metabolic phase (i.e.: transcription, replication). The DNA-associated protein of this finely suspended chromatin stains with hematoxylin, and thus imparts the *delicate* pale hematoxylinophilic chromasia to the parachromatin which is so characteristic of euplasia (Figs. VI-1, VI-2).

5. Chromatinic Net

Nearly all of the biologically *in*active heterochromatin, is condensed and aggregated into the chromatinic net and chromatinic rim. The net in euplasia usually is organized in fine threads and fine granules (Figs. VI-1, VI-2).

In altered general activity (retroplasia, proplasia, and malignant neoplasia) much of the small amount of the biologically inactive heterochromatin, which is finely suspended in the parachromatin in euplasia, condenses into the chromatinic net and the chromatinic rim. This yields abnormally cleared parachromatin, which is able to be detected by the experienced microscopist; however, the resulting increased chromasia of the chromatinic net and the chromatinic rim, is so slight that it is humanly imperceptible. The organization of the heterochromatin in the net and rim is altered from that of euplasia with the formation of degenerated clumps in retroplasia (Chap. VII), larger or coarser granules in proplasia (Chap. VIII), and large well-preserved and angulated clumps in malignant neoplasia (Chap. IX).

6. Chromatinic Rim

The chromatinic rim is the hematoxylinophilic structure visualized in light microscopy at the outer margin of the nucleus, and which for years has been called the 'nuclear membrane'. It is actually peripheral heterochromatin which has condensed and attached, by the lamina, to the inner surface of the true inner nuclear membrane (Chap. V). The latter, with the outer nuclear membrane, the perinuclear cisterna, and the nuclear pores, consti-

---

Fig. VI-3: Euplastic (normal) columnar cell, bronchial epithelium. Activity is moderate, as evidenced by the numerous nuclear pores (np) in the nuclear membrane, many of which exhibit a central granule. The chromatinic rim is of uniform thickness. While the outer nuclear membrane has numerous irregularities, the inner nuclear membrane is smooth and rounded, imparting that shape to the outer surface of the chromatinic rim. The chromatinic net pattern is mainly that of granular condensations which are uniformly dispersed throughout the parachromatin, as with light microscopy (cf.: Fig. VI-1). On very careful inspection, the parachromatin is seen to contain an extremely small amount of delicate, finely divided chromatin threads and granules, which are dispersed throughout its pale areas and which impart the pale hematoxylinophilia which is so significant in light microscopy (cf.: Fig. VI-1).

Osmium/methcrylate.

VI-3: × 25,000.

(Frost, Erozan, Donovan; unpublished electron micrograph.)

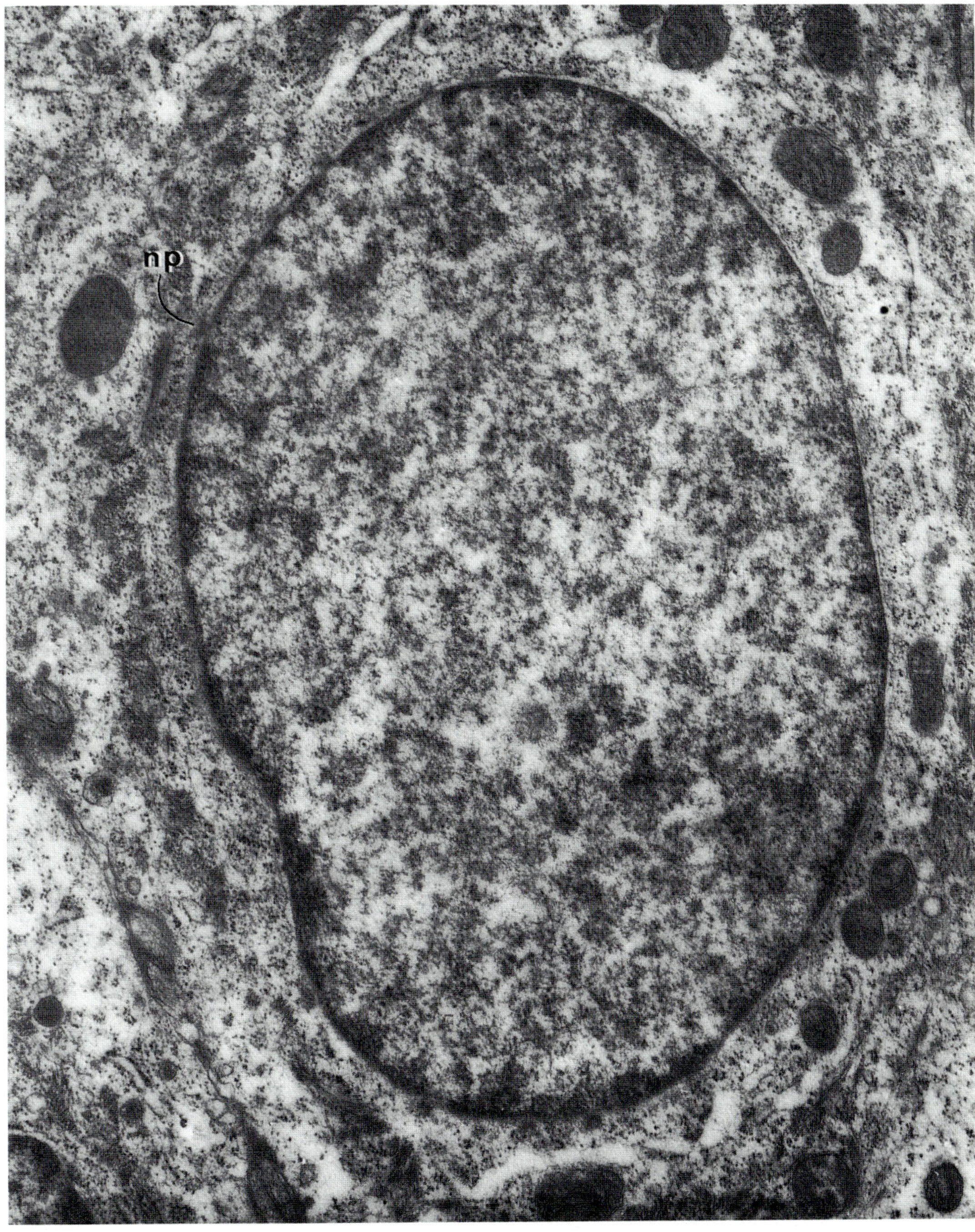
np

tute the nuclear envelope. The nuclear membrane is easily demonstrable under the electron microscope (Fig. VI-3), but is so thin that it is not visible by light microscopy (Figs. VI-1, VI-2). With the light microscope, therefore, the nuclear membrane of the euplastic cell appears as an *invisible limiting membrane* between two phases, the marginated chromatinic material and lamina against its inner surface with the cytoplasm against its outer surface, while its *shape* is that of the outer profile of the chromatinic rim. The amount of chromatinic material marginated with the lamina against the inner surface of the nuclear membrane, determines the thickness of the chromatinic rim (Figs. VI-1, VI-2). In euplasia it is of uniform thickness throughout a given nucleus, and from nucleus to nucleus of a diagnostic true tissue fragment (DTTF) (Table VI-1), or of a multinucleated cell.

7. Nuclear Membrane Shape

The shape of the nuclear membrane as evaluated in the light microscope, therefore, is the shape of the profile of the outer surface of the chromatinic rim (Figs. VI-1, VI-2). It is smoothly round in euplasia, unless other forces (e.g.: secretion, neighboring cells) indent it.

8. Nucleolus

The presence of nucleoli in euplasia is an indication of protein synthesis, either for intrinsic cellular use (e.g.: making organelles) or for extrinsic production (i.e.: making mucus for secretion). They are acidophilic with routine Papanicolaou staining, usually strikingly distinct from the hematoxylinophilic chromatin. At times, however, nucleolar acidophilia is *pale* from *poor counterstaining,* or will even stain with *hematoxylin* if that dye has been *too heavily applied or if the counterstain is weak.* Also, the nucleolus can be dwarfed and obscured by large amounts of hematoxylinophilic nucleolus associated chromatin in it and around it.

The number of nucleoli can markedly vary in euplasia between cell types of *different* functional differentiation, from none to many. Within a given cell type, however, the number of nucleoli is uniform among cells which were under the same influences (e.g.: DTTF), or among nuclei of a multinucleated cell. A useful rule of thumb is that the numbers of nucleoli in the nuclei of euplastic cells which were under the same influences (e.g.: multinucleated cells, cells in a DTTF) usually vary within plus or minus one (e.g.: none to two; one to three) (Fig. VI-1). At times they can vary as much as plus or minus two (e.g.: none to four; two to six). A greater variance than that, bears marked concern of malignant neoplasia (see Chap. IX).

9. Multinucleation

Multinucleation is not frequent in euplasia. When it does occur, however, each daughter nucleus is virtually identical with the others in size, shape, chromatin/parachromatin pattern, chromasia, and nucleoli.

C. Cytoplasmic Structure

The *amount* of cytoplasm in the euplastic cell tends to be abundant, yet it varies with the cell's functional differentiation. It is uniformly dispersed about the nucleus, with the nucleus being centrally placed, unless there are physiologic reasons (e.g.: functional differentiation activities) to account for an eccentric positioning of the nucleus or an out-of-round cytoplasm.

The *organelles* of the cytoplasm (e.g.: mitochondria, Golgi, endoplasmic reticulum) are not definitely identifiable *per se* in light microscopy with routine stain, but can be characterized by certain Papanicolaou staining characteristics, special stains and reactions. Mitochondria contribute to the fine vacuolation distributed generally throughout the cytoplasm (Figs. V-2, V-3), as does the Golgi, which is located mainly in the cytocentrum in the region of nuclear indentation [75].

The degree of cytoplasmic basophilia is related directly to the amount of ribonucleic acid and protein present as ribosomes. This is greater in the younger and more active cells (e.g.: germinal cells of a normal euplastic epithelium), and with increased general activity (proplasia, Chap. VIII). With maturation, basophilia generally decreases. Cytoplasm increasingly reflects the functional differentiation which the cell undergoes as it matures to play its part in the community of the body (Section D) and contains the structures of its function (e.g.: cilia, myofibrils, mucus, keratin, microvilli) (Figs. VI-4 through VI-7).

D. Multiple Cell Structures and Cell-to-Cell Relationships: Diagnostic True Tissue Fragments (DTTF)

It frequently becomes of great discriminating importance, therefore, to carefully and accurately analyze a group of cells as to their origin to identify with high *diagnostic* accuracy, whether they *definitely* had grown together before exfoliation as tissue – *diagnostic true tissue fragment* (DTTF) – or whether they had *possibly* come together secondarily (all *others*). There is widespread use of loosely used and differently interpreted terms (right column, Table VI-1), which is fraught with many interpretive and discriminatory dangers. When *diagnostically* recognized *true fragments* of *tissue* which have broken off and have exfoliated *intact* (Figs. V-1, V-2; VI-1), the exact cells which *grew* together and stayed together *still* lie side by side, attached to each other and in the same relationships with each other which they enjoyed in the intact tissue. As does the identification of a true *multinucleated* cell increase the number of valuable features of benignancy (Figs. VII-4; VIII-5, VIII-6; XII-3) and malignancy (Figs. IX-1, IX-6, IX-7) which are available to better determine biologic behavior from cellular morphology, so does the recognition of a *diagnostic true tissue fragment* (DTTF) add valuable criteria of benignancy (Figs. V-1, V-2; VI-1) and malignancy (Figs. XV-4, XV-13; XVII-4; XVIII-1).

For example, the *predictability* of the numbers of chromocenters or the numbers of nucleoli from nucleus to nucleus in a DTTF or in a multinucleated cell, is found in benignancy (Figs. V-1, V-2; VI-1); but the *lack* of this predictability to a significant degree (Chap. IX) can serve as a valuable criterion of malignancy (Figs. XII-19; XV-1; XVIII-1). Conversely, when a group of cells is actually not a true fragment of tissue, but is a conglomeration of primarily individual cells which secondarily came together, they can show non predictability in these key features simply because they arose from different areas, under different stimuli; or, in fact, they may not be of the same cell type.

There exist certain key criteria (Table VI-1) to help in the recognition of DTTF (Figs. V-2; VI-1; XII-19; XV-10, XV-13; XVII-10) from all other clusterings of cells (Fig. VII-7).

When a DTTF begins to degenerate, it may lose its diagnostic criteria; conversely, individual cells clustering together can 'mush' together as they begin to degenerate, with nuclear molding and loss of windows, mimicking a DTTF. Thus, to be diagnostically discriminable as a *DTTF,* it must be

---

Figures VI-4 through VI-7: Microvilli on mesothelial cells. All of the figures were *under*exposed in order to reveal the microvilli. Healthy microvilli (Figs. VI-4, VI-5, VI-7) are uniform. Degenerative cytoplasmic blebs blow up the microvilli (Fig. VI-6) or iron them out (Fig. VI-7).
Papanicolaou stain.
VI-4 through VI-6: $\times$ 1,000; VI-7: $\times$ 3,000.
(Figs. VI-4 through VI-6, courtesy: W. M. Howdon, M.D.)

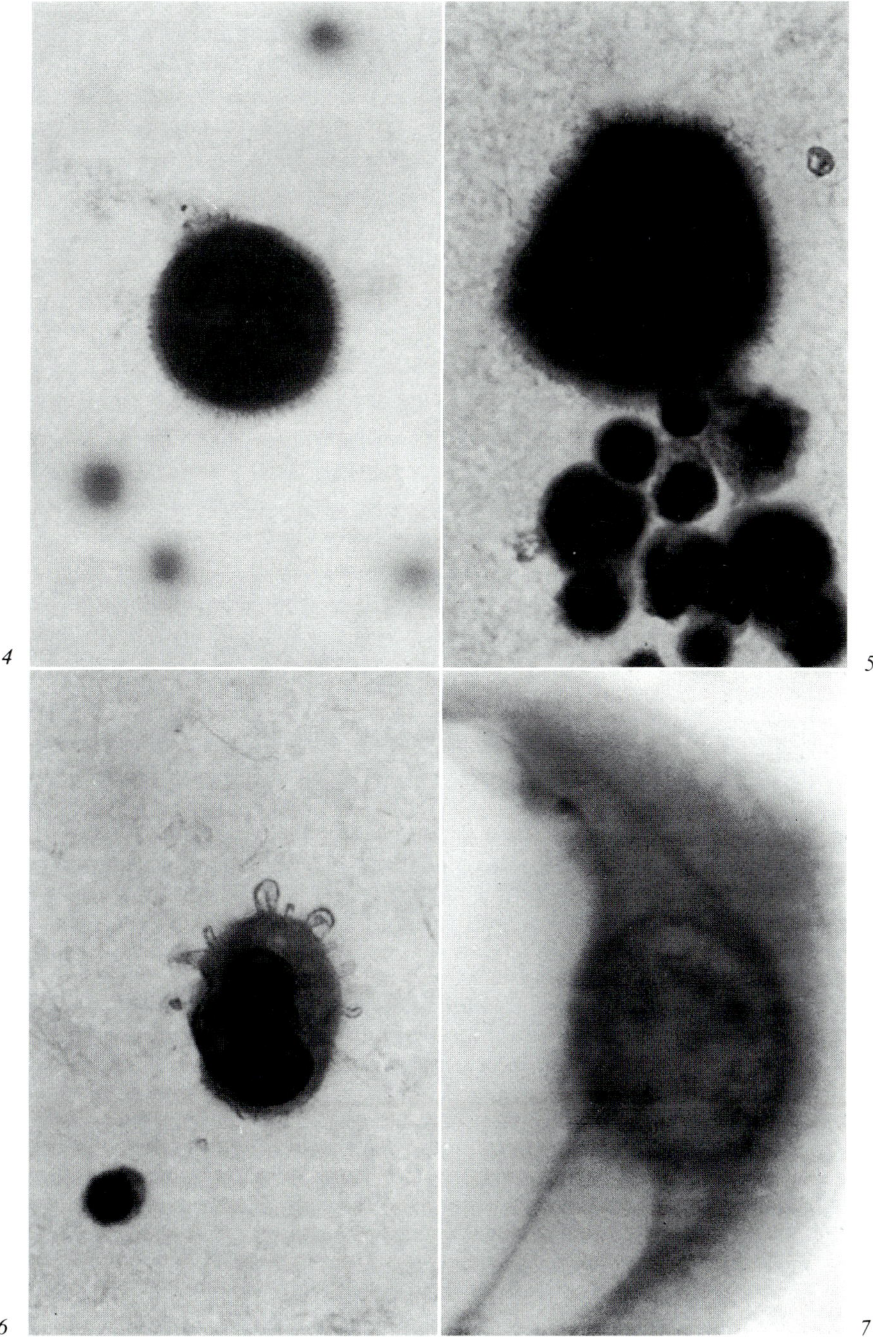

Table VI-1. Criteria to assist in identifying a *Diagnostic True Tissue Fragment (DTTF)*

Diagnostic True Tissue Fragment (DTTF) (e.g.: Figs. V-1, V-2; VI-1)	*Others,* e.g.: Poorly Preserved TTF, group, ball, aggregate, cluster, conglomeration, morula, pseudoacinus, pseudopapilla (e.g.: Fig. VII-7)
No 'windows' between adjacent cell borders	'Windows' between adjacent cell borders
Nuclear molding by adjacent nuclei in *well preserved cells*	No molding of nuclei by adjacent nuclei in *well preserved cells*
True epithelial luminal border – 'Fence-like' outer border of DTTF – Obtuse angles between cells – Intercellular tight junctions with terminal bars	No true epithelial luminal border – 'Loop-the-loop' outer border of cell mass – Acute angles between cells – No evidence of intercellular tight junctions or terminal bars
True acinus – Epithelial sheet (*en face* and ring) – Inner true epithelial luminal border – Inner lumen within DTTF	No true acinus – No true epithelial sheet – No true epithelial luminal border – No true lumen inside of cell mass
True villus, polyp, or papilloma – Epithelial sheet (*en face* and ring) – Outer true epithelial luminal border – Inner stroma, disorganized	No true villus, polyp, or papilloma – No true epithelial sheet – No true epithelial luminal border – No central stroma within epithelium

well preserved. When they show degeneration, they must be relegated to the 'wastebasket' of the right-hand column (Table VI-1), as they are diagnostically treacherous and are not to be used for crucial decisions. At times, therefore, it is not possible to make a sure distinction between a true tissue fragment falling apart *vs.* a more tightly fitting together conglomeration of primarily single cells. For the purposes of using discriminating features based upon it being a DTTF, such questionable structures should *not* be considered a Diagnostic True Tissue Fragment.

VII. Morphologic Characteristics of Retroplasia: Decreased General Activity

A. General Considerations

When the level of general activity has decreased from that which is baseline or normal for a particular cell type (Fig. II-4), morphologic changes of *retroplasia* occur. At this time multiple *retro*gressive physiologic changes exist, including those involving protein, fat, carbohydrate, mineral and nucleic acid metabolism.

There is alteration of cellular control over water content with impaired ability for the cell to maintain its water balance, with resulting increased retention or loss of this important substance. Many of these physiologic alterations yield morphologic changes of general structures or organelles which are useful in the study and diagnosis of biologic alterations (Figs. II-4 through II-6).

Morphologic changes of retroplasia depend upon numerous factors. These include *type* of stimulus and process(es) involved, *intensity, length* of action, *conditions* during the period *up to exfoliation, conditions* of the *subsequent* interval elapsing between exfoliation and fixation, *type* of fixative employed, and *preparation* of the specimen.

When cell death and fixation occur simultaneously in a euplastic cell, there are virtually no morphologic retrogressive changes (Fig. II-6). In most practical situations in cytopathology, however, sufficient time elapses for at least minimal morphologic alteration to occur (Figs. II-4, II-5); but, in the best and most diagnostically helpful preparations, this is kept to a bare minimum [100]. This critical feature is key to improving the quality and diagnostic accuracy of the specimen.

Degeneration is usually associated with morphologic changes which are fewer and milder when cellular death had not occurred before fixation [26, 140, 272, 280, 281, 296]. Increase of those morphologic characteristics which bespeak severe or irreversible physiologic alterations, indicates a higher probability of cellular *death. Necrosis,* occurring after death of the cells, is usually associated with the most severe cellular alterations of retroplasia.

B. Nuclear Structure

1. Nuclear Size and Shape

Extreme *variation* in *size* is the rule in degeneration, from a small pyknotic nucleus, lysed 'ghost' nucleus, or a disappeared nucleus, to an enormously swollen one (karyomegaly; up to 20 times). Water imbalance, either loss *or* increase, is the greatest single factor for determining nuclear size variance; however, the effect of degeneration can*not* be accurately assessed and then 'read out', because the degree of hematoxylinophilia varies markedly and unpredictably with denaturation of the DNA-associated proteins.

The *shape* characteristic of retroplasia is that of *wrinkling,* or multiple concavities (Figs. VII-1, VII-3, IX-9). This is a reflection of water loss and shrinkage of the nuclear contents, with inward collapse of the nuclear membrane. At times angular points in the membrane are formed where the concavities meet, which are *not* to be confused with the nuclear membrane irregularities indicative of malignant neoplasia (Chap. IX). While malignant cells often degenerate with resultant wrinkling as occurs in any other cell, *wrinkling* is *not* a criterion of malignant neoplasia but of retroplasia.

2. Hematoxylinophilic Chromasia

When the cell loses control over its water content, a number of major morphologic changes occur. As the nucleus takes in water and swells (Figs. VII-1, VII-2), its chromatin becomes pale, loses its pattern, and eventually lakes or marginates (Fig. VII-3). With water loss, the nucleus becomes small and dark with dehydration of its contents, resulting in hyperchromasia and eventual loss of chromatin pattern (i.e.: karyopyknosis) (Figs. XI-6; XII-2, XII-10).

Thus, a retroplastic nucleus usually becomes either *large and pale* (Fig. VII-2) or *small and dark* (i.e.: pyknotic, like the nucleus of a superficial cell) (Figs. XI-6; XII-2). When a degenerated cell paradoxically becomes

---

Figures VII-1, VII-2: Fast (vagino-pancervical) smears.

Fig. VII-1: Retroplasia of a degenerated, nonmalignant epithelial cell, probably columnar. The nucleus is markedly increased in size, compared to the entire well preserved neutrophil cell which closely approximates the size of a normal nucleus. The cytoplasm is dull, acidophilic and opaque. Much of it has been lost in the lower left and upper right, with a whisp extending out at about 7-o'clock and a disappearance of the cytoplasmic (cell or plasma) membrane to the left of that and at about 10-o'clock. This markedly increases the

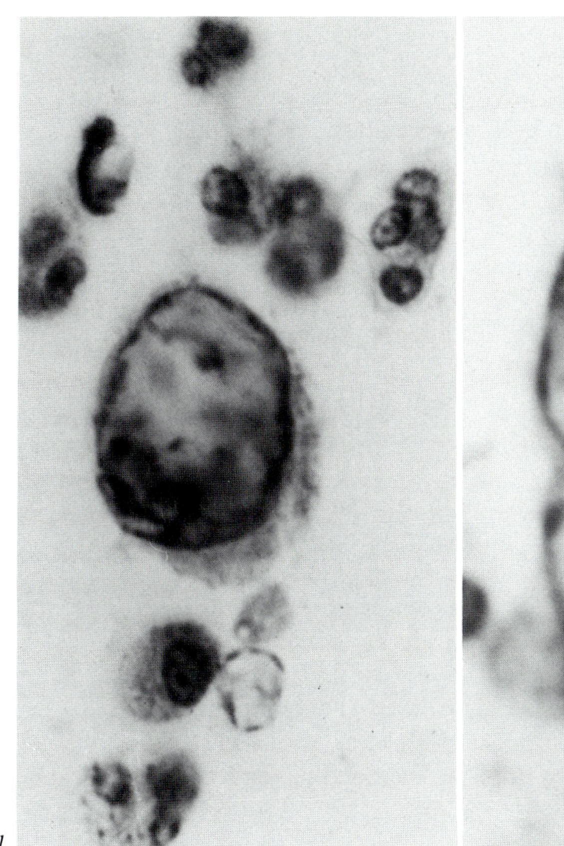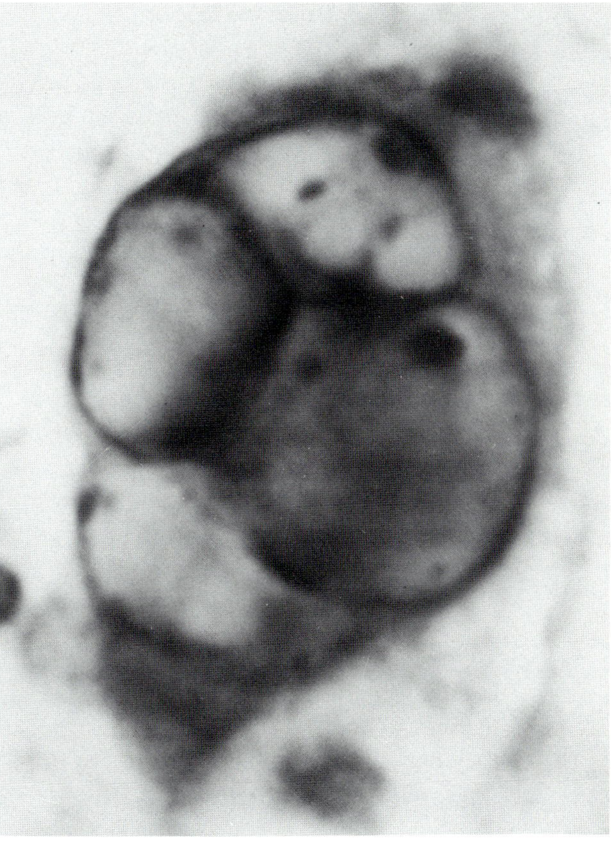

apparent N/C ratio which, of course, would be nearer to normal if cytoplasm had not been lost into the background. The parachromatin is cleared in many areas, with degeneration identified by the blurring of the chromatin/parachromatin interfaces throughout the chromatinic net and the inner surface of the chromatinic rim. This is in striking contrast to the crisply, well defined outer surface of the chromatinic rim, which is held sharply delineated from the cytoplasm by the true inner nuclear membrane. The latter is wrinkled in many areas, evidencing degeneration. Chromatin aggregates are large, dark, and irregular; but they are rounded and blurred, and most of them have marginated to the chromatinic rim.

Fig. VII-2: Proplasia followed by retroplasia. This very degenerated cell was first stimulated with proplastic *multinucleation, increase* in nuclear size, and granular chromatin (remnants still remaining in the right-hand nucleus). The cell then degenerated with marginated chromatin, clearing of the parachromatin, and blurring of the chromatin/parachromatin interfaces. Note the chromatin (particularly in the left-hand nucleus), which has a peculiar *gelatinous* degeneration. It is *not* ground glass, which consists of evenly dispersed and sharply defined *fine* granules as found in proplasia (cf.: Fig. XVI-6); but this is a jellied, degenerated, gelatinous DNA and its associated proteins. Herpes simplex virus infection.

Papanicolaou stain.

VII-1, VII-2: × 1,500.

large and dark, it usually was a large cell before degeneration (e.g.: proplasia, malignant neoplasia) which then became retroplastic (Figs. VII-1, VII-4, center; XII-10).

These alterations are due mainly to water loss. Variations in the degree of hematoxylinophilia of the DNA-associated proteins with degeneration, however, also plays a significant role in their genesis.

Such changes are usually diffuse. They may occur in localized regions, however, with clearing of areas, vacuole formation, condensation of cellular structures, and the formation of many types of inclusion bodies (e.g.: bismuth, lead [33, 309], virus) (Figs. VII-4 through VII-6).

3. Chromatin

In the *early* stages of cellular degeneration, *nuclear chromatinic* material provides some of the earliest and most sensitive signs of retroplasia. This is due mainly to denaturation and degeneration of the proteins associated with the DNA, starting near the very beginning of the process.

The *chromatin/parachromatin interfaces* are the borders of the chromatinic net and of the inner surface of the chromatinic rim, which are bathed in parachromatin. These are crisply defined and well preserved in the intact euplastic cell (Figs. VI-1 through VI-3). They *blur* and *smudge* as they denature or degenerate, and lyse into the surrounding parachromatin (Figs. VII-1 through VII-3).

4. Parachromatin

The *chromatin agglutination* which occurs with degeneration, causes many morphologic alterations. The parachromatin is swept clean of its fine and delicate chromatin by the DNA-associated proteins denaturing, by these chromatin fibers sticking together, and by their condensing onto the chromatinic rim and net (Fig. VII-3, compare with Fig. VI-3). Rather than the typical pale hematoxylinophilic chromasia of the parachromatin of the well-preserved euplastic nucleus (Figs. VI-1, VI-2), the parachromatin becomes very clear (Figs. VII-1, VII-2).

Fig. VII-3: Retroplasia. Bronchial columnar epithelial cell in early degeneration. The nuclear membrane is wrinkled as a result of loss of water from the nuclear contents. There is striking clearing of the interchromatin areas (parachromatin, by light microscopy) which are nearly void of the delicate chromatin fibers present in euplasia, proplasia, and malignant neoplasia (cf.: Figs. V-5 through V-8). There is increased density and thickness of the

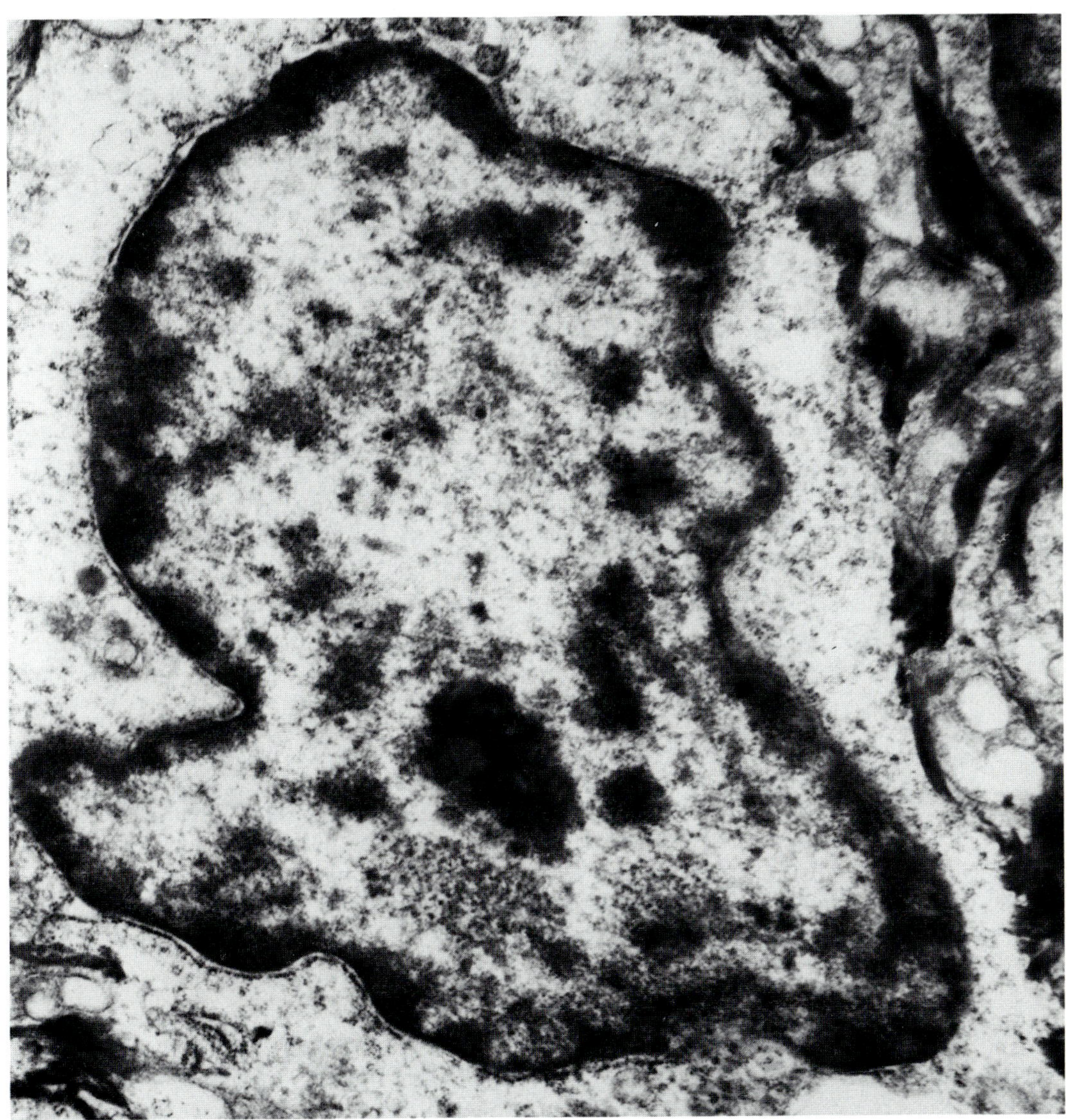

heterochromatin in the chromatinic rim and the chromatinic net. Note the decreased granularity and smudginess of these two structures (cf.: Figs. V-5 through V-8). There are still remains of the proplasia or euplasia which existed before degeneration set in (e.g.: nucleoli, nuclear pores), albeit structures are smudged.

Osmium/methacrylate.

VII-3: × 26,000.

(Frost, Erozan, Donovan; unpublished electron micrograph.)

This resulting *abnormally cleared parachromatin* of *retroplasia,* differs morphologically in light microscopy from the clearing found in *proplasia* and *malignant neoplasia only* in the rounded and smudged chromatin/parachromatin interfaces which it bathes (Figs. VII-1, VII-2, VII-4, VII-5). If morphology is to bespeak the markedly differing biologic behaviors of these three states, this is a critical and key feature to be closely noted and carefully evaluated.

5. Chromatinic Net

As chromatinic material degenerates, it agglutinates into larger structures and condenses into more densely hyperchromatic masses. The chromatin/parachromatin interfaces of these large chromatin clumps become blurred and rounded (Figs. VII-1, VII-6), usually in distinct contrast to the *angular* and *crisply delineated* large hyperchromatic chromatin clumps and macronucleoli which are present in well-preserved cancer cells (Chap. IX) (Fig. XVIII-1). Frequently, degenerating masses of the chromatinic net marginate outward to the inner surface of the nuclear membrane, presenting a 'beaded' appearance to the chromatinic rim (Figs. VII-1, VII-3).

When chromatin degenerates and becomes pale in retroplasia, it smudges and loses structure with eventual complete lysis into the parachro-

---

Fig. VII-4: Virus infection (probably herpes simplex), oral cellular spread. The two nuclei in the lower left became proplastic first with multinucleation, hyperchromasia, granular chromatin, karyomegaly, and undulating nuclear membranes due to initial cellular stimulation by virus; then, they have begun to degenerate with early smudging of the chromatin/parachromatin interfaces. The upper right nucleus shows initial proplasia which was followed by secondary retroplasia from viral injury and degeneration, with formation of a large stellate inclusion. The centrally placed viral inclusion is acidophilic with much admixed hematoxylinophilic and blurred (degenerated) chromatin. This is surrounded by a peri-inclusion halo of abnormally cleared parachromatin enclosed in a degenerating chromatinic rim, with thin transhalo bridges of degenerating chromatin stretching from inclusion to chromatinic rim. The large dark nucleus, lower left center, is undergoing karyopyknosis (condensed, dehydrated retroplasia).

Figs. VII-5, VII-6: Virus infection, sputum. In cells which were previously proplastic (multinucleation and some remaining granular chromatin), there is marked retroplasia (blurring and smudging of the chromatinic net and *inner* surface of the chromatinic rim, as their protein degenerates and dissolves into the parachromatin with a resulting *gelatinous* appearance). The *outer* profile of the chromatinic rim is held more crisply and sharply delineated from the cytoplasm by the invisible nuclear membrane.

Papanicolaou stain.

VII-4: $\times$ 2,000; VII-5, VII-6: $\times$ 1,400.

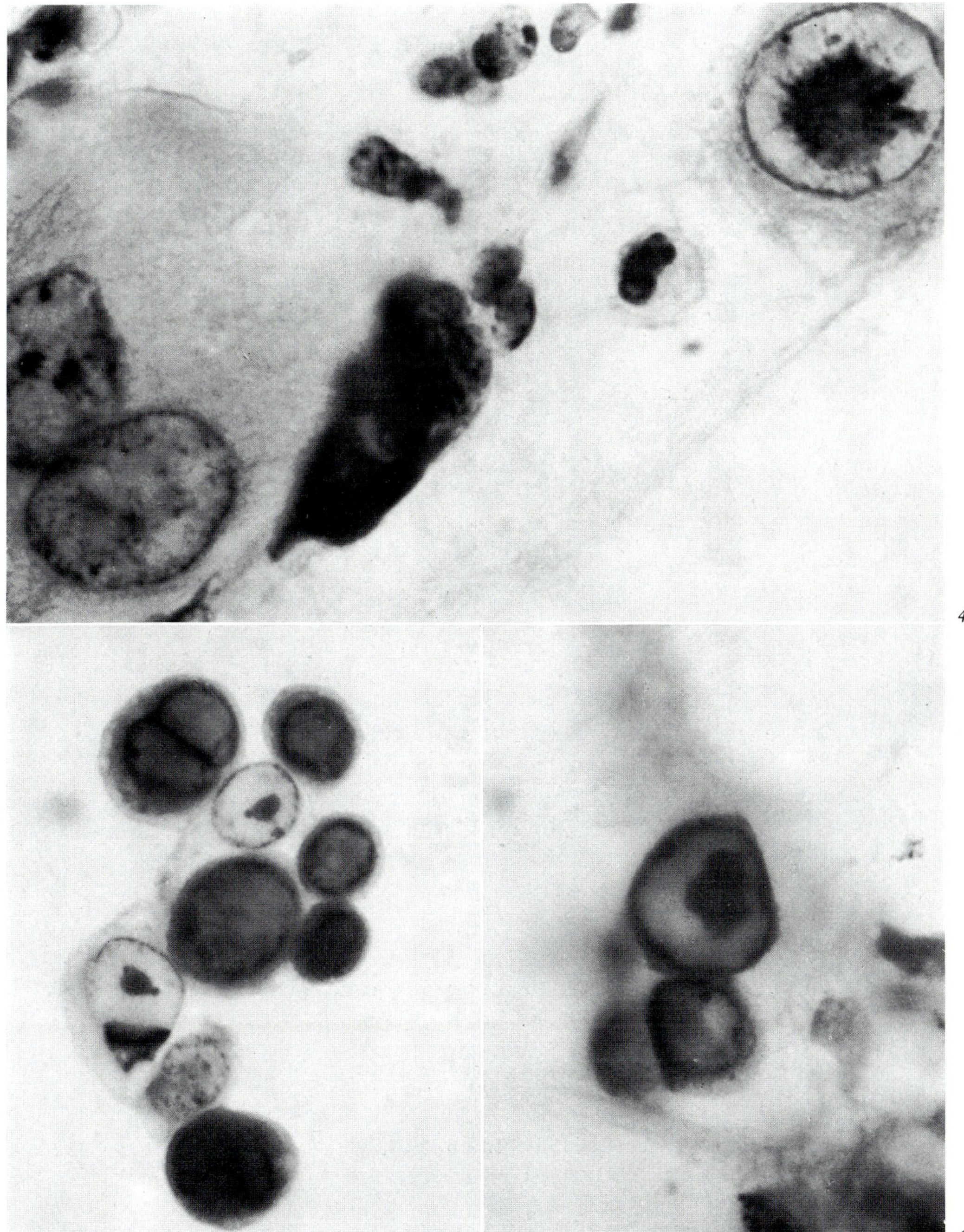

4

5

6

matin. This results in pale 'washed-out' appearing areas of the nucleus (Figs. VII-1, VII-2) which is frequently followed by rupture and loss by nucleolysis into the background.

At times degenerating chromatin appears *gelatinous* (Figs. VII-2, VII-5, VII-6), particularly in virus infection (e.g.: herpes, polyoma). This is not to be confused with the *'ground-glass'* chromatin pattern, which is formed by uniformly dispersed, *well preserved*, very fine granules of chromatin and is found in increased activity (Figs. VIII-5, VIII-6, XVI-6).

When chromatin becomes pale in retroplasia without further evidence of cell injury or degeneration, it can retain a fine, well-delineated chromatinic pattern. This bland, hypochromatic chromatinic structure is associated with certain longstanding retrogressive changes with decreased activity, but without death or significant degeneration (e.g.: aging, impaired nutrition, involutional hypoplasia). At times it is associated with foreign structures appearing in the nucleus and with longstanding or apparently minimally noxious agents, with which the cells have adapted to coexist (i.e.: bismuth, lead) [33, 309].

6. Chromatinic Rim

The chromatinic rim attached to the inner surface of the inner nuclear membrane, has a characteristic *biphasic* pattern in degeneration: Its inner margin, which is bathed in the parachromatin, appears blurred and smudged with degeneration; while its outer margin, which lies against the nuclear membrane, is retarded from degenerative lysing by the inner nuclear membrane, and retains its sharp delineation for a much longer period of time than do the chromatin/parachromatin interfaces (Figs. VII-1, VII-3, VII-6, VII-7). A comparison of these two borders of the chromatinic rim, the inner and the outer, provides valuable evidence concerning the degree of preservation *vs.* degeneration.

7. Nuclear Membrane Shape

With water *intake* during degeneration, a nucleus can swell and become more uniformly rounded. The nuclear membrane (as evidenced by the profile of the outer surface of the chromatinic rim) becomes stretched smooth and round, ironing out configurations which may be characteristic of the cell type (Figs. VII-2, VII-4) [265, 280, 281]. At times, blebs are evident in the light microscope between the nucleus and the cytoplasm. Many of these are localized blebs or swellings of degeneration in the nuclear cisterna, between the inner and outer nuclear membranes (Figs. V-6; VII-3).

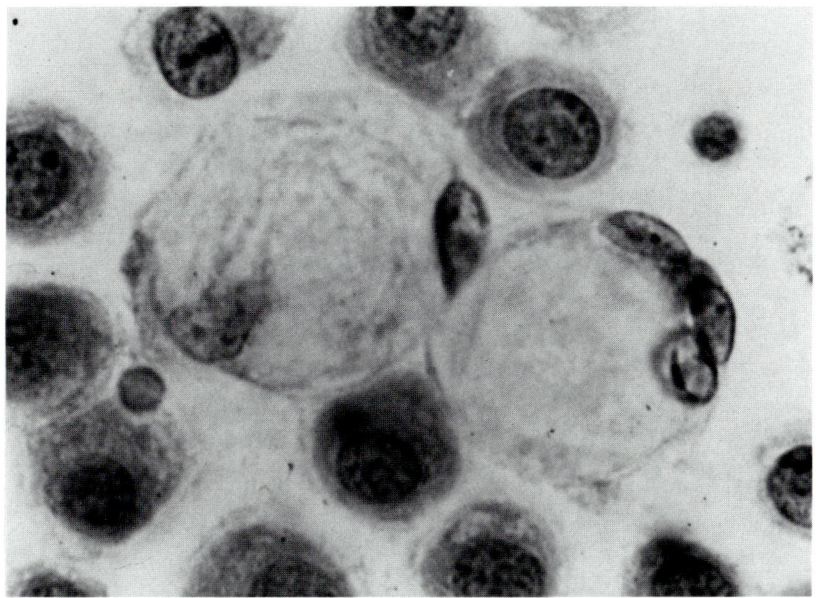

Fig. VII-7: Mild retroplasia in slightly degenerated mesothelial cells of peritoneal fluid. The nuclei have enlarged and there is multinucleation in one cell (right-center) indicating initial proplasia. In addition, and superimposed thereon, most cells show slightly blurred chromatin/parachromatin interfaces. The cytoplasm of most cells is filled with fine vacuoles of degeneration and has a characteristic ecto-endoplasmic biphasic degenerative change (e.g.: cells at 1- and 6-o'clock). While this is usually foamy and appears degenerated, at times it appears glassy and cannot be discriminated from the ecto-endoplasmic changes in atypical keratinization (Chap. XII). Two cells each contain a single large hyperdistended vacuole, so characteristic of degeneration in a watery fluid. This type of degeneration is common in watery fluid (e.g.: body cavity fluids) but rarely occurs in mucoid material (e.g.: respiratory, gynecologic), where such single hyperdistended vacuoles indicate atypical mucus secretion (Chap. XV). Note the window in the center between three cells, indicating that they are each primarily single cells, coming together secondarily, and are not a diagnostic true tissue fragment (Table VI-1). Multiple intercellular vacules of degeneration between cell-pairs at 1-o'clock and 8-o'clock slightly suggest intercellular bridges.

Papanicolaou stain.

VII-7: × 950.

With loss of water, the nucleus decreases in size, the nuclear membrane *wrinkles* and becomes a series of irregular concavities (Figs. VII-1, VII-3, IX-9). When the angle is sharp where two concavities join, this is not to be confused with the sharp angles and spicules in the shape of the nuclear membrane which can occur in a well-preserved malignant cell (Chap. IX). Likewise, when the angles are rounded where the concavities join, this can mimic undulation of a well-preserved proplastic cell (Chap. VIII). Cellular preservation is key in these determinations. The feature of greatest help here in determining the earliest or smallest change, is sharpness *vs.* blurriness of the chromatin/parachromatin interfaces, where the parachromatin bathes the chromatinic net and the inner surfaces of the chromatinic rim.

The chromatinic rim does not appreciably vary in thickness with degeneration, except where marginating degenerating chromatin will cause it to be thickened and to appear 'beaded' (Figs. VII-1, VII-3, VII-5).

8. Nucleolus

Nucleoli usually become small and insignificant in longstanding retrogressive processes, as very little proteins are being manufactured. In more acute degenerations, nucleoli can swell, vacuolate, fragment and, eventually dissolve. If such nucleolar swelling occurs before degenerative alterations become obvious in the chromatin, apparent nucleolar prominence can mislead into assessing biologic activity as that of increased protein production in proplasia, rather than that of degeneration of retroplasia. Very slight smudging of the chromatin may be the only evidence of the true degenerative nature of the process, or a slight pallid nuclear enlargement from decreased control of water.

C. Cytoplasmic Structure

In retroplasia, proteins denature and other constituents of the cytoplasm change, altering many morphologic characteristics. These are more marked with moderate to severe degeneration, death or necrosis than when degeneration plays a more minor role, such as in aging. In the latter, the end products of cell catabolism frequently appear as the highly refractile golden lipochrome granules (e.g.: lipofuscin, hemofuscin, ceroid).

General tinctorial qualities may be altered. Frequently the cytoplasmic structures lose their more characteristic staining reactions, and take on a somewhat uniform, drab acidophilia or nondescript gray. The brilliant transparency of the cytoplasm, which is so characteristic of the Papanicolaou stain and so useful in accurate evaluation of the biologic behavior and diagnosis, is frequently lost to a dull opacity.

Submicroscopic structures aggregate and, thus, become visible. The cytoplasmic texture becomes granular, flocculent, cloudy, foamy, or vacuolated (Figs. VII-1, VII-2, VII-7). Invisible lipids become visible, accumulate and appear as clear droplets (i.e.: fatty phanerosis) [112] and fat increases in amount.

A glassy quality to the cytoplasm, hyaline (*hyalos:* Gr. = glass), may appear when the protein alteration occurs in watery fluids (Fig. VII-7), especially in the outer rim of the cytoplasm. It frequently occurs in tissues associated with high temperatures (e.g.: burns) or particular toxins (i.e.: typhoid). While it can be of any color, it may be strikingly acidophilic (e.g.: Zenker's hyaline degeneration) and mimic a high degree of mature keratinization; like atypical keratin, however, it can be of virtually any color (Chaps. XI, XII).

Degenerating cytoplasm may appear condensed, either diffusely or concentrically. In watery fluids (e.g.: body cavity fluid, cyst fluid, urine, tissue culture) the outer rim of cytoplasm may condense concentrically, taking on a character entirely different from the more homogeneous or flocculent inner cytoplasm about the nucleus (Fig. VII-7). At times this is difficult to discriminate from the ecto-endoplasmic formation of atypical keratinization in squamous cell maturation (Figs. XII-7, XII-16, XII-17), so that this important feature of functional differentiation loses much value in such watery fluids.

When a cell loses control over water balance, increasing quantities of water enter the cytoplasm. Submicroscopic structures swell and the cytoplasmic texture becomes cloudy, foamy, and vacuolated. Eventually it ruptures into the background.

Such retrogressive vacuoles are usually multiple and uniformly dispersed throughout the cytoplasm, without appreciable shifting of the nucleus. This is to be distinguished from a foamy histiocyte with an eccentric nucleus, and from a secretory cell with multiple vacuoles or with a single, pouched-out, hyperdistended vacuole characteristic of secretion which forces the nucleus to the side. When cells undergo retrogression in an extremely watery milieu (e.g.: body cavity fluids, tissue culture, urine) single, large, hyperdistended vacuoles of degeneration may develop which mimic those of secretion, even compressing and forcing the nucleus laterally (Fig. VII-7).

As the degenerating cell continues to increase its intracytoplasmic water, it increases in size and becomes pale. Eventually, the cell membrane ruptures and spills forth cytoplasmic contents as debris in the background (Chap. IV).

The *bare* nucleus or *partially bare* nucleus which results from such cellular degeneration and cytolysis, frequently has alarming degenerating features (e.g.: karyomegaly, hyperchromasia, chromatin clumping, parachromatin clearing) (Fig. VII-1) which can be easily mistaken by the unwary for a cancer cell having an extremely thin rim of cytoplasm (Figs. IX-3; XIII-9; XVIII-5). The crucial difference between the bare or partially bare nucleus and the diagnostic malignant cell with an extremely high nucleo-cytoplasmic (N/C) ratio (e.g.: small cell undifferentiated) is that the former does *not* have *intact cytoplasm* which the latter must have in order for the cell to be diagnostic of cancer (Chap. IX, B). Oil immersion, 100× objective, and precise Koehler illumination (Table IV-1) may have to be utilized when this is a crucial point, in order to obtain sufficient resolution to determine *without doubt* the presence or absence of an intact cell membrane.

The *truly bare* nucleus, therefore, around which *intact* cytoplasm cannot be definitely identified, must be considered a *degenerated* cell and not be used for diagnostic purposes (Fig. VII-1). By extremely bizarre features, it may arouse great suspicion and demand more thorough investigation; however, it should *not* form the *basis* for a *diagnosis* of cancer.

As the cytoplasmic border degenerates and softens, it can become markedly irregular and grotesque before it ruptures. Spidery cytoplasmic projections are formed. Even false pseudopods can occur in a cell in its death agony [40]. In chronic irritation with prolonged degeneration, grotesque structures develop. The *multiple* cytoplasmic tails of *degeneration* are not to be confused with the bizarre cytoplasmic tail formations (e.g.: tadpole, spindle cell) which a cell can form in atypical cytoplasmic thinning in *active* functional differentiation of stratified squamous epithelium (Figs. XII-1, XII-3, XII-10, XII-12) (Chap. XII, D, 2).

Cilia and other structures of the cell membrane (i.e.: microvilli, brush border) (Figs. V-3, VI-4, VI-6) are sensitive to degeneration and show morphologic changes early. Usually cilia are among the first structures to disappear, followed by loss of their terminal plate [108]. In longstanding, mild retroplasia (i.e.: chronic irritation, cigarette smoking), the cilia first decrease their motility and then entirely lose their ability to beat [25], they then shorten their length and the number of cilia per cell is markedly reduced [17, 60, 147].

Retroplastic cytoplasm will at times pinch itself off. This is most striking in viral *infection* of columnar cells of the respiratory tract and of the endocervix (ciliocytophthoria) [50, 203], and in the *physiologic* monthly transition of ciliated endocervical cells to nonciliated cells, shedding detached ciliated tufts (DCT) (Fig. IV-12) [144]. Both of these situations produce a bit of anucleate cytoplasm which curiously retains its actively beating cilia (Fig. IV-12), and leaves behind a nucleus with the remaining cytoplasm retaining its tail.

In chronic retroplasia with minimal degeneration, such as in aging or in an adverse nutritional milieu, the cytoplasm may be only decreased in amount. Cytoplasmic organelles become fewer and less obvious, and there is decreased basophilia of the cytoplasm.

Some cells develop 'wear-and-tear' or 'senility' pigments, lipochromes, in their cytoplasm (e.g.: hemofuscin, lipofuscin, ceroid) with chronic retroplasia. These appear to represent lipid-containing residues of lysosomal digestion or to be associated with oxidation of unsaturated lipids [16, 266]. With age, other lipid droplets and mucoprotein particles also appear in the cytoplasm, as do increased cross-linkages in proteins which cause altered cytoplasmic physical-chemical properties (e.g.: reduced solubility and elasticity, increased viscosity, tinctorial changes).

D. Multiple Cells

As cell borders degenerate and become indistinct, tissue fragments appear to be syncytia. By recognizing these as degenerating pseudosyncytial masses rather than truly multinucleated single cells, much confusion is avoided in arriving at a diagnosis regarding cancer (Chap. IX). Likewise, degenerating *individual cells* of any type (e.g.: histiocytes, epithelial cells, stoma) which come together can lose their cell borders and form pseudosyncytial masses. These all can resemble single multinucleated cells, and the nonpredictability between nuclei may be confused with cancer (Chap. IX) unless careful note is made of their retroplastic changes.

Two useful distinguishing features which can be of great help in identifying the true process(es), should be recognized (ref. Table VI-1). One is that the outer boundary of such pseudosyncytial masses may retain much of the shape of each individual cell's border, giving a 'loop-the-loop' appearance to that boundary (Fig. VII-7). This is in contradistinction to the uniform, linear 'community' epithelial luminal border formed by a fragment of epithelium at its luminal surface (Figs. V-1, V-2), or to the uniform cell membrane of a single, multinucleated cell (Figs. IX-1, IX-6; XII-7, XII-11).

Another useful feature is that single cells coming together frequently retain intercellular spaces or 'windows' between their cytoplasmic membranes (Fig. VII-7, Table VI-1). These intercellular 'windows' occur mainly at the angles where three or four cells come together. As the cellular borders become increasingly degenerated, however, these distinguishing features are lost [94].

VIII. Morphologic Characteristics of Proplasia: Increased General Activity

Most of the significant *morphologic characteristics* of proplasia are present in the *nuclear* structures. The *degree of maturation* attained by the *cytoplasm* yields additional information regarding the type and severity of the proplasia. Both of these reflect the increased level of general biologic activity which is present.

There is a wide range of proplasias, in type and in severity. Most are normal physiologic increases in biologic activities, such as preparation for replication (Fig. II-7), increased secretion, response to physiologic stimuli (e.g.: endocrine, nervous), response to external stimuli (Fig. II-11), healing and repair (Figs. II-8, II-9) [59]. At times, however, proplasia is an abnormal response to repeatedly and chronically encountered injurious agents (Figs. II-10, II-12) which can bear connotation of neoplastic development (i.e.: dysplasia) and even, when *extreme,* of premalignant neoplasia (i.e.: *in situ* preinvasive cancer) [94, 99, 113, 229].

A. Nuclear Structure

1. Nuclear Size and Shape

Karyomegaly is striking in most proplasias. The increase is usually about *four* times the size of a 'normal' nucleus (i.e.: the size of an average well preserved neutrophil, Figs. VI-1, VI-2), but it can enlarge up to *six* times. Enlargement of *more than six times,* is *not* usually due to proplasia, but indicates that one is probably dealing with another process(es) (e.g.: cancer, degeneration, chemotherapy, radiation, folate deficiency, pernicious anemia).

Even with enlargement, however, the proplastic nucleus is *not* pale, as it usually is in retroplasia. On the contrary, it is usually quite hyperchromatic.

The degree of nuclear increase in size, though, is of *no* help in determining the *severity* of the increase of biologic activity. It takes only a *slight* stimulus to increase the nuclear size to *maximum,* and it is unable to become larger with further increased activity. Nuclear size increase, therefore, is not as good a gauge of activity as is hyperchromasia, parachromatin clearing, chromatin granule size, chromatin rim thickness and nuclear membrane undulation.

2. Hematoxylinophilic Chromasia

The hematoxylinophilic chromatinic material becomes hyperchromatic with increased general activity. This is *not* the blurred, smudged hyperchromasia of degeneration (Chap. VII); instead, in the *well-preserved proplastic* cell, the *degree of hematoxylinophilia* is directly proportionate to the amount of DNA-associated protein which is present in the nucleus and, thus, to the *amount of DNA* (Figs. VIII-1, VIII-2, VIII-3).

3. Chromatin

Increased biologic activity places a higher demand upon the nucleus for both increased DNA transcription of RNA and increased replication of DNA. Both of these activities occur in the parachromatin (i.e.: interchromatin areas) where the decondensed or minimally condensed biologically active euchromatin is located (Chap. V).

There is also some inactive heterochromatin in the parachromatin; but, by far, most of the heterochromatin is highly condensed in the chromatinic net and the chromatinic rim. The small amount of euchromatin and heterochromatin in the parachromatin areas during euplasia, impart the pale hematoxylinophilia to the parachromatin in the Papanicolaou stain (Fig. VII-2).

In retroplasia, the degenerating chromatin in the areas of parachromatin, irrespective of type (i.e.: euchromatin or heterochromatin; biologically active or inactive), stick together and move to adhere to the large masses of heterochromatin in the chromatinic net and rim (Fig. VII-1). This *physical* clearing out from the parachromatin of *both eu*chromatin and *hetero*chromatin and *reducing* biologic activity is, *biologically,* diametrically different from the parachromatin clearing in proplasia which facilitates *increased* biologic activity.

In *proplasia,* highly selective amounts of mainly *hetero*chromatin leave the parachromatin and condense in the chromatinic net and rim. Thus, while there is *less total* amount of chromatin left behind in the parachromatin, which results in 'cleared parachromatin', its active *eu*chromatin is proportionately *higher.*

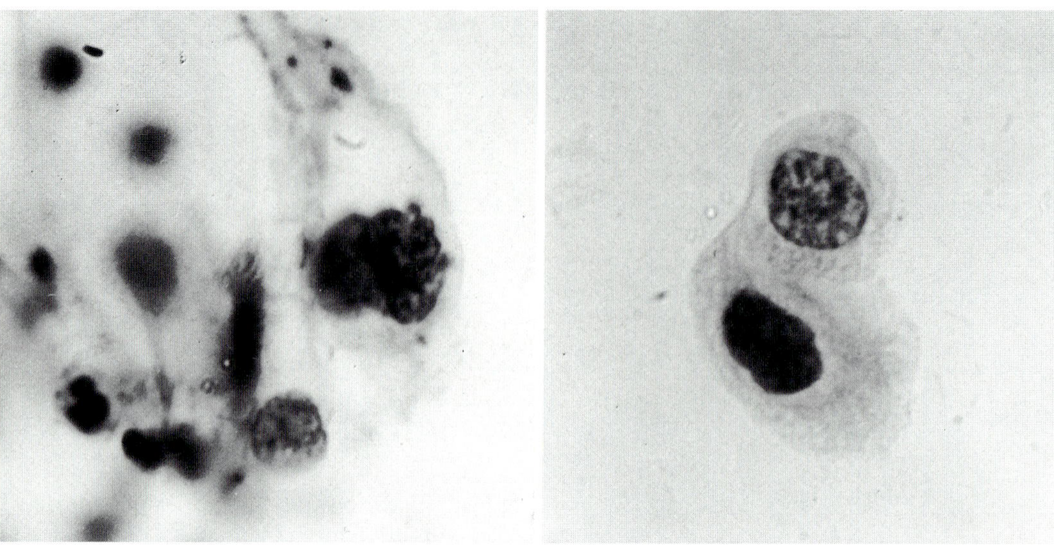

Figures VIII-1, VIII-2: Fast (vagino-pancervical) smears.

Fig. VIII-1: Mild proplasia with mature squamous cytoplasm (mature squamous dys-karyosis) with superimposed degeneration. Slight inflammatory atypia of the cervix. Fine chromatin granules with minimal hyperchromasia of a single nucleus (below); compare its slightly increased size (2 ×) over the neutrophil. Multinucleation of the upper two nuclei is due to initial proplasia of stimulation. Most of their hyperchromasia is due to slight degen-eration (note slight blurring of most of the chromatin/parachromatin interfaces) superim-posed on top of proplasia. While some of the nuclear membrane irregularities may be wavy from the stimulation, they are masked by the marked wrinkling of degeneration.

Fig. VIII-2: Slight to moderate proplasia. There is quite a bit of degeneration super-imposed on the lower dark nucleus, with blurred hyperchromasia, shrinkage and wrinkling (karyopyknosis) in a moderately mature dyskaryotic cell. The upper cell has scanty, imma-ture cytoplasm, but only *moderate* proplastic nuclear changes which Papanicolaou termed dyskaryosis. There is undulation of the nuclear membrane, uniform thickness of the chro-matinic rim, fine chromatin granules with a few moderate granules, moderate hyperchro-masia, and parachromatin clearing uniformly dispersed about the chromatinic net.

Papanicolaou stain.

VIII-1, VIII-2: × 1,000.

Morphologically, the parachromatin clearing of *retroplasia* is recog-nized by the *blurriness* of the chromatin/parachromatin interfaces sur-rounding the cleared areas (Figs. VII-1, VII-3). Conversely, the parachro-matin clearing of *proplasia* is distinguished from that of retroplasia, by the *well-preserved, crisply defined,* 'cookie-cutter' chromatin/parachromatin in-terfaces of a well-preserved nucleus (Figs. VIII-3, VIII-4).

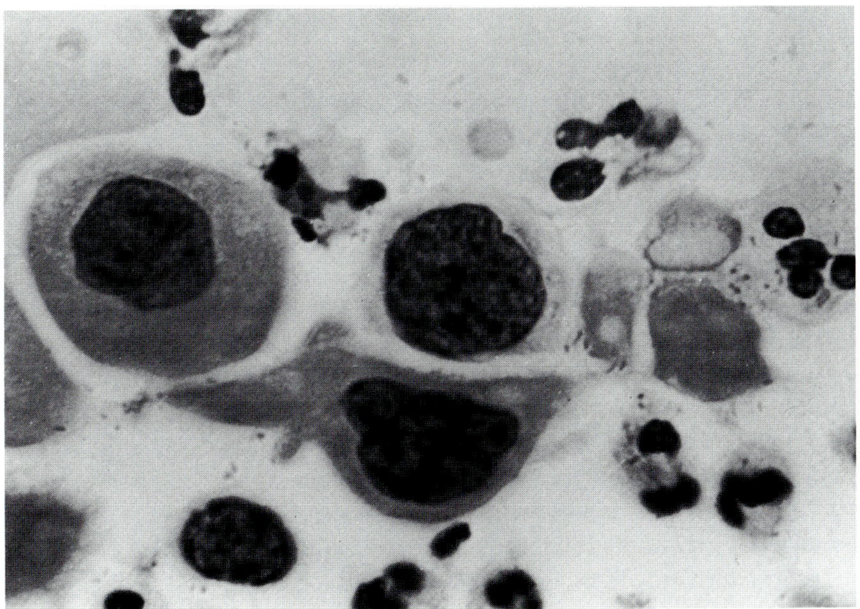

Fig. VIII-3: Fast (vagino-pancervical) smear. *Marked proplasia.* The upper nucleus in the center is a characteristic prototype of severe dyskaryosis, which could have been shed from either an *in situ* carcinoma or a marked dysplasia, depending upon the maturity of its cytoplasm. The nuclear membrane is severely *undulated* with uniform *thickening* of the chromatinic rim. The chromatin is finely granular throughout, with many moderate granules superimposed and a few coarse granules (up to the size of nuclear lobes of a neutrophil). There is marked hyperchromasia (note, it is *not* overstained as there is good nuclear detail in the neutrophil on the right) and the markedly cleared parachromatin is uniformly dispersed about the chromatinic net. The cytoplasm appears very scanty; but careful inspection reveals that it runs as a tail to the lower left, and across a bridge on the right to include the first small ovoid drop of cytoplasm; thus it is an immature dyskaryotic cell and is not diagnostic of *in situ* carcinoma. The other two cells have moderately mature cytoplasm (adequate amount and thick). The nucleus of the lower one has quite a bit of degenerative blurring, which could account for most of its hyperchromasia.
Papanicolaou stain.
VIII-3: × 1,500.

In proplasia, chromatin is condensed or aggregated into *granules.* Their size (i.e.: small, medium, coarse) increases with increased activity.

Uniformity of the chromatin/parachromatin pattern, is an additional morphologic characteristic of proplasia. The two most important features to evaluate in assessing the significance of pattern uniformity, is the uniform *dispersion* of the chromatinic granules within the parachromatin, and a uniform *degree of clearing* of the parachromatin throughout virtually the entire nucleus.

4. Parachromatin

Thus, in contrast to the hematoxylinophilic *hyper*chromasia of the *chromatinic net* and the *chromatinic rim* in proplasia, the *parachromatin* becomes *hypo*chromatic, or 'cleared'. With increased cellular activity of proplasia, greater amounts of the finely dispersed inactivated *hetero*chromatin, which normally remain in the parachromatin of the resting euplastic cell, move from the parachromatin to the larger masses of heterochromatin in the chromatinic net and rim (Figs. VIII-3, VIII-4). This allows the remaining parachromatin to be *richer* in *eu*chromatin, and for the latter to be more *highly dispersed* in the nucleoplasm of the parachromatin, apparently facilitating its biologic activities.

This parachromatin clearing is directly proportionate to the degree of biologic activity of the cell (cf.: Figs. VIII-2, VIII-3). In extreme proplasia, parachromatin may become markedly cleared, or *hypo*hematoxylinophilic, in contrast to the markedly increased *hyper*hematoxylinophilia of the chromatinic net and rim. However, the granules of the chromatinic net still tend to be *uniformly dispersed* throughout this cleared parachromatin.

5. Chromatinic Net

The most frequent shape of chromatin structures in proplasia, is that of *granules*. These *increase* in chromasia, in number, and in *size* (i.e.: fine, moderate, coarse) proportionate to the degree of increased general activity (cf.: Figs. VIII-1 through VIII-3).

Fig. VIII-4: A proplastic cell of squamous metaplasia of the bronchus. Note the markedly *wavy* nuclear envelope. It is the *inner* nuclear membrane which determines the shape of the outer profile of the chromatinic rim. The outer nuclear membrane (om) is thrown into marked irregularities in numerous areas. Where there are deep folds in the inner nuclear membrane (thus the profile of the chromatinic rim) note the numerous nuclear pores (np), and the accumulation of ribosomes (r) in the cytoplasm of the trough in the wave. This organization is very reminiscent of a secretory gland. In one of the areas of deeply furrowed membrane, the pore is cut tangentially (arrow), providing a cross-section of its protein complex. Chromatin is in fine, moderate, and coarse granular accumulations with areas of extreme interchromatin (parachromatin) clearing.
Osmium/methacrylate.
VIII-4: × 25,000.
(Frost, Erozan, Donovan; unpublished electron micrograph.)

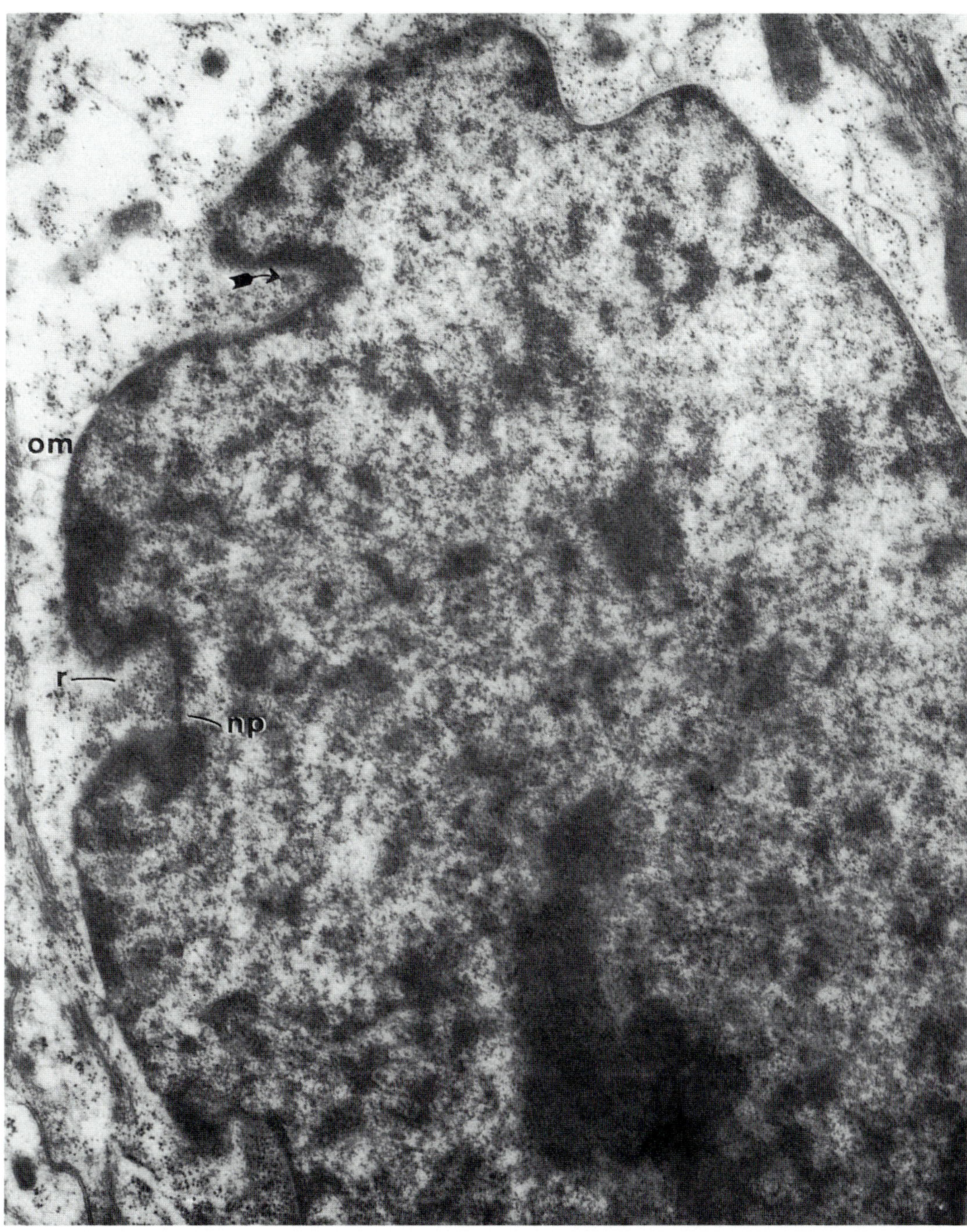
om
r
np

When all granules are *fine* and uniformly dispersed, this pattern is frequently referred to as 'ground glass' (Fig. VIII-1). As activity increases, they increase in size to *moderate* granules (Fig. VIII-2).

When there is markedly increased activity, *coarse* granules appear throughout the nucleus, of about the size of a lobe of a neutrophil nucleus (Fig. VIII-3). Conspicuous by their absence, are the large clumps or aggregate masses of chromatin, which frequently characterize invasive cancer (Figs. IX-1, IX-8).

There is a *uniform pattern of dispersion* of these chromatin granules within the parachromatin throughout the nucleus of proplasia. This is similar to the uniformity of dispersion of the fine chromatin threads and granules in euplasia, and is in contrast to the marked irregularity of chromatin dispersion encountered in discriminable nuclei of malignant neoplasia (Figs. IX-1, IX-8, IX-10, IX-13).

6. Chromatinic Rim

The chromatinic rim is of uniform thickness throughout its extent in a *given* cell, with very little tendency for it to vary in thickness except for a somewhat granular appearance. These granules of the chromatinic rim tend to be approximately the same size as the granules of the chromatinic net.

Some proplastic nuclei have extremely thin chromatinic rims, while nuclei of neighboring cells from the same lesion and in the same biologic situation, have thicker rims. The chromatinic rim of *each* nucleus, however, is uniformly thick throughout that given nucleus (Figs. VIII-3, VIII-4).

This *uniformity* of thickness is a hallmark of *euplasia* as well as *proplasia*. It is a valuable differential characteristic from the marked *variation* in chromatinic rim thickness in the same cell which occurs in some discriminable cells from malignant neoplasia (Figs. IX-8, IX-13).

7. Nuclear Membrane Shape

The nucleus of proplasia is generally *rounded* or *vesicular* (*vesicula:* L. = small bladder, small blister). It frequently gives the appearance that it is filled with fluid under pressure.

The nuclear membrane becomes undulated (wavy) with increased biologic activity (Figs. VIII-3, VIII-4, IX-9). This wavy morphologic alteration to the membrane, appears to represent a structural response of the nucleus to a biologic need for increased nucleo-cytoplasmic exchange [295]. The nuclear membrane is the interface across which this nucleo-cytoplasmic activity takes place [142, 223]. It encloses what is essentially an incompressible, but deformable substance. By forming such a waved contour, the surface area of the nuclear membrane is enlarged for increased nucleo-cytoplasmic exchange without changing nuclear volume [295].

In the *trough of a wave* in the nuclear membrane of a very active proplastic nucleus, there is frequently to be noted by electron microscopy an increased number of nuclear pores concentrated in that portion of the membrane (Fig. VIII-4). In addition, there is a congregation of free ribosomes in the cytoplasm of the trough, somewhat reminiscent of the formation of a gland in a mucous membrane to facilitate secretion. This structural arrangement may well allow the proteins of the nuclear pore a sheltered area for them to complete their complex task of monitoring and altering the substances (e.g.: ribosomal RNA and associated proteins) passing through the pore (Chap. V) before they are released into the cytoplasmic pool (e.g.: of ribosomes).

This undulation of the nuclear membrane increases in direct proportion to the increase of cellular activity. At times these undulations of increased biologic activity are localized to one area of the nuclear surface. If this is the case, it is usually the area adjacent to the region of the center of the cell, the cytocentrum and where there is centriolar activity.

It may result in nuclear indentation on the side near the cytocentrum, and is more common in the monocyte-macrophage cell types forming the hof (Fig. IV-1). It also occurs in secretory cells at the luminal end of the nucleus (Figs. V-1, V-2), while the other epithelial and connective tissues characteristically have generalized nuclear membrane undulation.

Undulation is not a characteristic of euplasia (Chap. VI), where the nucleus is *round or rounded* (Fig. VI-1). In retroplasia (Chap. VII), nuclear membrane *wrinkling* occurs with loss of water (Fig. VII-1). The gentle *waves* of proplasia (Fig. VIII-3), with a uniformly thick chromatinic rim, are in striking contrast to the macabre and extreme *irregularities* (Fig. IX-4, IX-9) which are characteristic in some discriminable nuclear membranes of cells from malignant neoplasia (Chap. IX).

8. Nucleolus

Nucleoli serve as a hallmark of protein production, either for *intrinsic* cell use or for *extrinsic* secretion. They can become exceptionally large and extremely *prominent* (even one-third of the nucleus) in nuclei of cells undergoing *non*malignant greatly increased metabolic activity, such as in rapid cytoplasmic growth and cellular replication or in marked secretion.

In cells functionally differentiating toward *stratified squamous epithelium,* however, the *nucleoli disappear,* in even rapidly growing epithelium, as each cell matures and nears the surface (Figs. XVI-3, XVI-7). Thus, proplastic cells which are shed from stratified squamous epithelium virtually never have nucleoli (Figs. VIII-3; XVI-6) (the rare exceptions include: severe atrophy, postradiation atypia, extreme atypical repair [114]).

The presence of prominent nucleoli in cells appearing to be from a proplastic *squamous* lesion, therefore, should evoke caution that one may be dealing with an invasive cancer (Fig. XII-13). In proplasia, in all cell types other than squamous (e.g.: active columnar cells, regenerating muscle, growing fibroblasts, immunoblasts), nucleoli increase in size and number, and become very conspicuous. Thus, prominent nucleoli are frequent in exfoliated proplastic columnar cells.

Nucleolar prominence is thus a characteristic morphologic feature of proplasia when either growth or secretory production, or both, is actively occurring. The one *exception* is *stratified squamous epithelium.*

An important characteristic of extreme proplasia is that both chromatin aggregates and prominent nucleoli have a relatively *rounded* contour (Fig. V-2). This is in contrast to the extremely sharp, pointed, jagged, angular features of the same two structures which are noted in some discriminable invasive malignant neoplasms (Fig. XVIII-1).

9. Mitosis

Increased general activity can bring forth *increased numbers* of *normal* mitoses. Not only are *metaphase* figures increased, but also *impending prophase, prophase,* and *telophase* figures.

In *impending prophase,* there is condensed beading of a *well-preserved* chromatinic rim (i.e.: marginated chromatin granules becoming a part of the chromatinic rim) as well as prominent granules in the chromatinic net. The chromatinic rim remains *intact* between the marginated granules (i.e.: an *intact* true nuclear membrane). This may be difficult or impossible to discriminate from a *markedly* dyskaryotic nucleus of severe proplasia or *in situ* carcinoma and may, in fact, represent the same phenomenon.

In *prophase,* the nucleus appears essentially as it does in impending prophase, *except* for the disappearance of segments of the chromatinic rim between the marginated granules. This signifies the disappearance of the true nuclear membrane with its lamina and marginated chromatin.

In *metaphase* the shape of the nucleus is lost, as is all semblance of chromatinic rim or evidence of true nuclear membrane (Fig. IX-14). The individual chromosomes are recognizable individually, or in pairs. They are spread throughout the cytoplasm (*en face* view of metaphase plate) or are lined up in two rows (profile view of the plate) often with the mitotic spindle and centrioles.

Telophase appears as two small, adjacent, identical cells with small, hyperchromatic, well-preserved, identical nuclei (Fig. XVI-7). They may appear to be karyopyknotic, but their identical features and their small, rounded cytoplasm discriminate them. The tight, dark nuclei begin to loosen up and enlarge, and the nuclear membrane again becomes visible as the outer profile of the newly-formed chromatinic rim.

10. Multinucleation

Multiple nuclei are common in proplasia (Figs. VIII-1, VIII-5, VIII-6). Such daughter nuclei within the same cell, however, retain characteristics which are markedly similar to each other. This is valuable in contrasting and recognizing the anisokaryosis present in some discriminable multinucleated invasive cancer cells (Figs. IX-5 through IX-7). In the proplastic multinucleated cell, nuclear size and shape may vary somewhat; but other significant characteristics (i.e.: parachromatin clearing, chromatin pattern, significant difference in hematoxylinophilia, numbers of nucleoli) retain the sameness of a 'mirror image' from one to another, while they can vary markedly in malignant neoplasia (Chap. IX).

B. Cytoplasmic Structure

While the morphology of the nucleus imparts most of the information regarding the degree and significance of increased activity, cytoplasmic features do have value. The *degree* of *cytoplasmic maturation* yields important information regarding the *severity* of the proplasia.

As a general rule, the *higher* the level is of typical cytoplasmic maturation, which a proplastic cell attains: the more *normal* is its *character,* the *lower* is the degree of *proplastic* activity, and *less likely* is the probability of *neoplastic* behavior.

Conversely, *more immaturity* and *atypical functional differentiation* in the cytoplasm of a proplastic cell, bespeak: a *greater* degree of increased *activity, more atypical behavior,* and *greater* probability of *neoplastic initiation.*

In evaluating cellular characteristics, it is convenient to adjudge the degree of maturation of the cytoplasm as being *immature, moderately mature,* or *mature* (Fig. VIII-8). The typicality or atypicality of functional differentiation is discussed in Section C where those differentiating toward stratified squamous and columnar epithelium will be considered in more detail, but the same general principles apply to virtually all cell types.

C. The Cell as a Whole

Papanicolaou recognized the *prominent role* which the *nuclear* structure plays in evaluating the morphologic picture of proplasia. He applied the term 'dyskaryosis' (*dys-:* N.L., Gr. = ill, difficult, faulty, abnormal; *-karyon:* Gr. = nucleus) to the condition of *cells* with this abnormal pattern of proplasia. He then applied it to cell types, according to their cytoplasmic functional differentiation (e.g.: intermediate cell dyskaryosis, parabasal cell dyskaryosis, endocervical cell dyskaryosis) [201, 300].

Dyskaryosis is a *cytologic* term and is *not* to be used as a clinical term. It is most valuable to denote the degree of severity of the proplastic nuclear change

---

Figures VIII-5 through VIII-7: Vaginal pool smear of a 61-year-old female. The specimen had been allowed by the clinician to slightly air-dry before fixation. Note the 'bland' nuclear artifacts from air-drying.

Fig. VIII-5: Parabasal cell atrophy (i.e.: teleatrophy; M.I.: 100/0/0 [99]) with atrophic vaginitis. Only one cell, at the extreme lower left, has a normal nucleus the size of an average well-preserved neutrophil. The other five are proplastic cells, shed from an area injured from severe chronic irritation from the atrophic vaginitis and from harsh douches which the patient had self prescribed. Nuclei are four to six times the size of the average neutrophil. There is frequent multinucleation and the chromatin is in fine granules uniformly dispersed throughout a slightly cleared parachromatin ('ground glass' chromatinic net) with only an occasional moderate-sized granule. There is undulation of the nuclear membrane. The chromatinic rims are slightly thickened and vary from nucleus to nucleus, but each is of uniform thickness throughout its own nucleus. There is slight hematoxylinophilic hyperchromasia.

Fig. VII-6: The *moderately mature dyskaryotic cell* (upper left) shows some degenerative retroplastic changes (e.g.: chromatin/parachromatin interfaces more blurred than the outer surface of the chromatinic rim) so that size, chromatin clumping, nonuniform parachromatin clearing, and hyperchromasia cannot be accurately assessed. The large immature dyskaryotic cell is better preserved.

Fig. VIII-7: Two weeks following estrogen therapy. Minimal dyskaryosis with atypical maturation remains. This disappeared following a few more weeks of continued estrogen therapy.

Papanicolaou stain.

VIII-5: × 800; VIII-6: × 2,000; VIII-7: × 500.

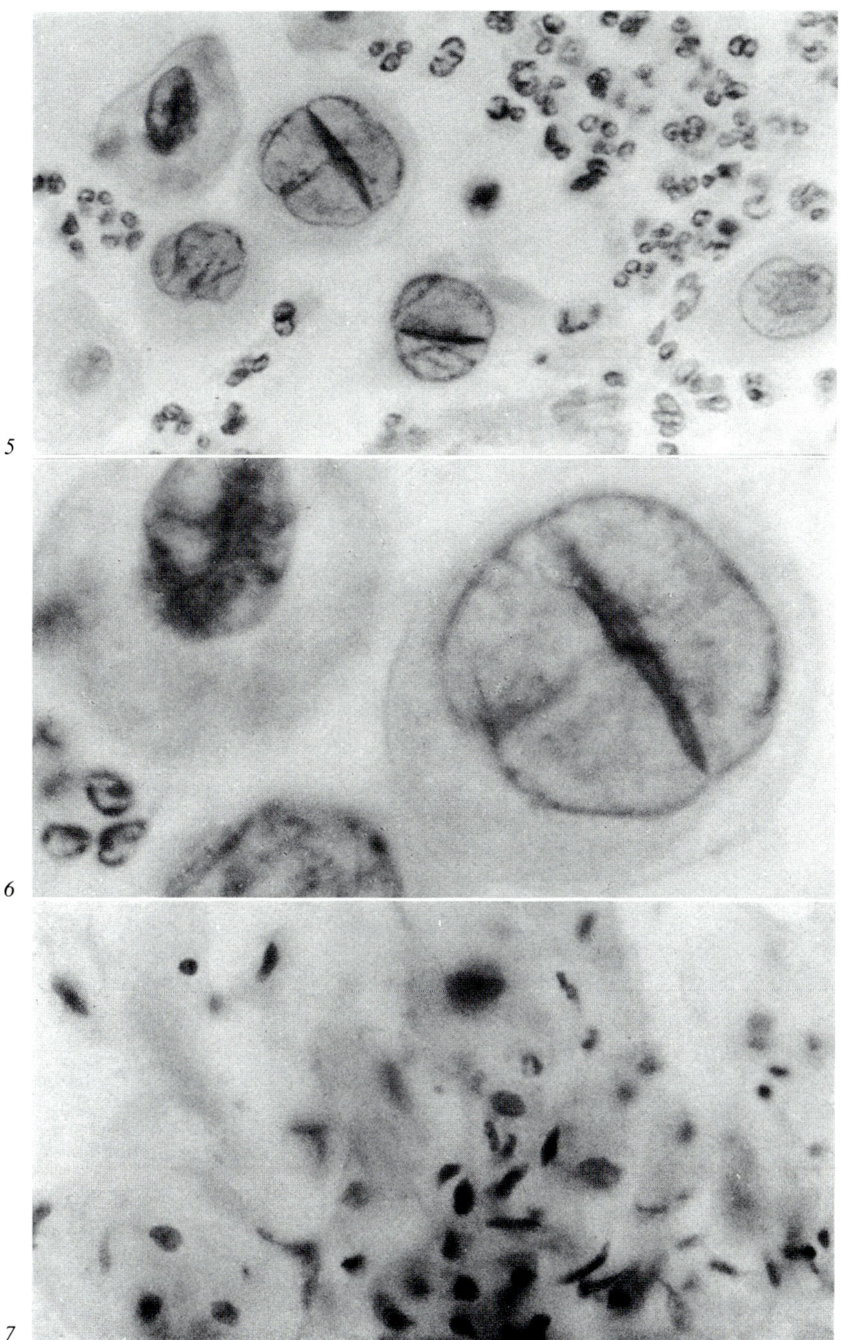

5

6

7

(i.e.: slight, moderate, marked dyskaryosis) and the degree of maturity of the cytoplasm (i.e.: mature, moderately mature, immature) (Fig. VIII-8) to help assess the *level* and *atypicality* of the cellular and tissue activity.

When only one part of the picture is present, at times only that structural finding or feature is referred to, *alone* (e.g.: karyomegaly, 'ground glass', hyperchromasia). It usually is most informative and valuable, however, to recognize the possibility of its being a part of a proplastic process, rather than an isolated and merely 'interesting' morphologic finding. In *marked* atypia, marked dysplasia, and CIN III the morphologic picture of proplasia is full-blown (Figs. VIII-5 through VIII-7; XVI-4 through XVI-9) [69, 94, 99, 155] with marked nuclear dyskaryosis and immature cytoplasm in many cells (Fig. VIII-3). With *in situ* carcinoma the cytoplasm is even scantier (Figs. XIII-1, XIII-3, XIII-9, XIII-13).

Dyskaryosis has been greatly misunderstood; at times it has been discarded, misused and abused. It has even been employed, for a short period of time, as a connotative clinical diagnosis. It is of greatest value, however, when its use is reserved entirely as a denotative, *morphologic* term identifying this family of cellular structural changes.

1. Proplastic Epithelium with Stratified Squamous Maturation

When the surface cells of a proplastic epithelium are *stratified squamous* as they mature and *lie parallel* to the lumen and to the basement membrane, *squamous dyskaryotic cells* of varying degrees of maturity exfoliate therefrom. If it is native keratinizing stratified squamous epithelium (e.g.: oral, vaginal), one proper term is *atypical squamous epithelium*. If it is squamous metaplasia, the preferred term is *atypical squamous metaplasia*. If the exact character of the source is not clear (e.g.: sputum) or it is immaterial, the preferred term to cover both is simply *squamous atypia*.

Nomenclature for essentially the same atypias, actually varies greatly. It includes the three just mentioned with modifiers (e.g.: slight, moderate, or marked) to indicate varying degrees of atypia, as well as many more connotative terms moving in and out of favor and usage (e.g.: reactive or regenerating squamous epithelium, atypical hyperplasia, basal cell hyperactivity, spinal cell hyperactivity, dysplasia, intraepithelial neoplasia).

The *mature* squamous dyskaryotic cell (Figs. VIII-1, VIII-8) has abundant cytoplasm, which matures in essentially the same typical fashion as does the cytoplasm of the normal squame (intermediate or superficial cell). Frequently, there is less cytoplasm in these atypical squames, so that it is a

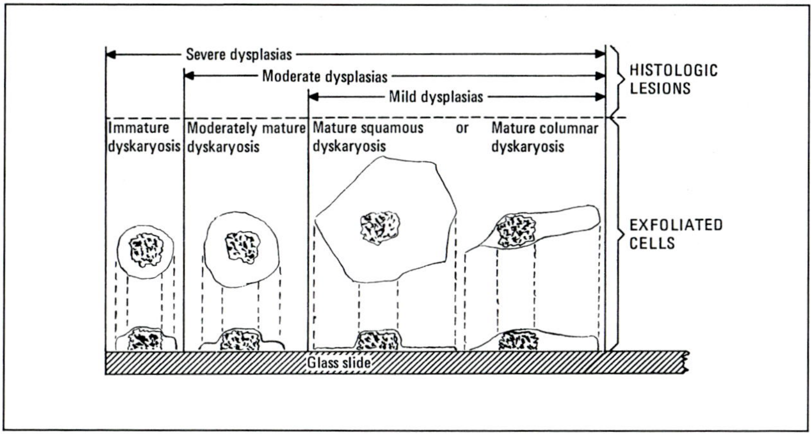

Fig. VIII-8.

smaller wafer than the normal squame. Thus it is symmetrically thinned into a polygonal wafer, or squame, with the nucleus centrally located. In severe irritations, some atypical functional differentiation may be present (Chap. XIII), which includes bizarre asymmetrical thinning of the cytoplasm with tail formation (i.e.: tadpole cell; spindle cell) (Figs. XII-1 through XII-4), atypical stratification (i.e.: pearl formation) (Fig. XI-6), and varying degrees of keratinization.

The centrally placed nucleus exhibits degrees of nuclear changes of proplasia which reflect the intensity of the cell's biologic activity (Figs. VIII-6; XII-3; XIII-5). The nucleus is enlarged, the nuclear membrane undulated, and the chromatinic rim is of uniform thickness. The chromatinic net is granular, hyperchromatinic, and uniformly dispersed within cleared parachromatin.

Nucleoli are *absent* in squamous atypia or dyskaryosis, in contrast to the prominent nucleoli in the mature *columnar* atypical or dyskaryotic cells (Fig. VIII-8) and dyskaryotic cells shed from other cell types (e.g.: muscle, fibrocytes, mesothelium, macrophages). Furthermore, this is in sharp contradistinction to invasive squamous cell carcinoma shedding nucleoli in its maturely differentiating invasive squamous carcinoma cells. The latter may be morphologically identical to the mature squamous dyskaryotic cell, except for nucleoli (Fig. XII-13).

The *moderately* mature squamous dyskaryotic cell (Figs. VIII-2, VIII-8) usually sheds from an epithelium which is maturing to stratified squamous and which is *moderately* increased in activity. It has moderately abundant cytoplasm which is unable to thin or squamify before exfoliation; thus, it retains thick cytoplasm from nuclear border to cell border.

As the proplastic lesion becomes more severely active and atypical, a higher *percentage* of the dyskaryotic cells which are shed from it are only able to mature their cytoplasm to this level of *moderate* maturation (Fig. VIII-8) [70]. Furthermore, their nuclear features are more marked than those of the mature squamous dyskaryotic cell (i.e.: more hyperchromatic, more coarsely granular chromatin, thicker chromatinic rim, increased nuclear membrane undulation). Nucleoli are not prominent, in contrast to the prominent nucleoli of moderately mature *columnar* dyskaryotic cells.

The *immature* dyskaryotic cell (Figs. VIII-3, VIII-8) has much *less* cytoplasm, which is *thick* and is not squamified. While very scanty, it is not the extremely thin rim of cytoplasm present in the discriminable cells shed from *in situ* carcinoma (Figs. XIII-1, XIII-9, XIII-13). The high nucleo-cytoplasmic ratio of the immature dyskaryotic cell, is a hallmark of its immaturity and bespeaks the *inability* to *mature* of the cells comprising the epithelial lesion from which they are shed (e.g.: marked squamous atypia; markedly atypical squamous epithelium, markedly atypical squamous metaplasia; severe dysplasia; severe intraepithelial neoplasia; CIN-III). The more severely proplastic and dysplastic the lesion becomes, the higher is the percentage of the *immature* dyskaryotic cells among the dyskaryotic cells which are shed from the lesion (Fig. VIII-8) [70].

2. Proplastic Epithelium with Columnar Maturation

When the surface cells of a proplastic epithelium mature to *columnar* shape and orient *perpendicularly* to the lumen, they shed as *atypical columnar cells* or *columnar dyskaryotic cells* (Fig. VIII-8) of varying degrees of atypia and of maturity. Terms employed for this proplastic epithelium include regenerative columnar epithelium, reactive columnar epithelium, atypical columnar epithelium, reserve cell hyperplasia, subcolumnar cell hyperplasia, subcylindrical cell hyperplasia, presquamous squamous metaplasia, and transformation zone.

The *mature columnar dyskaryotic cell* (Fig. VIII-8) has abundant cytoplasm which is *thick* and assumes a typical *columnar* shape. A tail of cytoplasm is frequently recognizable at the basement membrane end. The luminal end has the bulk of cytoplasm. The luminal border usually is crisp but may be indistinct. It may have cilia, a brush border or other microvillous processes, maturing much as a euplastic columnar cell of that particular anatomic area except for its nucleus.

The dyskaryotic nucleus is eccentrically located toward the basement membrane end of the cell. In contrast to the nucleus of the mature squamous dyskaryotic cells, *nucleoli* are *prominent* in these mature columnar dyskaryotic cells (Fig. VIII-8).

The *moderately mature* and the *immature dyskaryotic cells* are shed from the luminal surface or from the subluminal areas (e.g.: subcylindrical cells, reserve cells, parabasal cells) of proplastic epithelium which is maturing toward columnar. Their cytoplasm is thick and of decreasing amount with increasing immaturity. They are usually round or rounded, but each may have a small tail from its attachment to the basement membrane. They also have *prominent nucleoli,* like the mature columnar dyskaryotic cells, but otherwise are similar to the squamous dyskaryotic cells of the same degree of cytoplasmic immaturity (Fig. VIII-8).

3. Proplastic Tissues with Other Maturation

Virtually all living tissues undergo progressive, altered behavior with increased general activity. For example, reaction to injury with regeneration in *skeletal muscle* (e.g.: X-ray, trauma, irritants) produces cells having the same large, hyperchromatic, vesicular, dyskaryotic nuclei with undulated nuclear membranes, in cytoplasm with varying degrees of rhabdomyocyte maturation. They too have prominent *nucleoli,* as with columnar dyskaryotic cells.

Macrophages can become extremely proplastic in the presence of irritants and other 'calls to duty' for their phagocytic function. Many nuclear features of dyskaryosis are present to bespeak this increased activity. Even mitoses can be numerous in this cell type, when the need for additional phagocytes is great. The chromatin dispersion pattern throughout the parachromatin can be less uniform than it is in dyskaryosis of epithelial cells, and there can be one or two large chromatin agglutinations in addition to the granules so typical of dyskaryosis. Such 'active' macrophages characteristically have blurred chromatin/parachromatin interfaces mimicking retroplasia.

These alterations in biologic activity are reflected in the morphology of the cells in the tissues, as well as in the exfoliated cells. As the stimulant is removed from the milieu, or healing is complete, proplasia gradually changes to euplasia if irreversible changes (e.g.: initiation, mutations) have not occurred (Figs. II-11, II-15).

IX. Morphologic Characteristics of Malignant Neoplasia: General Activity of Invasive Cancer

Most morphologic changes associated with cancer, which are valuable for correctly adjudging a cell's malignant behavior, are contained in the *nucleus* [92, 94, 99, 103, 109]. At times, an atypical *relationship* of *cytoplasm to nucleus* assists in determining malignant behavior; but most of the information which the cytoplasm has to give, is in determining cellular functional differentiation. *Nucleus-to-nucleus relationships* within *diagnostic true tissue fragments* (DTTF) (Table VI-1) and within *multinucleated cells,* can be made to yield additional valuable information for determining malignant behavior.

There are three general trends of greatest diagnostic value in the morphologic characteristics of malignant neoplasia (invasive cancer): *sharp angularity* of otherwise rounded or oval prominent structures (e.g.: nucleoli, chromocenters, nuclear membrane); *unpredictable irregularity* of otherwise uniform and predictable characteristics or structures (e.g.: degree of hyperchromasia, degree of parachromatin clearing, size of cleared parachromatin areas, thickness of chromatinic rim); and *extremes* of cellular features which in euplasia are moderate and used as reference, or are subtle and inconspicuous (e.g.: chromatin clumping, parachromatin clearing, nuclear size, amount of cytoplasm).

A. Nuclear Structure

Nuclear structures contain most of the valuable malignant criteria. They involve, chiefly, alterations in the structure of chromatin, parachromatin, nuclear membrane, nucleolus, and nuclear relationships to the cytoplasm (Figs. IX-1 through IX-8).

1. Nuclear Size and Shape

Extremes in either nuclear size or shape, or both, give some indication of neoplasia in the proper set of circumstances. There is frequently great size difference in retroplastic and proplastic processes, which weaken this criterion; however, extremely large nuclei, which are well preserved and have scanty, well-preserved cytoplasm, are very suggestive of malignancy.

Nuclei change their shape from round, according to physiologic need (e.g.: indentation at the cytocentrum (i.e.: hof) of a histiocyte (Fig. IV-1), molding from secretion, undulation of the nuclear membrane from increased nucleo-cytoplasmic metabolic interchange). But when shape alteration is extreme and grotesque (e.g.: off round, elongated, sharp, pointed, jagged, angular) for no apparent reason, it greatly favors malignancy (Fig. IX-9) (more at 7, below).

2. Hematoxylinophilic Chromasia

Most malignant cells display hematoxylinophilic *hyper*chromasia. When the cell is well preserved, this finding directly reflects *increased DNA and associated proteins* (Figs. IX-1 through IX-8) [110, 111, 189, 194, 305].

It is to be carefully noted, however, that hyperchromasia, *per se,* is *not* discriminating and can be present in all three: retroplasia (Chap. VII), proplasia (Chap. VIII), and malignant neoplasia. Careful morphologic differentiation must be made between these three, in order for this feature to be of the great and distinctive value of which it is capable.

The degree of nuclear staining by hematoxylin (hematoxylinophilia) *can* be made proportionate to the amount of chromatinic material present; but significant factors and artifacts must be controlled. The greatest of these are degeneration and reproducibility in staining and preparation [7–14, 100, 124].

Hematoxylin stains the proteins associated with the DNA, not the DNA, itself. With degeneration, the DNA-associated proteins denature and alter their hematoxylinophilia. The result of this is that the stainability and proportion (of this degraded protein *to* DNA) becomes *un*predictable and, therefore, unusable as a quantifier.

A saving factor for the careful observer, however, is that with this degeneration the chromatin/parachromatin interfaces become blurred (Chap. VII) and signal the inaccuracy. In well-preserved, properly stained, and properly prepared cells [7–14, 100, 124], the morphologic finding of *crisp* chromatin/parachromatin interfaces (i.e.: the 'cookie-cutter' borders of well-preserved chromatin bathed in parachromatin) is valuable evidence that any hyperchromasia which may be present is *not* due to degeneration, but signifies *increased* biologic activity.

The hyperchromasia of both proplasia and malignant neoplasia in well-preserved cells, is due to this increased DNA and its associated proteins. This is identifiable and able to be evaluated in cells with well-defined, crisp chromatin/parachromatin interfaces.

3. Chromatin

A major characteristic of most cancers is that the pattern of their chromatinic material can be markedly abnormal in malignant neoplasia (Figs. IX-1 through IX-8). Usually the overall nuclear pattern is greatly altered. The chromatin condenses and *clumps* into large angular masses, unpredictably distributed in the parachromatin in a most irregular pattern (Figs. IX-1 through IX-8, IX-13).

Figures IX-1 through IX-8: Criteria of malignant neoplasia. There is marked *parachromatin* clearing in many of these malignant cells (Figs. IX-1, IX-3, IX-4, IX-6 through IX-8). In some (Figs. IX-3, IX-4, IX-6, IX-8) there is variance in the *degree* of clearing throughout a given nucleus, and marked *variation* in the *size* of the areas of clearing (Fig. IX-8). In two, however (Figs. IX-1, IX-7), cleared parachromatin is uniformly dispersed; this feature does not discriminate malignant neoplasia from extreme proplasia in these cases. The *chromatinic net* has sharp angularity of massive chromocenters (Figs. IX-1, IX-8). The *chromatinic rim* is irregular in thickness (Figs. IX-1, IX-5, IX-6, IX-8) with heterochromatin amassed in some areas, and with these masses having sharp angularities (Figs. IX-1, IX-6: lower right); and other areas are devoid of chromatinic rim, so that the invisible nuclear membrane appears merely as a faint boundary between nucleus and cytoplasm (Figs. IX-6: 4-o'clock; IX-8: 7- and 9-o'clock). The *nuclear membrane* (i.e.: profile of the outer surface of the chromatinic rim) is bizarre in shape (Figs. IX-1 through IX-4, IX-6) with many small irregularities which are *not* from wrinkling (i.e.: *not* degeneration) (Fig. IX-1), outward pointed spicules (Fig. IX-2), right-angle 'bites' (Fig. IX-3: lower nucleus, 1-o'clock), sharply angled indentations (Fig. IX-4: 12-o'clock, lower nucleus) and deep irregularities (Fig. IX-6) for no apparent reasons. There is a large *nucleolus* (Fig. IX-8: 9-o'clock) with a sharp irregularity pointing downward. Most of the nuclei are markedly increased in *size* (Fig. IX-6: compare with the neutrophil, lower left which is the approximate size of a normal epithelial cell nucleus). *Multinucleation* (Figs. IX-1, IX-5 through IX-7), *per se,* does *not* discriminate malignant neoplasia from proplasia, even though frequently present in cancer; however, the marked *difference* in *size* of the sister nuclei (Figs. IX-6, IX-7), shape (Fig. IX-1), chromasia and chromatin pattern (Fig. IX-5), are helpful criteria. A high *nucleo-cytoplasmic ratio* in a *well-preserved* cell is a valuable criterion (Figs. IX-3, IX-6, IX-8), especially in the cells with huge nuclei.

Papanicolaou stain.

IX-1 through IX-4, IX-6, IX-8: $\times$ 2,000; IX-5, IX-7: $\times$ 1,320.

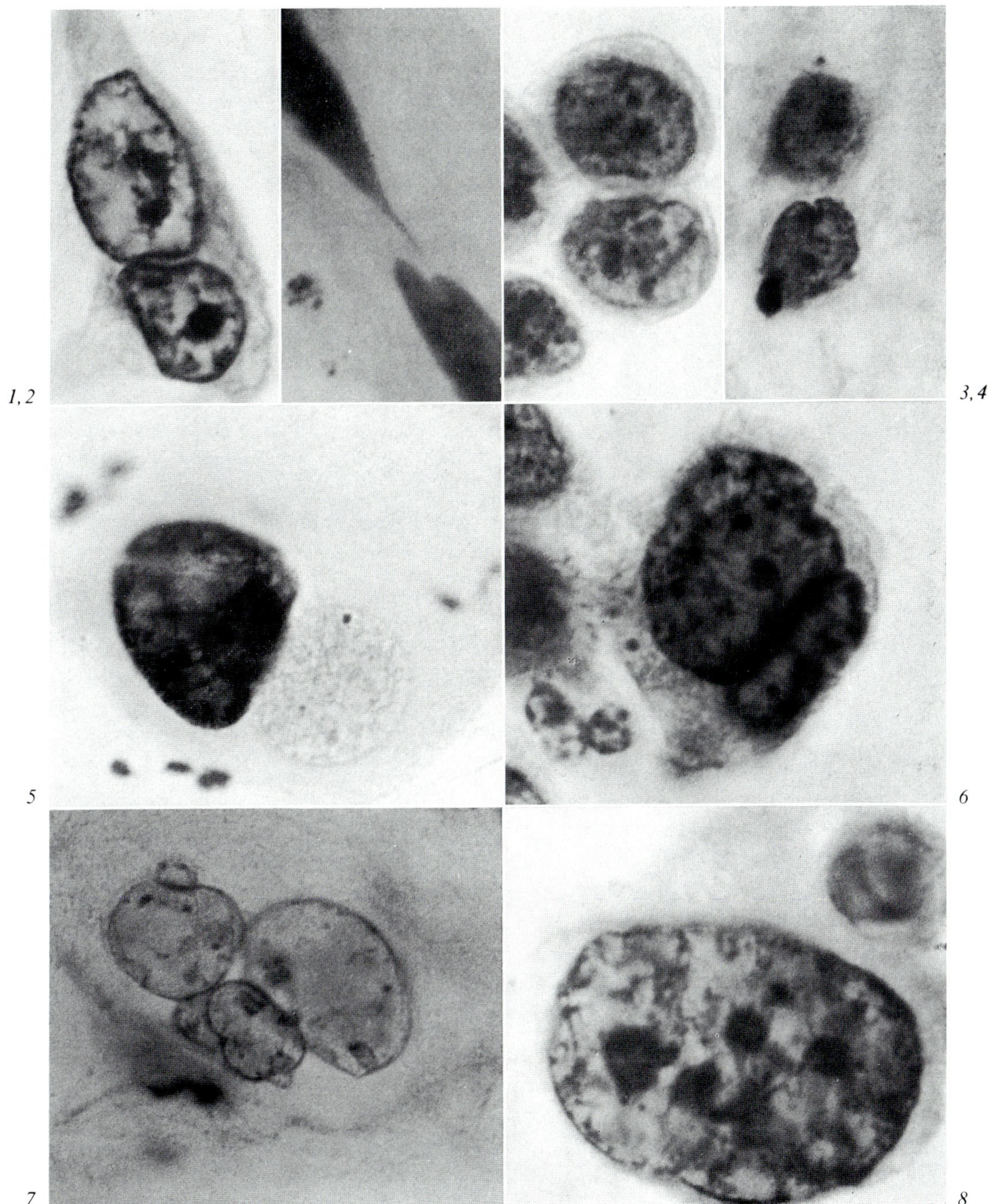

1,2
3,4
5
6
7
8

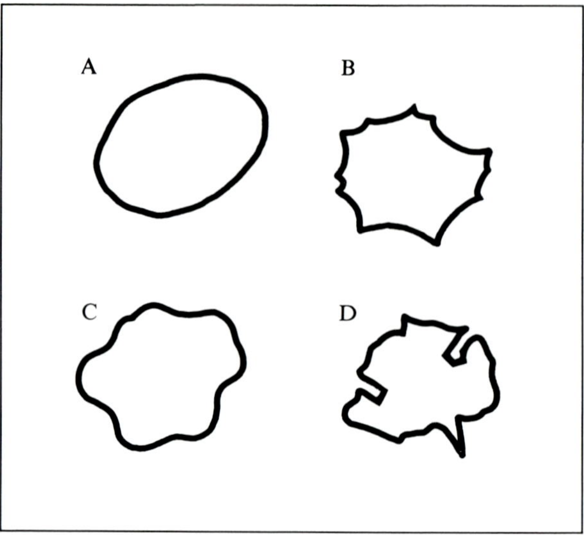

Fig. IX-9: Nuclear membrane shape. While the inner nuclear membrane is too thin to be visualized in the light microscope, its shape is the shape of the profile of the outer surface of the chromatinic rim (Figs. V-5 through V-8). In *euplasia* (A) the nuclear membrane tends to be *round or oval,* as a result of natural surface forces of a membrane around a plastic body, unless distorted by obvious factors (e.g.: molding by an adjacent nucleus, vacuole; indentation (hof) toward the cytocentrum of a macrophage). In *retroplasia* (B) the nuclear membrane tends to *wrinkle,* with water loss and shrinkage, as a series of concavities, between which can occur either sharp or rounded corners – these are *not* the angles of malignancy. In *proplasia* (C) the nuclear membrane is thrown into *waves* or *undulations,* as with any plastic body attaining maximal surface area for its volume, when there is increased need for more transport activity across the membrane. In *malignant neoplasia* (D) the nuclear membrane has unpredictable and unexplainable irregularities (e.g.: sharply pointed spicules, deeply angled clefts, sharp 90° angles) without obvious reasons (e.g.: molding from an adjacent nucleus indicates that the cells grew together in tissue (Table VI-1) and is thus not a malignant criterion).

4. Parachromatin

Extreme abnormal *clearing* of the parachromatin occurs in malignant neoplasia; at times, it is as clear as the clearest portion of the background (Figs. IX-1, IX-3, IX-6, IX-8, IX-13). As in proplasia, this appears to be due to a movement of much of the biologically inactive heterochromatin, which is in a more finely divided form in the parachromatin of a euplastic cell, into the condensed chromatinic net and chromatinic rim. This leaves less chromatin in the parachromatin, but it is richer in the biologically active euchromatin. This decreased hematoxylinophilia of the pale parachromatin of euplasia, results in this extreme clearing.

To tell the extreme parachromatin clearing which can also be found in proplasia from that of malignant neoplasia, there are two additional features of great discriminating value in identifying malignant neoplasia. One is *variance* in the *degree* of parachromatin clearing in different areas of the same nucleus (Figs. IX-3, IX-4, IX-8), in contrast to the uniformity of the parachromatin clearing in proplasia (Figs. VIII-2, VIII-3, VIII-7) (cf.: Figs. VIII-4 *vs.* IX-13). The second is variance in the *size* of the *areas* of parachromatin clearing, where there are large and small areas scattered at apparent random throughout the same nucleus (Figs. IX-3, IX-8) in contrast to their more uniform size in proplasia (Figs. VIII-2, VIII-3) and *in situ* carcinoma (Fig. XIII-1).

In retroplasia, varying degrees of nonuniformity of the parachromatin clearing occur as the degenerated chromatin (both heterochromatin and euchromatin) indiscriminately agglutinates into the chromatinic net and rim. However, the poor preservation with smudged outlines of these masses of degenerated chromatin bathed in this parachromatin (blurred chromatin/parachromatin interfaces) (Fig. VII-1) bespeaks that the hyperchromasia may well be due to the degeneration. Conversely, well-preserved cells with crisp 'cookie-cutter' interfaces of chromatin/parachromatin borders, of the chromatin clumps bathed in abnormally cleared parachromatin (Figs. IX-3, IX-6, IX-8), give strong evidence of malignant neoplasia.

5. Chromatinic Net

Patterns of the chromatinic net can vary, in malignant neoplasia, from a bland normal-appearing nucleus (lower right, Fig. IX-5), to a most macabre pattern. Of great help are the large, well-preserved aggregates of heterochromatin whose outlines are extremely *irregular* with sharp, pointed, jagged, angular projections and indentations (Fig. IX-8). In the well-preserved cell, outlines of these chromatin masses are *crisply defined,* as if they had been formed by a cookie cutter. With degeneration, this sharp 'cookie-cutter' outline is lost due to smearing, blurring, and smudging of the chromatin margin (left nucleus, Fig. IX-5). As any cell degenerates, its chromatin can agglutinate into clumps and become hyperchromatic, frequently closely mimicking the chromatin masses found in malignancy *except* that the chromatin/parachromatin margins blur (Fig. VII-1). A *degenerated* hyperchromatinic cell can raise concern, but should *not* be the basis for a *diagnosis* of malignancy. A well-preserved cell with distinct chromatin/ parachromatin interface borders, as if sharply cut out with a cookie-cutter, is important for accurate evaluation of malignant behavior (Figs. IX-1 through IX-8).

This chromatin clumping associated with malignant behavior does *not* occur uniformly throughout the nucleus. Thus the overall nuclear chromatin *pattern* is greatly altered, uniformity is lost, and marked *irregularity* of pattern becomes a striking feature (Figs. IX-3, IX-6, IX-8 *vs.* cf.: Fig. VIII-3). This is a key feature in assessing the degree of tumor advancement along neoplastic progression (e.g.: *in situ* through *invasive* through *metastatic*) [70].

In noninvasive or *in situ* cancers, chromatinic material tends to be arranged in coarse *granules,* smaller masses which are *rounded,* and are *uniformly* dispersed throughout the nucleus (Figs. XIII-1, XIII-9, XIII-13). This is the most severe form of the pattern of dyskaryosis found in proplasias, dysplasias, and progressive lesions of cellular unrest, and neoplasias less than malignant neoplasia or invasive cancer.

While cells characteristic of *in situ* carcinoma are also found shed from invasive carcinoma, as are normal-appearing cancer cells, these are not diagnostic of the most severe lesion. Therefore, one must diligently search for those cells which indicate the most severe lesion present, those which are more characteristic of invasive cancer, with chromatin clumps which are *massive, angular* or pointed, and which are distributed in a *nonuniform* pattern throughout the unevenly cleared parachromatin.

6. Chromatinic Rim

While many cancer cells have uniformly thin chromatinic rims, indistinguishable from those in euplasia, this feature does not help in identifying them as malignant. Usually the rim thickens in neoplasia; but if it remains uniformly thickened, this feature does not discriminate it from proplasia.

When the chromatinic rim offers the greatest help in identification, there are large irregular clumps of heterochromatin amassed, with the lamina, against the inner aspect of the inner nuclear envelope (Fig. IX-13). In addition, in small but significant nearby areas of the nuclear membrane, there is no chromatin massed against it at all.

This gives extreme thickness to the chromatinic rim and, in other small but significant areas of the same nuclear membrane, there will be seen virtually no chromatinic rim against it (Figs. IX-6, IX-8). In these areas, the bare nuclear membrane alone is so extremely thin that it cannot be discerned with the resolution of the light microscope, so that it appears merely as a boundary between the nuclear contents and the cytoplasmic contents (Figs. V-8; IX-6, IX-8). The resulting *extreme thickness* alternating with *extreme thinness* of the same nuclear membrane, can be a valuable and discriminating criterion of malignancy.

7. Nuclear Membrane Shape

This structure is too thin to be resolved by the light microscope. The *shape* of the inner nuclear membrane, however, is clearly outlined by the shape of the profile of the outer surface of the chromatinic membrane.

In malignant neoplasia, this *shape* of the nuclear membrane can be extremely irregular (Fig. IX-9). It can project into the cytoplasm as a sharp angle (Figs. IX-4, IX-10, IX-11, IX-13) or as a pointed pinnacle or spindle (Fig. IX-2). It can indent into the nucleus as a sharp 'bite' with angled corners (Figs. IX-3, IX-10), as an unpredictable, deep, angular nuclear fold or indentation (Figs. IX-4, IX-10, IX-13, IX-14).

These significant, *unpredictable* irregularities are extremely valuable malignant criteria, when critically identified. They are not to be confused with degenerative wrinkling of the nuclear membrane (Figs. VII-1, VII-3, IX-9). Neither are they to be confused with nuclear membrane irregularity which is explainable by mechanical or physiological forces, such as secretion (Figs. XV-13; XVI-2), granules (Figs. XI-2, XI-4), or nuclear molding from adjacent nuclei in multinucleation (Figs. VII-5; VIII-1; IX-1, IX-6), in tissue (Figs. XVI-2, XVI-7) in a diagnostic true tissue fragment (Fig. XVIII-1) or degenerated cells compressing and mushing together. The most macabre and unpredictably irregular shapes are good hallmarks of malignant neoplasia.

In most cell types, the location of the nucleus within its cytoplasm and its orientation are *not* happenstance, but reflect the cell's biologic behavior. Nuclear orientation about the cytocentrum (i.e.: in macrophages, *hof:* Ger. = courtyard) is a prominent feature of metabolic activity. In these cells, nuclear flattening faces the cytocentrum (Fig. IV-1). With more extensive activity, indentation and lobulation occurs which also orients toward the cytocentrum and partially enfolds this cellular zone of activity.

While *malignant cells* frequently also exhibit this orientation, they may carry the tendency of nuclear folding to excess, with extreme pointing and unpredictable angularity in this region of the hof, facing the cytocentrum. Furthermore, in malignancy the area of the nuclear membrane irregularity is not always strictly oriented toward the cytocentrum, as it is in the benign cell. It is strikingly grotesque to find, in what appears to be a histiocyte, an unexpected 'anti-hof' area of nuclear membrane irregularity facing the cellular membrane away from the expected region of the cytocentrum, rather than, or in addition to, its expected flattening or indented irregularities oriented toward the hof (Fig. IX-12). This discrepancy can be of great discriminating value as a malignant criterion.

8. Nucleolus

Excessive nucleolar *irregularity* (e.g.: in shape, size, numbers) is frequently present in malignant neoplasia, but is also present in extreme proplasia. This can be dangerously confusing in diagnosis, so that great care is necessary in identifying nucleolar changes. Findings in a nucleolus which help discriminate cancer fall into two major areas: nucleoli which are *large and sharply irregular,* and significant *variances in nucleolar numbers per nucleus.*

a. Large and Sharply Irregular Nucleoli. Nucleoli can be discriminating for cancer when they are *large, irregular and sharply cornered or spiculated.* The *corners* of the discriminating large nucleolus are *not rounded,* but are *sharply jagged, angled, pointed* and *indented* (Fig. XVIII-1). This characteristic of nucleoli, therefore, is like that found in chromatin clumps, where *sharp* projections outward and indentations inward in large structures, are significantly discriminating in identifying malignant neoplasia [99, 103, 202].

Figures IX-10 through IX-12: Nuclear membrane irregularities in cancer.

Fig. IX-10: Needle aspiration, adenocarcinoma, liver. There is a large indentation ('bite') in the nuclear membrane (outer profile of the chromatinic rim) at about 11-o'clock of the larger cell. A sharp fold overhangs irregularities in the indentations. There are two rounded but prominent nucleoli in each nucleus, which do not provide criteria for malignancy but do indicate increased protein production in this secretory cancer. Note hazy products of secretion (i.e.: bile) lie in a multiply-vacuolated cytoplasm. Parachromatin clearing is extreme and unpredictable.

Fig. IX-11: Gastric aspiration, adenocarcinoma, stomach. The nuclear membrane is extremely irregular, extending with angles and folds into the large amount of cytoplasm. This uneven distribution of cytoplasm is abnormal as the nuclear membrane is compressed into, and closely approximating, the cellular membrane for over half of its extent, in spite of the abundance of cytoplasm extending off to the upper left. The upper cell has an extremely high nucleo-cytoplasmic ratio.

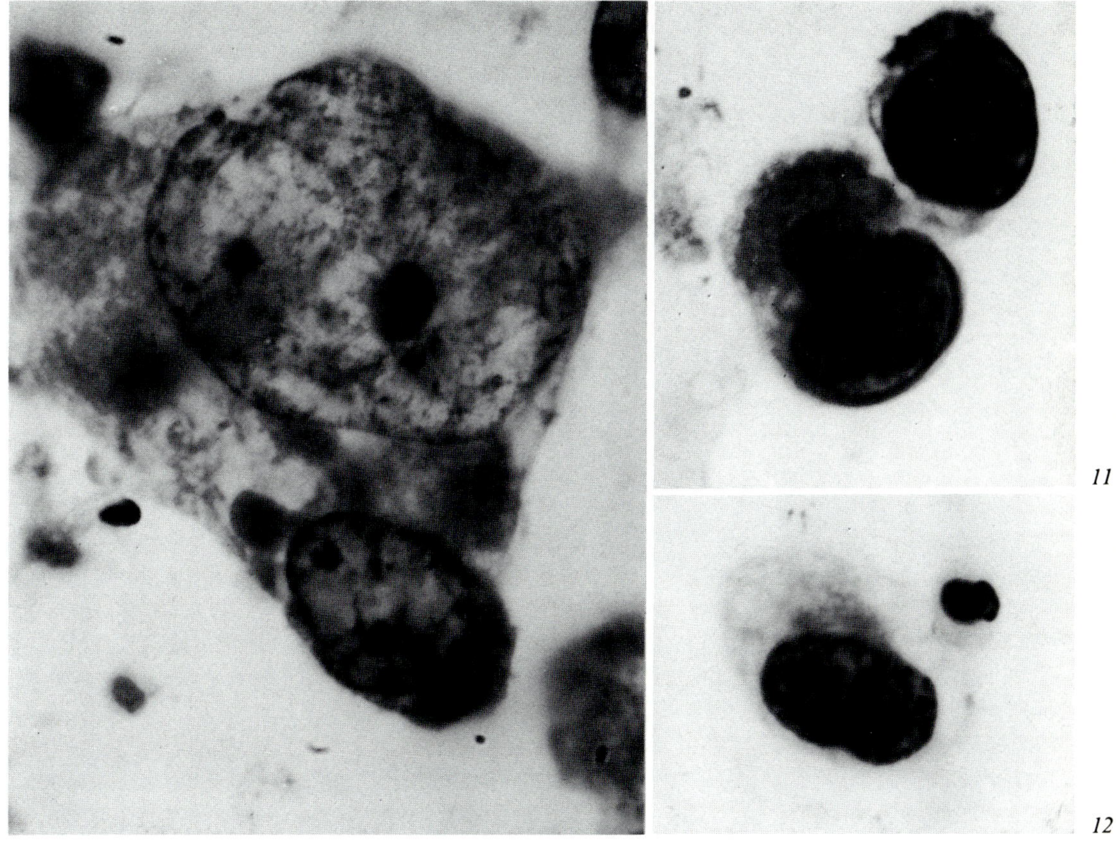

Fig. IX-12: *Sputum, adenocarcinoma, lung, with cytoplasmic secretion.* This cell closely resembles a macrophage, with its kidney-bean shaped nucleus and foamy cytoplasm containing fuzzy particles which could be phagocytosed particles undergoing digestion. The nuclear irregularity, however, does *not* face the cytocentrum, as does a macrophage indentation in forming a hof. Instead, the nuclear membrane irregularity is *opposite* the *center* of the cell, in an 'anti-hof' position – a good malignant criterion. Note the coarse granules of the chromatinic net and the chromatinic rim, the outer profile of which indicates the shape of the nuclear membrane, and the irregular clearing of the parachromatin.

Papanicolaou stain.

IX-10 through IX-12: $\times$ 1,250.

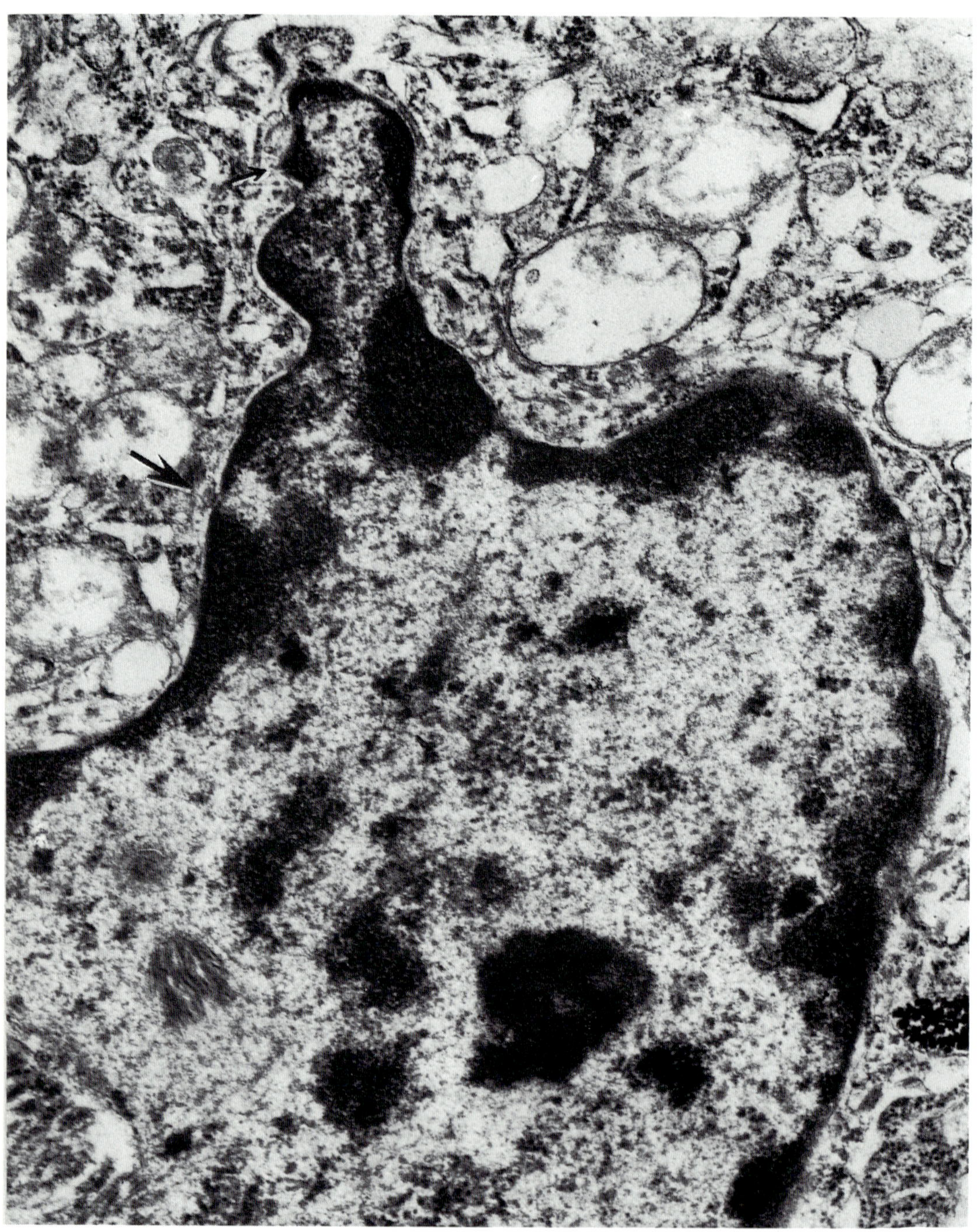

13

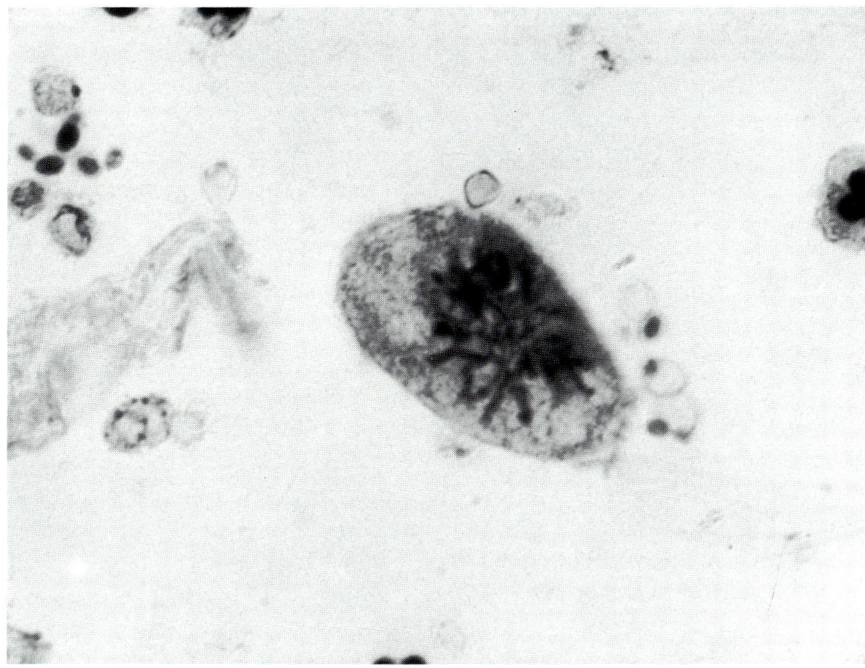

Fig. IX-13: Squamous cell carcinoma of the lung. Numerous sharply angulated irregularities are present in the inner nuclear membrane where, moving down along the left-hand side of the nucleus from the top, there is a sharp right angle outward (thin arrow), followed by a sharp right angle inward. At least two of the involuted irregularities contain active nuclear pores. Nuclear pore activity is high throughout the nucleus. They are numerous, and many have a well-developed central granule and prominent fibers trailing inward, into the nucleus, and outward, into the cytoplasm (heavy arrow). The chromatinic rim varies markedly in thickness, with prominent and irregular aggregations of well-preserved heterochromatin marginated to it.

Osmium/methacrylate.

IX-13: × 40,000.

(Frost, Erozan, Donovan; unpublished electron micrograph.)

Fig. IX-14: Abnormal mitosis, transitional cell carcinoma, voided urine. This is an *en face* view of the metaphase plate of a dividing cancer cell. There were many more than the normal number (46) of human chromosomes (23 pairs in the female; 22 pairs plus one X- and one Y-chromosome in the male). The paired nature of the chromosomes are best discriminated in this cell between 6- and 8-o'clock of the plate.

Papanicolaou stain.

IX-14: × 1,250.

Large size, *per se,* is not a discriminating criterion between proplasia and malignant neoplasia; but a *large* nucleolus *with* sharply *angled irregularity* is of great value. Macronucleoli are associated with increased protein production (i.e.: secretion; severe proplasia) (Figs. V-1, V-2). When such activity is *non*neoplastic, the large nucleoli are *round* or *rounded;* and if irregular, their projections, corners, and indentations are *rounded* and *blunted* rather than sharp and angular.

Varying amounts of nucleolus associated chromatin can be identified about the periphery of the nucleolus. When it is disproportionately great, this basophilic nucleolus associated chromatin is so prominent that it may mask the central core of acidophilic nucleolar material.

For the assessment of malignant neoplastic activity, it makes little difference whether such a mixed body is mainly a chromocenter or a nucleolus, as *either* is of discriminating value when it is *large,* excessively *irregular,* and *pointed.* However, for the assessment of protein production, either for secretion or for building cellular structure, it is important to recognize the nucleolar component and to assess its biologic significance.

b. Variances in Nucleolar Numbers per Nucleus. Numbers of nucleoli per nucleus, can give valuable evidence of benignancy or malignancy. Lack of *predictability* of the *numbers* of nucleoli per nucleus is a valuable discriminating finding in a *multinucleated cell* or in a *diagnostic true tissue fragment* (DTTF, Table VI-1).

Increase in the *number* of nucleoli per nucleus is merely proportionate to the degree of cellular activity or protein production. The number appears to depend upon the type of activity, changing as a cell's function changes (e.g.: endometrium, liver). Thus, merely nucleolar number, *per se,* does not distinguish well between cancer and noncancer.

But in these instances, even though their numbers increase significantly they usually do not vary from nucleus-to-nucleus, in a multinucleated cell or in a diagnostic true tissue fragment, more than $\pm$ 1 (e.g.: from 1 to 3) or even $\pm$ 2 (e.g.: from 1 to 5). Discrimination is good, however, when there is an appreciable *difference* in the numbers of nucleoli present *either* in sister nuclei within a multinucleated cell, or in different nuclei of like cells (e.g.: diagnostic true tissue fragment) (Fig. XVIII-1). To favor malignancy, however, such variation should be $\pm$ 3 (e.g.: *one* nucleolus in one nucleus, *seven* in another) or more.

c. The Presence per se of Nucleoli. The mere presence of a nucleolus, in itself, does *not* constitute a malignant criterion. There are too many benign situations where large, prominent nucleoli are present (e.g.: rapid growth, repair, secretion) which can*not* be differentiated from cancer by the mere presence of nucleoli, for it to be a safe and dependable criterion. There are two exceptions, however, worthy of further consideration.

Lymphomas and leukemias. Very active lymphocytes (e.g.: immunoblasts) can have small or moderate-sized nucleoli. Otherwise, leukocytes tend to be without them, or with very inconspicuous ones.

Malignant leukocytes, on the other hand, frequently have large, prominent, multiple nucleoli. At times they are also irregular and pointed, and other good malignant criteria will be present; however, huge nucleoli may be the only or the most striking finding and, thus, can be considered a malignant criterion (Figs. XVII-5 through XVII-13).

Invasive squamous cell carcinoma. With invasion, nucleoli appear as evidence of elevated protein production. This can be the first and only morphologic change indicative of an acquired biologic ability to invade. Its other morphologic features may be entirely compatible with nonmalignancy, or what we may recognize today as premalignant conditions (e.g.: dysplasia, dyskaryosis, proplasia) (cf.: Figs. VIII-3 *vs.* XII-13). In this situation, therefore, the mere presence of nucleoli should serve as the impetus for further studies to rule in or out the presence of invasive cancer and, in this sense, may also be considered a malignant criterion.

9. Mitoses

The presence of *normal mitoses* in exfoliated material does not discriminate between proplasia and malignant neoplasia. Actually, in the relatively nonphysiologic materials (e.g.: pulmonary secretions, vaginal fluid, urine), single cells in mitosis are nearly always of the monocyte-histiocyte-macrophage lineage. These nonphysiologic fluids supress cell division, except in the macrophage which, because of the 'garbage-to-collect', readily undergoes replication.

On the other hand, fragments of tissue recently broken off from healthy and rapidly growing tissue, either malignant or nonmalignant, can contain cells caught in mitosis. Thus, in the more physiologic or tissue culture-like material (i.e.: body cavity fluid) and in needle aspirations, malignant as well as nonmalignant cells can be caught in mitosis. Thus, the mere presence of *normal* mitoses, *per se,* offers virtually no support for malignancy.

The presence of *abnormal mitoses* is another matter, however, and offers valuable evidence of probable malignancy. Such bizarre mitoses, when the metaphase plate is viewed in *profile* (Fig. XVIII-1), as multipolar figures (e.g.: tripolar, quadripolar) and lag chromosomes or, when the plate is viewed *en face* (Fig. IX-14), as numbers of chromosomes which are definitely greater or less than 46, are good indications of malignancy. They are relatively rare in exfoliated material, however, except for body cavity fluids and needle aspirations.

The mere *presence* of *normal* mitoses, therefore, is poorly-discriminating evidence of malignant neoplasia. The abnormal mitotic figures, however, give evidence of neoplasia.

10. Multinucleation

Multinucleation, *per se,* like the presence of mitoses, is *not* a malignant criterion. It indicates increased activity and, thus, does *not* discriminate between *malignant neoplasia* and *extreme proplasia,* where it can be most common and striking (cf.: Chap. VIII).

Multinucleation, however, does provide an unequaled opportunity to compare those features, nucleus-to-nucleus, which are critical in accurate discrimination of cancer. Multinucleation with identical or mirror image nuclei, can be from a nonmalignant process(es) and thus nondiscriminating. Great *variation* between the daughter nuclei, occurs in malignant neoplasia and aids in discriminating it from the proplasias.

The principal features to be assessed nucleus-to-nucleus, include variation in nuclear size and shape, chromatinic net, chromatinic rim, parachromatin, chromatin/parachromatin pattern, nucleoli, nuclear membrane, and nuclear chromasia. Of least assistance is nuclear shape and size variation, as they at times occur in the extreme proplasias, so that their variation must be extreme to be significant (Fig. IX-7). To be of greatest value in assessing malignancy, therefore, the daughter nuclei must show significant variation in the other key morphologic features (Figs. IX-1, IX-5, IX-6; XII-7, XII-11).

B. Cytoplasmic Structure

The cytoplasm bears only a few discriminating criteria of malignant neoplasia, as it is mainly the site for functional differentiating characteristics (see the Fourth Section). These key features for malignancy which do involve the cytoplasmic, concern the cytoplasmic *relationship* to the nucleus in *amount* (ratio of nuclear size to cytoplasmic size, N/C ratio), and the *relationship* of *nuclear membrane to cell membrane.*

1. Ratio of Nuclear Size to Cytoplasmic Size
(N/C Ratio)

The *amount* of nucleus and of cytoplasm, is usually expressed as the *nucleo-cytoplasmic ratio* (N/C). A high N/C ratio in a large cell, is a valuable malignant finding (Figs. IX-3, IX-6, IX-8; XVIII-5) if the cell is *larger* than an immunoblast or histiocyte up to very large cells. In very small cells, the size of leukocytes, it loses most of its significance due to the high N/C ratio of lymphocytes. In cells the size of macrophages or parabasal cells, and larger, it is valuable.

For accurate and valid evaluation, a cell must have *intact* cytoplasm. A truly bare nucleus is the remnant of a degenerated cell. It is susceptible, unpredictably, to morphologic changes of retroplasia which can danger-ously mimic malignant criteria and thus discriminate poorly from malig-nancy. This accounts for many reported false positive cancer identifica-tions, as well as false negatives.

Highest resolution (100 × oil immersion objective with Koehler illu-mination, Table IV-1) is necessary at times to confidently determine that a scant rim is, in fact, an intact cytoplasm rather than residual degenerated fragments (Fig. XVIII-5). A truly bare nucleus, or a nucleus with small wisps of degenerating cytoplasm (Fig. VII-1) are able to warn of possible abnormality and indicate need for further specimens to be examined, but should *not* be used to establish a diagnosis. For greatest accuracy in discrim-inating cancer, cells must be *whole* and have *intact* cytoplasm and cell membrane.

Extremely *scant,* intact cytoplasm constitutes an excellent malignant criterion, especially in a *large* cell (Figs. I-1; IX-8). A thin rim of cytoplasm around a severely dyskaryotic nucleus, is of great diagnostic significance in cells shed from *in situ* squamous cell carcinoma (Figs. XIII-1, XIII-3, XIII-9, XIII-13). This constitutes the most valuable, single criterion outside of nuclear structures in differentiating them from cells of extremely severe dysplasia.

When an intact, thin rim of cytoplasm is of such crucial diagnostic significance, pitfalls in its evaluation must be carefully ruled out. First, the cytoplasm is to be established as being intact.

Secondly, it is to be determined that a significant amount of cytoplasm does *not* lie *above* or *below* the nucleus in such a way that it does not appear in the cellular profile. When cellular spreads are made of unfixed speci-mens, a cell seeks the surface of the slide, spreads laterally and thins from surface tension, and displays most of its cytoplasm lateral to its nucleus.

Thus, when cells are isolated and have spread their cytoplasm to the limit, a high N/C ratio with intact cell border is significant (Fig. XVIII-5, upper left and lower right cells).

In material which is *fixed before* spreading, however (e.g.: specimens collected in fixative), a cell rounds up into a firm ball. Like a hard-boiled egg, it does not flatten completely when a spread is made, retaining greater amounts of cytoplasm than usual above and below the nucleus and displaying a proportionately smaller amount of cytoplasm laterally than it actually possesses.

Furthermore, even an unfixed cell, when it is *embedded* in thick mucus or nestled among other cells or debris, can*not* spread its cytoplasm, which appears scantier than it actually is (Fig. VI-1). Significant amounts of cytoplasm can thus be hidden above or below the nucleus. Careful examination under good resolution and using through-and-through focusing, is necessary to rule this out, wherever scanty cytoplasm is a critical criterion. At times, however, this is extremely difficult to determine with certainty, and doubt may still remain; in such a circumstance, the finding of scanty cytoplasm is diminished in significance.

2. Relationship of Nuclear Membrane to Cell Membrane

The relationship of nuclear membrane to cell membrane (plasma membrane) at times is bizarre in cancer, one membrane appearing to have a peculiar affinity for the other. This situation is rarely found in normal cells.

This is evidenced in a number of ways. In cells with a completely adequate amount of cytoplasm, the membranes may 'seek' each other over a long stretch of nuclear surface, and maintain a strikingly uniform distance apart, as if measured by an engineer's calipers. When cytoplasm is scanty, this close relationship is expected; however, this abnormality is striking when the abundance of cytoplasm allows for a large amount of it to flare off and accumulate at one side of the cell, while in another area the cell membrane closely and uniformly 'hugs' the nuclear envelope (Fig. IX-11). Grotesque *nuclear* outlines may be followed faithfully by equally grotesque *cell* outlines, with the two membranes running strictly parallel to each other.

This affinity of the two membranes for each other, at times is carried past and around the corners of nuclei. The cell membrane appears to hug and pull itself around the nucleus, before flaring out to enclose the abundant mass of remaining cytoplasm or to bend back out along the tails (Figs. IX-11; XIII-17).

C. Multiple Cells and Cell-to-Cell Relationships
(Organelle Structures, Diagnostic True Tissue Fragments)

When 'sibling' cells in tissue or diagnostic true tissue fragments are compared cell-to-cell, many features in euplasia are predictable from cell-to-cell. In malignant neoplasia, however, *unpredictable variability* of previously mentioned key malignant criteria from cell-to-cell, can be striking. Further information regarding malignant behavior, can be gained from key · morphologic changes in cells which have been caused by other cells neighboring them in the tissue.

1. Nuclei
There can be great *variability* in chromatin pattern (i.e.: chromasia, parachromatin, configuration and number of heterochromatin clumps, chromatinic rim), numbers of nucleoli, nuclear size, nuclear shape, and numbers of multiple nuclei.

In many rapidly growing *tissues, both* proplastic and malignant neoplastic, there is molding of nuclei by the *nuclei* of adjacent cells. This is evidence of their rapid, expanding growth together in tissue and assists in recognizing a diagnostic true tissue fragment (Table VI-1) , but not to discriminate cancer from noncancer. However, *nuclear molding* by adjacent *soft cytoplasm* (not by a vacuole, granule, etc.), is an abnormal finding and, while infrequent, constitutes good evidence of malignancy.

2. Cytoplasm
The variability in the amount of cytoplasm and in the amount of the total cell, of one sibling cell to the next, can be accurately determined in tissue fragments. This variability (anisocytosis) is increased in malignancy. The extreme scantiness of cytoplasm, which is so common in cancer, gives the appearance of marked 'crowding'.

3. Foreign Structures
The finding of tissue structures which are foreign or extremely rare to the area of the specimen, may provide valuable evidence of malignant neoplasia. Thus, well-formed papillary fragments, polyps, stalks, acini, and ducts which are essentially foreign (e.g.: in body cavity fluids, urine), suggest involvement by malignant neoplasia. Certain disease states (e.g.: rheumatoid, collagen vascular) can mimic these, however, by forming papillary or

acinar structures of the mesothelium, and irritations of the urinary tract (e.g.: stones, blood clots, cysts) can form papillary structures of the urothelium.

D. Caveat

Let it be reemphasized again that there is *no perfect single* malignant criterion which, when present, means cancer is present; or, when absent, means there is no cancer. Neither is there a perfect single benign criterion.

It is by *recognizing* those morphologic changes which are most frequently present in cancer and rarely in noncancer, evaluating these malignant criteria carefully and, based upon knowledge and experience, *adjudging* them and determining what biologic process(es) is involved, that accurate diagnoses are reached.

Section D
Functional Differentiation:
Its Morphologic Characteristics

X. Introduction: Morphologic Characteristics of Functional Differentiation

The capabilities for cells to *differentiate*, in order to perform their *functions,* are myriad. While they usually occur in well-defined functional packages or 'syndromes', recognized as cell types, they can produce virtually an uninterrupted spectrum of morphologic alterations.

Functional differentiation occurs in typical and atypical forms. *Typical* functional differentiation and its morphologic features, occur in tissues and cells of all levels of general activity, especially of euplasia (Chaps. XI, XIV). *Atypical* forms occur chiefly in the states of altered general activity: retroplasia (Chap. VII), proplasia (Chap. VIII), metaplasia (Chap. XV), developing neoplasia, and malignant neoplasia (Chaps. XII, XV, XVI).

In contrast to the *nuclear* alterations which bespeak mostly general activity, functional differentiation is reflected mainly in *cytoplasmic* structures (e.g.: mucus production, keratinization, myofibril development, pigment formation) [94]. In addition, there do exist some important nuclear characteristics (e.g.: karyopyknosis in stratified squamous maturation; nucleoli in protein production), as well as formations of organs by cells in diagnostic true tissue fragments (e.g.: papillary fronds, glands) which bespeak functional differentiation.

A. Cell Types

In euplasia, functional differentiation is usually clear-cut [5, 235, 252]. This has allowed for classical categorizations into specialized cell types (e.g.: mucus secreting columnar cells; keratinizing stratified squamous cells) as set forth in classical and standard works on histology [42, 135] and cytology [46, 47, 53, 61, 204, 205, 241, 253, 269, 270].

For orderly thought, it is classical to relate to 'tissues of origin' and 'cells of origin' as though they are born into a certain cell type and are incapable of change. Yet, it has been well documented for many years that, under proper conditions, apparent differentiation of tissues and cells can be altered [306].

Mononuclear cells of the blood are capable of changing to reticulum cells and fibroblasts in a subcutaneous filter membrane chamber [105]. Keratinized ectoderm in tissue culture becomes mucified under high Vitamin A effect [77] and, conversely, stratified squamous epithelium develops from columnar epithelium (i.e.: metaplasia) under chronic irritation in many sites, e.g.: tracheo-bronchial tree, endocervix uteri (Chap. XV).

Cellular functional differentiation is complex, and its mechanisms are many. The latter are unfolding at a rapid rate to investigation in cellular biology [3, 165, 254].

B. Transcription and Protein Synthesis

Only a very few specific sections of the total chromatin DNA sequences are available for RNA transcription as biologically active euchromatin [89, 294]. The bulk of a cell's genetic complement is the biologically inactive heterochromatin, which is repressed or masked by proteins [121]. A small amount is dispersed in the parachromatin, and the rest is highly compacted in the chromatinic net and the chromatinic rim.

Those portions of the genome which are unmasked and are available for transcription [210], vary with cell type and with a given cell's functional state.

The ten nucleolar sites on the five pairs of satellite chromosomes (i.e.: number 13, 14, 15, 21, 22) make ten individual nucleoli only in some cell types, and then only in late telophase and early anaphase [15]. During interphase (metabolic phase), they coalesce into one or more nucleoli for the rest of the cell's intermitotic activities of protein production. In the nucleoli, the ribosomal RNA (rRNA) is manufactured, associated with its *acidophilic* protein, transported through the proteins of the nuclear pores into the cytoplasm where the rRNA-associated protein is then *basophilic* and either attaches to the outer nuclear membrane and endoplasmic reticulum (ER) producing the rough ER, or lies loose in the cytoplasm as free ribosomes (Chap. V) [34, 240, 271, 288, 291].

The acidophilic rRNA-associated proteins of the proribosomes in the nucleus, produce the acidophilia (i.e.: red, orange, yellow) of the nucleoli of a cell which is involved with protein production. Likewise, the basophilic rRNA-associated proteins of the ribosomes in the cytoplasm, cause a deep basophilia of the cytoplasm (blue, green) to be associated with an actively metabolizing cell (i.e.: producing protein for growth, repair, secretion). These structures thus become the morphologic basis for recognizing protein production.

The other RNAs (e.g.: messenger RNA, mRNA; transfer RNA, tRNA; diffuse nuclear RNA, dnRNA) do not accumulate conspicuously in the nucleus, as does the ribosomal RNA of the nucleolus. Instead, they pass into the cytoplasm, where they associate with the ribosomes and take part in the synthesis of proteins (e.g.: keratins, mucins, myosins, actin, tubulin, insulin) (Chap. V) [214].

Innumerable other intermediate steps intervene between transcription of RNA in the nucleus and its translation with protein synthesis in the cytoplasm for cellular functional differentiation. This abbreviated model, however, is direct and facilitates understanding of the mechanisms involved in altering functional differentiation (i.e.: metaplasia, modulation) and in producing its atypical forms which can be encountered in retroplasia, proplasia, and malignant neoplasia.

It has been said that 'the nucleus directs and the cytoplasm toils'. As a cell differentiates to perform these toilsome functions, its cytoplasmic structures bespeak such biologic behavior.

C. Degree of Maturation and Functional Differentiation

The *degree* of maturation and of functional differentiation which is attained by cells of *proplasia* and of *malignant neoplasia,* is generally *less* than that attained by nonmalignant cells. This 'dedifferentiation', as it has been referred to over the years, is frequently confused with malignancy. It is a *separate* phenomenon which is tied closely with the cell's level of activity and its 'preoccupation for other tasks', or its difficulty of differentiating to perform tasks other than surviving in a hostile environment. Thus, it is not to be considered a malignant criterion, and it does not discriminate proplasia from malignant neoplasia.

The degree of maturation attained by cells of a cancer tends to be *inversely* proportionate to the *aggressiveness* of the cancer, and *directly* proportionate to the *favorableness of the prognosis* of the host's ability to combat the new growth. This is by no means always the case, but it forms the basis for numerous useful classifications (e.g.: Broders [48], Wentz-Reagan [297]) and other morphologic indicators designed and utilized to prognosticate biologic potential, therapeutic susceptibility, and overall outcome of a given tumor.

All of the types of cellular differentiation found in cancer are much too numerous to be covered adequately. It is felt that the major concepts concerning the cancer cell's functional differentiation are best conveyed, in a work of this size and intent, in the few types chosen for this Section. Thorough familiarity with the basic principles, as applied to these cell types, permits application to other states of functional differentiation.

Section C considers *typical* differentiation principally of the euplastic cell (Chaps. XI, XIII) and *atypical* forms of this differentiation. The latter is mainly associated with the states of biologic 'unrest': retroplasia, proplasia and, especially, malignant neoplasia.

D. Accuracy of Cell Typing of Cancer

The *accuracy* of cell typing can be of great practical importance in atypias and, particularly, in *cancer*. In the latter, when therapy is based entirely upon a cytopathologic interpretation of cell type, accuracy becomes of significant consequence (e.g.: small cell carcinoma *vs.* nonsmall cell carcinoma of the lung).

Typing accuracy is directly proportionate to the *quality of the specimen* (e.g.: clinical procurement, preservation, preparation, staining) and the *adequacy of significant material* within the specimen. This is at least as important in *cellular* specimens as it is in *tissue* specimens, where a miserable biopsy has a far lower chance of identifying tumor type of the final totally removed lesion in the surgical specimen, than does an adequately sampled and prepared biopsy.

The *accuracy* of cytologic cell typing has a tremendous advantage over tissue biopsy *and* a tremendous disadvantage. For the latter, even when sampling is adequate, cytology does not have tissue pattern to utilize, and must rely virtually upon individual cell characteristics of functional differentiation and a few, small, diagnostic true tissue fragments (DTTF) (Table VI-1). Conversely, however, cellular specimens cover a wider surface area in their sampling, so that dissimilar differentiations in separate areas of a heterogeneous tumor have a better chance of being represented within the small cytologic material evaluated than in a small biopsy.

Achievable accuracy of cell typing from a cytologic sample and the *importance* of that accuracy, vary as to tumor type and as to anatomic site.

As an example, critically examine the data from a series of *464 consecutive* specimens (241 sputum; 223 bronchoscopic) from our hospital *routine* diagnostic laboratory files, which were *all diagnosed as cancer,* and which were from *consecutive* patients who had *tissue confirmation* of lung cancer [68].

Table X-1. Cytologic classification of lung cancer

Cytologic diagnosis		Histologic diagnosis			Agreement
SQ	166	166	SQ		100%
SM	67	65	SM		97%
AD	30	30	AD		100%
All nonSM, n.o.d.	126	14	LG	11%	96%
		54	SQ, p.d.	43%	
		48	AD, p.d.	38%	
		5	A/S, p.d.	4%	
		5	SM	4%	

SQ = Squamous cell carcinoma, WHO I; SM = small cell carcinoma, WHO II; AD = adenocarcinoma, WHO III; NonSM = nonsmall cell carcinoma; n.o.d. = not otherwise diagnostic; LG = large cell undifferentiated carcinoma, WHO IV; A/S = adenosquamous carcinoma; p.d. = poorly differentiated.

There were 263 specimens which were diagnosed as the patient having a *definite* cell type (i.e.: SQ, AD, SM). An additional 126 were felt to have nonsmall cell cancer, but it was not possible to diagnose a further functional differentiation from the specimen (Table X-1).

Of the 166 diagnosed as *squamous cell carcinoma* (SQ), all were proven by histology (*100%* agreement) (Table X-2). Likewise, all of the 30 diagnosed as *adenocarcinoma* (AD) were proven by histology (*100%* agreement) (Table X-3). Of the 67 diagnosed as *small cell carcinoma* (SM), 65 were proven by histology (*97%* agreement) (Table X-4). Of all 263 diagnosed as a definite cell type (Table X-1), 261 were proven by histology (*99%* agreement).

Of the 126 called *nonsmall cell cancer* (nonSM) but without evidence of functional differentiation sufficient for a more specific diagnosis of cell type, 123 were proven by histology to be nonSM (*96%* agreement) (Table X-5). Therefore, of *all* 322 called *nonSM* (i.e.: SQ, AD, nonSM) (Table X-1) 319 were proven by histology (*99%* agreement).

It is important clinically to differentiate with accuracy between SM and nonSM lung cancer for therapy. When adequate tissue cannot be obtained for histopathologic diagnosis, cytopathologic diagnosis must be made *but with greatest care and accuracy* so that it can be *relied upon with confidence.*

Table X-2. Squamous cell carcinoma (SQ) of the lung

Cytologic diagnosis	Histologic diagnosis							
	SQ		AD		SM	LG	A/S	total
	well diff.	poorly diff.	well diff.	poorly diff.				
SQ	131 (79%)	35 (21%)	–	–	–	–	–	166
NonSM, possibly SQ	19 (33%)	22 (38%)	–	8 (14%)	3 (5%)	4 (7%)	2 (3%)	58

Table X-3. Adenocarcinoma (AD) of the lung

Cytologic diagnosis	Histologic diagnosis							
	SQ		AD		SM	LG	A/S	total
	well diff.	poorly diff.	well diff.	poorly diff.				
AD	–	–	27 (90%)*	3 (10%)	–	–	–	30
NonSM, possibly AD	2 (3%)	8 (13%)	10 (16%)	29 (48%)	2 (3%)	7 (12%)	3 (5%)	61

* 5 bronchioloalveolar.

Table X-4. Small cell carcinoma (SM) of the lung

Cytologic diagnosis	Histologic diagnosis							
	SQ		AD		SM	LG	A/S	total
	well diff.	poorly diff.	well diff.	poorly diff.				
SM	–	–	–	1 (1½%)	65 (97%)	–	1 (1½%)	67

Table X-5. Nonsmall cell carcinoma, not otherwise diagnostic, of the lung

Cytologic diagnosis	Histologic diagnosis							
	SQ		AD		SM	LG	A/S	total
	well diff.	poorly diff.	well diff.	poorly diff.				
NonSM, n.o.d.	–	3 (43%)	–	1 (14%)	–	3 (43%)	–	7
NonSM, some SQ features	19 (33%)	22 (38%)	–	8 (14%)	3 (5%)	4 (7%)	2 (3%)	58
NonSM, some AD features	2 (3%)	8 (13%)	10 (16%)	29 (48%)	2 (3%)	7 (11%)	3 (5%)	61
All NonSM, n.o.d.	21 (17%)	33 (26%)	10 (8%)	38 (30%)	5 (4%)	14 (11%)	5 (4%)	126

It is essential for the diagnostician to strive to *know* when he is *sure,* and to *know* when he is *not* sure – and to report *only* the former as being *diagnostic.* Suspicions have *no* place being represented as diagnostic, in accurate modern day medical diagnosis and therapy.

The importance of this determination varies in different sites, cell types, and clinical situations. Careful attention to the valuable details of functional differentiation, discussed in this Section B, is key to accuracy.

E. Caveat

A final general note. Features of functional differentiation are *not* features of malignancy, no matter how atypical they may be. A Cardinal Rule for proceeding to diagnosis, is:

First, determine that cancer is present (mainly by *nuclear* features, Chaps. V–IX);

Second, determine the cell type of cancer present by its functional differentiation (mainly by *cytoplasmic* features, Chaps. XI–XVIII).

Reversing this order, by using functional differentiating features as malignant criteria, no matter how atypical, is not only usually meaningless but frequently is misleading – and asks for diagnostic disaster, rather than the *diagnostic confidence* which *careful evaluation* and *reporting* are able to ensure.

XI. Typical Functional Differentiation of Keratinizing Stratified Squamous Epithelium

A. General Considerations

Certain key morphologic features characterize the functional differentiation of euplastic tissues to form keratinized stratified squamous epithelium. They include a *central* nucleus, *thread-like* chromatin pattern (interphase), *karyopyknosis* with maturation, *intercellular bridges, stratification, keratinization, thinning,* orientation *parallel* to basement membrane and lumen, and exfoliation as *single* cells (Fig. VI-2).

Many classifications based upon the morphology of the cellular layers of this epithelium, have been proposed and employed [42, 135]. By an international agreement on nomenclature [301, 302], three major cell types from keratinizing stratified squamous epithelium are now recognized, based upon the state of maturation attained by the cell at the time of its exfoliation from the epithelium: *parabasal* cell, *intermediate* cell, *superficial* cell. As these major categories are based upon the degree of typical normal maturation attained at the time of their exfoliation from the surface, they also represent levels of maturation attained in the tissue, but not necessarily pure zones of the tissue. While these categories can be further subdivided, if needed or desired clinically or for investigational purposes, they generally are of greatest value when simply used *per se* [99].

B. The Nucleus in Keratinizing Stratified Squamous Differentiation

Nuclear features are of less help than cytoplasmic structures, in providing evidence of keratinizing stratified squamous cell maturation. Nevertheless, nuclei do reflect functional differentiation, particularly in keratinizing stratified squamous epithelium of higher maturation (i.e.: superficial cell maturation). Nuclear morphologic features, which are of varying degrees of assistance, include: placement and orientation within the cell, shape, chromatin pattern, pyknosis, and disappearance.

1. Placement and Orientation within the Cell

The nucleus of a cell shed from keratinizing stratified squamous epithelium, virtually always seeks a *central* location within the cytoplasm (Figs. VI-2, XI-5). This is regardless of the cell's degree of maturation (i.e.: parabasal, intermediate cell, superficial cell).

While keratohyaline granules may occasionally distort the nucleus (Figs. XI-1 through XI-4), the nuclei are with*out* distortion from any recognizable center of activity or cytocentrum, such as many other types of cells demonstrate (e.g.: monocytes, macrophages, columnar cells). There is no directional or planar orientation of the nonround or oval nucleus within the cytoplasm, is symmetrically distributed about it, as viewed from above.

2. Shape

It retains a round or oval shape. Through the parabasal cell and intermediate cell stages, the nucleus appears *vesicular,* with the nuclear membrane drawn tightly about its semiliquid, moldable contents (Figs. XI-1 through XI-4).

3. Chromatin Pattern

The nucleus of this epithelial cell in the interphase (i.e.: metabolic, resting, intermitotic) has a characteristic chromatin pattern. The chromatinic net consists of a uniformly dispersed fine network of thin fibers and small granules (Fig. VI-2). Chromocenters, or prominent granules, are few and inconspicuous. As the cells mature, this delicate chromatin pattern becomes slightly coarser (Fig. XI-5).

The chromatinic rim is also delicate and uniformly thin over the nuclear surface (Figs. VI-2, XI-5). In the human female, one prominent

---

Figures XI-1 through XI-4: Squamous cells from keratinizing stratified squamous epithelium. Oral smears. The nuclei are centrally placed, round or oval, and vesicular. Chromatin in the chromatinic net is mainly in delicate threads and small granules. The cytoplasm is orange, and contains many keratins and precursors uniformly dispersed throughout. In addition, there are keratinization particles of the keratohyaline type. No eleidin granules are present. Some of the keratohyaline granules are in extremely close association with their nucleus. The multiple nuclei (Fig. XI-3) are virtually identical. There is beginning chromatin blurring, margination and hyperchromasia of early karyopyknosis in all nuclei.

Papanicolaou stain.

XI-1, XI-2: × 2,000; XI-3: × 3,500; XI-4: × 6,300.

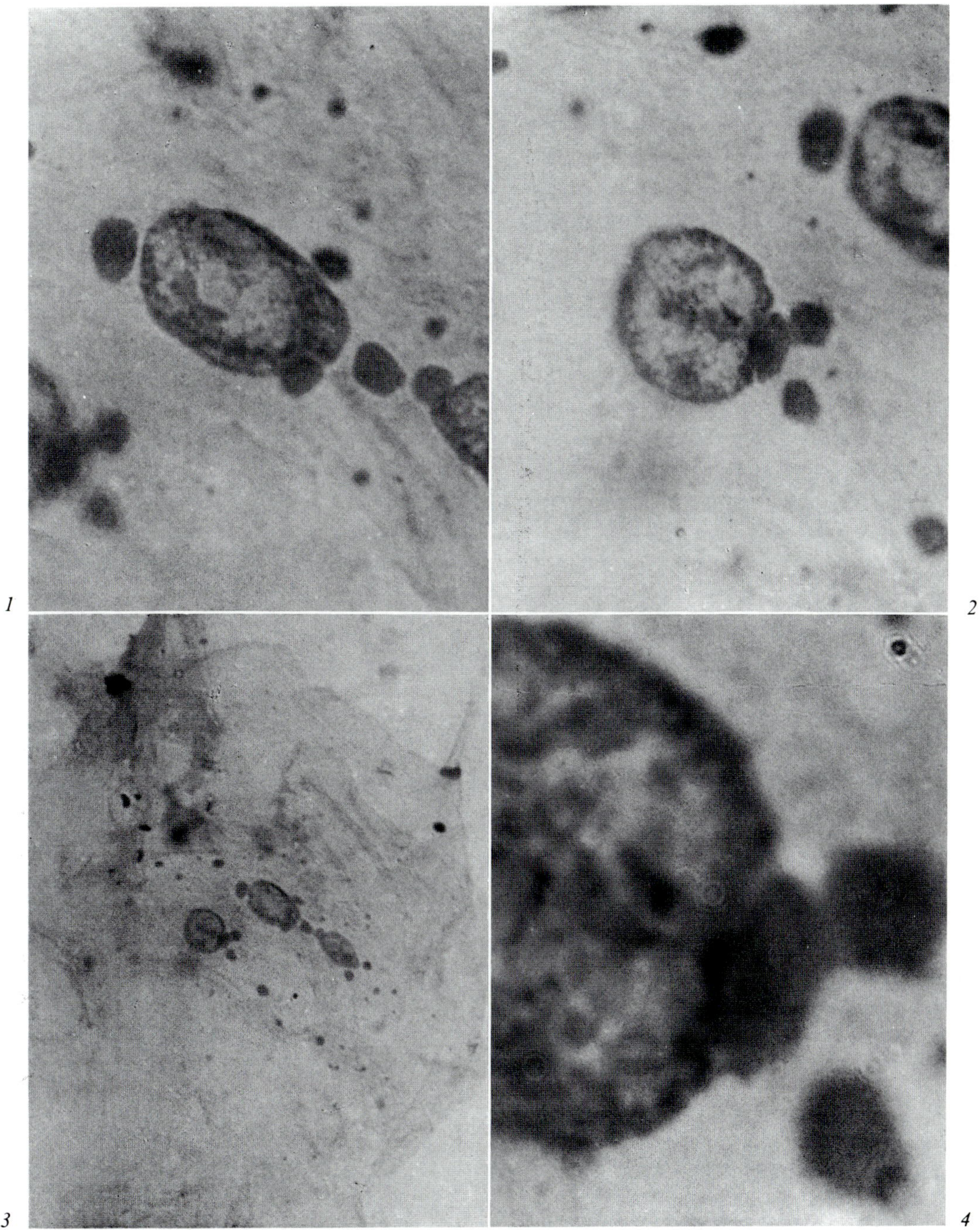

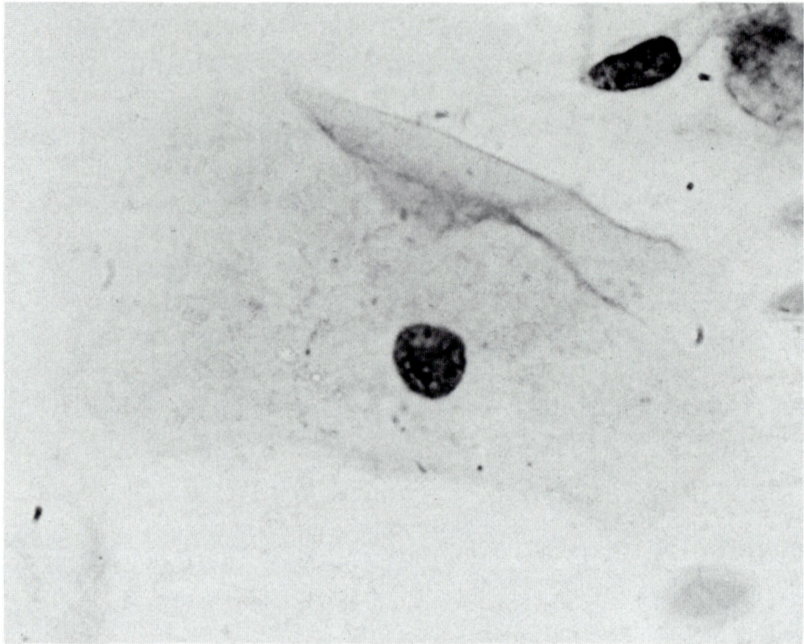

Fig. XI-5: Fast (vagino-pancervical) smear. This *intermediate squamous cell* is lying flat and singly in a clean background, as is found in an unopposed estrogen effect. There is a prominent chromocenter, with a striking thread-like chromatinic net. The thin, wafer polygonal cytoplasm (blue) has longitudinal wrinkles, and one corner has folded back upon itself. This is *not navicular* folding [99] but merely mechanical wrinkling of a thin wafer squame.
Papanicolaou stain.
XI-5: × 1,160.

chromocenter is condensed into the chromatinic rim and is absent in the male. This female X-sex chromocenter, X-chromatin, or Barr body [29, 30, 99] represents one of the two X-sex chromosomes which is biologically inactive, heterochromatic and condensed in the chromatinic rim against the inner nuclear membrane. The Barr body measures approximately 1 micron in diameter [99].

The other X-sex chromosome is needed in its euchromatic and biologically active form for cell viability, and is thus not identifiable in the interphase nucleus of either man or woman. The Barr body can occur at any part of the chromatinic rim on the nuclear membrane, as we view it in the interphase nucleus; however, for *definite identification* in the human, it is recognized *only* in those nuclei which are oriented in the cell so that the Barr body lies in the lateral equatorial chromatinic rim as it is viewed in profile [99]. In this way it is identifiable as lying on the inner surface of the nuclear membrane, as a part of, and melding into the chromatinic rim.

For definite *identification,* it is important that the nucleus be *well preserved;* furthermore, the chromocenter must be a part of the chromatinic rim and, by using a $100 \times$ oil immersion objective, is unable to be resolved as being separate therefrom. This X-chromocenter is identifiable by these strict criteria [97, 99, 190] in about 20 to 60 percent of exfoliated euplastic squamous cells of the normal human female. Its presence indicates: 1) a heterochromatic X-sex chromosome in the chromatinic rim at the nuclear envelope *and,* 2) a second euchromatic X-sex chromosome unidentifiably dispersed within the chromatinic net and parachromatin of the metabolic nucleus. As the latter is necessary for cell survival, its presence is presumed. Thus the *minimum* number of X-sex chromosomes which are present in a given cell (X_{min}) is represented by the *maximum* number of Barr bodies recognized per cell (B_{max}) plus one, or: $X_{min} = B_{max} + 1$.

4. Karyopyknosis and Disappearance from the Cytoplasm

As the cell matures (specif.: hypermatures) from intermediate to superficial type, karyopyknosis abruptly occurs. Two major processes are involved. There is *denaturation* of the nuclear contents (internal smudging of chromatin/parachromatin interfaces, alteration of chromasia, loss of pattern), and *water* loss with nuclear dehydration (nuclear membrane wrinkling, shrinkage below 6 microns, hyperchromasia). Under phase microscopy, or lowered condenser lighting, the pyknotic nucleus has a red sheen [299].

Three distinct morphologic phases in the development of karyopyknosis are thus to be observed: chromatin blurring with eventual *loss of chromatin pattern, shrinkage, hyperchromasia.* These do not occur in any given order, and usually occur virtually simultaneously. Thus, about 90 percent of the karyopyknotic cells have all three of these key recognition features. When only one feature does occur in a *well*-preserved specimen, however, that is usually considered sufficient to identify the fact that karyopyknosis is occurring (e.g.: in determining estrogenic effect; Maturation Index) [96, 99, 299].

A pyknotic nucleus may fragment (karyorrhexis). As maturation continues, with or without fragmentation, the nucleus eventually lyses and disappears from the cell. At times the resulting anucleate superficial cell retains the nuclear 'ghost', a small pale area within the cytoplasm [99].

C. The Cytoplasm in Keratinizing Stratified Squamous Differentiation

In the less mature cells (i.e.: germinal cells, parabasal cells) the cytoplasm is more elastic. On exfoliation it rounds up into thick rubbery balls. With maturation, the cytoplasm reduces this general elasticity of immaturity and becomes thinner when placed upon a glass slide.

Additionally, and specifically in keratinizing stratified squamous epithelium, this cytoplasm becomes increasingly rigid as the hard and tough keratin precursors and, eventually, keratin appear in the cytoplasm [3, 49, 81, 182, 183]. With such increasing maturity, the exfoliated squamous cell deviates more from this thick and rounded prototype of youth. It more rigidly retains the structure it held in tissues before exfoliation, with virtually no alteration in the highly keratinized cells.

The more mature keratin, as present in the upper layers of the skin (horny layer, stratum corneum, stratum lucidum, eleidin layer), appears glassy and acidophilic. By electron microscopy, it is found to consist of tonofibrils embedded in a tough matrix. The latter appears to be derived from the material of the keratohyaline granules [184].

The material of the *keratohyaline granules* forms in the cytoplasm of the intermediate cells of the outer Malpighian layer. It is probably formed in the rough endoplasmic reticulum (rER) and, perhaps, even in the perinuclear cisterna (Figs. XI-1 through XI-4) from the outer nuclear membrane. It is then packaged by the cell into keratohyaline granules, where it appears in the outer intermediate cells and superficial cells, particularly in chronic irritation. It appears to eventually contribute to form the matrix between the tonofibrils of maturing keratin.

The *eleidin granules* appear to represent masses of tonofibrils with enough matrix to hold them together as a granule. They appear more crystalline than the more lipid appearing keratohyaline granules.

1. Keratinization

The word in the title of this class of epithelium is keratiniz*ing,* rather than keratiniz*ed.* As inferred thereby, the process of keratinization is occurring throughout its layers, even though keratinized cells are not recognizable on the surface of the nonkeratinized or noncornified members of this class of epithelium by usual hematoxylin and eosin and Papanicolaou stains.

There are many keratins and prekeratins being identified in increasing numbers by immunodiagnostic techniques [88, 184, 267]. For the most part, these appear to be varying levels of maturation of the protein molecules, and alterations in their structure for differing functional needs. Keratins and their less mature precursors (prekeratins) typically occur *uniformly* throughout the cytoplasm. Even though some precursors appear as keratinization particles (i.e.: keratohyaline, eleidin) (Figs. XI-1 through XI-4), most precursors are generally distributed uniformly throughout the cytoplasm.

Keratohyaline granules are essentially smooth and rounded, appearing to be lipid in character, while eleidin is more irregular in outline and has some crystalline morphologic features, Keratohyaline tends to be basophilic, from deep black-purple (India-ink) to golden-brown (Figs. XI-1 through XI-4). Eleidin tends to be acidophilic (yellow-orange-red), like the eleidin layer of the skin, from which it derives its name. They both can appear from dull to gleaming, but at times the eleidin has a definite hyaline (i.e.: glassy) appearance. They both appear in the upper layers of the more mature keratinizing stratified squamous epithelium, with irritation, and in irritational atypias; but it is usually only the eleidin keratinization particle which is found in cancer and the severe atypias.

While the definitive identity of keratins is by special stains and, particularly, immunodiagnostic techniques [88, 132, 267], there are a number of morphologic features present in the routine Papanicolaou stain which facilitate its detection and identification. Those with the greatest discrimination involve *hyalinization* and *color*.

a. Hyalinization. This *glassy* appearance to the cytoplasm (*hyalos:* Gr. = glass) constitutes the most discriminative characteristic of the presence of the more mature keratins. This extremely hyaline character frequently imparts to the cytoplasm the impression of viewing a brilliantly *gleaming*, colored glass of a church window, with the sun beaming in from behind.

The term hyaline has come to be frequently used in pathology to connote an acidophilia, such as Zenker's hyaline degeneration of muscle fibers in toxic febrile illnesses (e.g.: typhoid), rather than simply its more denotative use for glassy. Regardless of the tinctorial shade of the cytoplasm (e.g.: blue, green, gray, yellow, red, orange), this extreme glassy, hyaline character is a good discriminating hallmark of keratinization.

b. Color. As the process of keratinization develops with cellular maturation, the *tinctorial* properties of the cytoplasm under Papanicolaou stain, change from basophilia (i.e.: green to blue) toward acidophilia (e.g.: yellow, orange, red). The mature keratins stain strikingly acidophilic.

Other substances and conditions also cause the cytoplasm to stain acidophilic, of which the most common and important to recognize are *degeneration, inflammation,* and *air-drying.* They can produce an intense acidophilia, most frequently in the yellow or the red-orange. For this reason, the presence of a deep brilliant *orangophilia,* or hyaline 'Halloween orange' (a useful characterization coined by S. T. Shutt, CT(ASCP)), is a useful distinguishing characteristic for the tentative identification of mature keratin.

In the earlier stages of keratinization, however, tinctorial characteristics are neither well developed nor stable. They also are extremely sensitive to artifactual influences (i.e.: inflammation, degeneration, pH change). In epithelia not normally containing mature keratins (i.e.: the noncornified stratified squamous epithelium of the vagina), the tinctorial property of cytoplasm containing the keratin precursors is nonspecific; in fact, it is frequently misleading. In such noncornified stratified squamous epithelium, neither the extreme orangophilia nor the hyaline qualities of keratinization occurs. More complete keratinization does arise in these areas under pathologic situations (e.g.: hyperkeratosis, leukoplakia) (Chap. XV).

Due to great sensitivity to artifacts, therefore, color alone is a poor discriminating feature of keratinization. The best tinctorial morphologic evidence in the routine stain for mature keratins is a *hyalinized, deep orangophilia* of the cytoplasm.

2. Thinning

The maturing cells layer-out into a *thin* squame (L. = scale), which is oriented *parallel* to the basement membrane and to the lumen (Fig. XVI-7). It uniformly squamifies into an extremely fine plate (Fig. XI-5), thinning symmetrically around a centrally-placed nucleus. The thinning is *uniform* throughout the *whole* cytoplasm, from the region of the nuclear membrane to the cell membrane.

In tissue, the upper and lower cell membranes are usually held apart by fluid cytoplasmic contents (e.g.: glycogen, mucopolysaccharides), especially in the intermediate cells. On exfoliation of the cell, most of the contents are secreted into the background. The two cell membranes approach each other with great force into close apposition, being held out taut by the tough keratin precursors in typical wafer shape.

In contrast, the cells of *simple squamous epithelium* are not to be confused with these of *keratinizing stratified squamous epithelium.* The former (e.g.: pulmonary alveolar lining cells, renal Bowman's capsule epithelium, mesothelium, endothelium) are neither keratinizing nor keratinized. Even though they are also flattened squames, simple squamous cells are soft and malleable, round up when shed, and appear thick in cellular spreads (Fig. VII-7). Functionally and morphologically, therefore, the squamous cells of simple squamous epithelium are to be clearly separated from those of keratinizing stratified squamous epithelium.

D. The Exfoliated Cell of Keratinizing Stratified Squamous Differentiation

1. Parabasal Cell
When the epithelium does not mature past that of the parabasal cell (e.g.: complete atrophy, parabasal atrophy, teleatrophy), the cells exfoliated are of parabasal cell type. The cytoplasm is rubbery and retains its thickness, even in cellular spreads. It is usually small, rounded, and basophilic (blue, green, gray); but size, shape, or color do *not* discriminate for accurate identification.

The nucleus seeks a central position. When it is moved off to one side, either from mechanical force or degeneration, the cell may mimic a histiocyte or columnar cell. Characteristically, the nucleus has an active metabolic, interphase pattern. In atypical situations, however, it can be proplastic (Chap. VIII) and even pyknotic.

The dyskaryotic cell (Chaps. VIII, XV) must be recognized by its nucleus [99] and *not* be included in this normal cell series (e.g.: in cytohormonal evaluation) [96]. A dyskaryotic cell with moderately mature or immature cytoplasm (Figs. VIII-2, VIII-3) is frequently mistakenly identified for a parabasal cell, whose cytoplasm is of similar maturity; however, when it is proplastic and when the dyskaryotic nucleus is large and hyperchromatic, it is distinguishable and should be properly identified. Only cells with euplastic or small pyknotic nuclei are included as normal parabasal cell type.

2. Intermediate Cell
When epithelium matures to the level of the intermediate cell (e.g.: progestogen effect in vaginal mucous membrane), that cell type is exfoliated. Its cytoplasm is a thin wafer, abundantly spread to an attenuated polygonal squame. Usually basophilic, its tinctorial qualities are not constant enough to help in accurate identification. The central nucleus is vesicular with a metabolic interphase chromatinic pattern. Mature squamous dyskaryotic cells and cells from metaplasia, must be recognized and

excluded from these normal squames (e.g.: in cytohormonal evaluation) [99]. The large, hyperchromatic nucleus helps to identify the dyskaryotic cell. The wafer-thin cytoplasm of the squame shed from metaplasia is frequently considerably smaller than that of the normal intermediate cell.

3. Superficial Cell

a. Nucleated Superficial Cell. When maturation of the stratified squamous epithelium is advanced to a degree where the nucleated superficial cell lies on the surface (e.g.: estrogenic effect in the keratinizing stratified squamous epithelium of the vaginal mucous membrane), that cell type is exfoliated. This is the classical superficial cell [99].

Its cytoplasm is a thin polygonal wafer, basically identical to that of the intermediate cell. Usually eosinophilic and large, its color and size are not accurate identifying features.

The nucleus is centrally located and is karyopyknotic. This and the anucleated superficial cell are actually hypermature.

b. Anucleated Superficial Cell. As maturation continues (e.g.: keratinized stratified squamous epithelium of the outside skin), the nucleus of the cell lying on the surface of the epithelium fragments, lyses and, eventually, disappears. Evidence of continuing keratinization appears in the cytoplasm (e.g.: keratohyaline and eleidin granules) as hypermaturation continues to the level of the granular layer and stratum lucidum [99].

Eventually, hyaline keratins fill the complete cytoplasm, with its deep glassy orangophilia, of those cells shed from the stratum corneum. Along with this, the pyknotic nucleus fragments and lyses, so that cells shed from the stratum lucidum and stratum corneum are typically anucleate superficial cells.

To recapitulate identification of the three major cell types of keratinizing stratified squamous epithelium:

Parabasal cell:
    cytoplasm – thick;
    nucleus – metabolic or pyknotic, not dyskaryotic.

Intermediate cell:
    cytoplasm – thin (a squame);
    nucleus – metabolic (vesicular with preserved chromatin pattern).

Superficial cell:
    cytoplasm – thin (a squame);
    nucleus – pyknotic or absent.

E. Tissue Fragments in Keratinizing Stratified Squamous Maturation

1. Stratification

Squamous cells tend to shed singly. However, when a fragment of keratinizing stratified squamous epithelium sheds, the functional feature of *stratification* is readily discerned. Typically this appears as a multiple-celled, thick sheet of squames viewed en face in a mosaic, pavement-like pattern. This nondescript grouping, however, usually is not able to be accurately identified as a true tissue fragment which is discriminable from merely a clump of squames secondarily piled on top of each other.

In some normal sites (e.g.: tonsillar plugs), as well as in *atypical* situations (e.g.: acanthotic plugs in irritated areas), tissue fragments exfoliate as concentrically arranged, stratified cells, or *pearls* (cell 'nests', or 'pearly bodies') (Fig. XI-6). In this arrangement, the cells lying on top are frequently nested into concavities of the underlying cell, as a baseball in a catcher's mitt (Chaps. XII, XV).

This superficial nesting has been erroneously referred to as 'cannibalism' [202]. While this was an interesting connotative observation, it is actually untrue, as cytophagocytosis does *not* occur in well-differentiated keratinizing stratified squamous epithelium, either benign or malignant, while it *can* occur in undifferentiated cancers. Use of the term, therefore, has been discontinued and replaced by 'pearl'.

Such pearl formation represents merely an extreme in the maturation features of stratification, those of cells lying on top of each other in *strata* and *holding together* to stay there (Fig. XVI-7). The nuclei of these benign pearls appear nonmalignant, usually somewhat proplastic and retroplastic (Fig. XI-6), but not malignant neoplastic as with malignant pearls (Fig. XII-5).

2. Intercellular Bridges

Intercellular bridges are prominent and commonly found in tissue, mainly in the middle and lower third of the epithelium (Figs. XVI-5, XVI-7, XVI-9). After exfoliation, however, their *identification* is *infrequent* as they are usually ripped apart and represented by only a few thin strands of cytoplasm between cells.

They represent areas of extremely strong intercellular attachment. The cytoplasmic processes which extend across between cells, occur with a fairly uniform periodicity along the cell boundaries. While present in virtually all levels of stratified squamous epithelial cell maturation, intercellular bridges are best retained and demonstrable in the parabasal cells. They become less

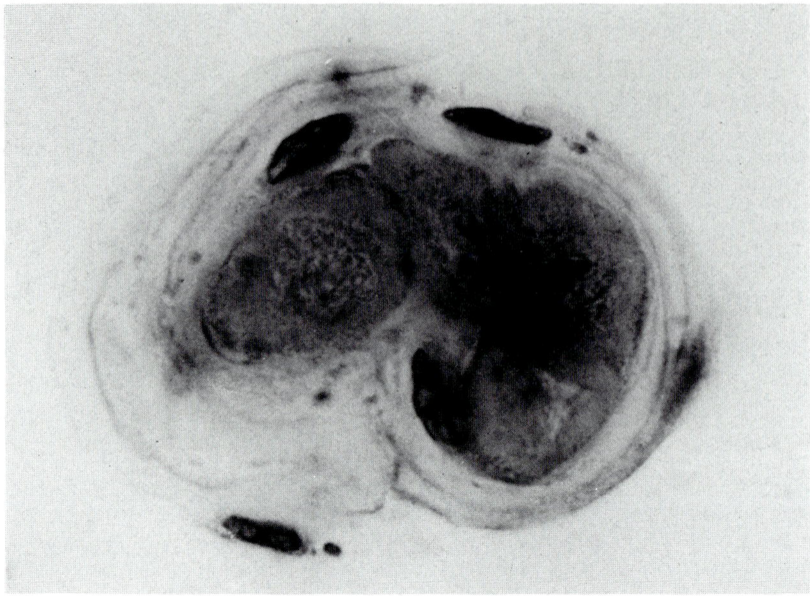

Fig. XI-6: Squamous cells, shed as a benign pearl. The four anucleate, thick, dark cells are lying on top of the four lighter stained squames; successively, one on top of the other in a truly stratified configuration. This diagnostic true tissue fragment (DTTF) is basically euplastic; however, it has somewhat proplastic stimulated nuclei which, secondarily, are retroplastic (i.e.: wrinkled, hyperchromatic, blurred chromatin/parachromatin interfaces).
Papanicolaou stain.
XI-6: × 1,160.

distinctive in the intermediate cells and, in the superficial cells, degenerate into microvilli of nondiscriminating, nonuniform periodicity.

Intercellular bridges are discriminatingly identified when well preserved by their fairly *uniform periodicity* and the *uniform parallelism of the cell borders* (Figs. XII-8, XII-9; XVI-9). They must carefully be distinguished from intercellular vacuoles of degeneration (see section in Chap. XII, E, 2) (Figs. XII-18, XII-19).

With degeneration, cells with *or* without intercellular bridges can form vacuoles of degeneration between their cell walls. Slight intercellular vacuolation can dangerously mimic intercellular bridges. Variations in size of these vacuoles, however, give these *pseudo*bridges *non*uniform periodicity; furthermore, the adjacent cell walls alongside of the vacuoles, are *scalloped* rather than *straight and uniformly parallel.*

XII. Atypical Functional Differentiation: Keratinized Stratified Squamous Epithelium and Squamous Cell Carcinoma

A. Atypical Stratified Squamous Differentiation

In states other than euplasia, and even occasionally in euplasia to a much less extent, functional differentiation of keratinizing stratified squamous epithelium is frequently *atypical*. The probability for these atypicalities to occur increases in retroplasia, is greater in proplasia, and is greatest in malignant neoplasia. Atypical functional differentiation is extremely frequent in tissues which are in transition from one state to another (Chap. XVI).

The *features* of these atypicalities of functional differentiation, however, do *not* differ essentially when found in retroplasia, proplasia, and malignant neoplasia. They do, however, differ in their frequency and severity.

They are *not*, therefore, to be considered as malignant criteria. In fact, their discriminating value is not only so poor that it is virtually nonexistent, using them as malignant criteria is extremely treacherous and unpredictable. All too frequently they have been the cause for *false identification* and *diagnosis* of cancer when they have been allowed to alter one's opinion regarding the presence or absence of malignant neoplasia.

A Cardinal Rule of great value to observe in approaching a diagnosis of cancer and its cell type, is to proceed in this order:

First, determine that cancer is present (mainly by *nuclear* features, Chaps. V–IX);

Second, determine the cell type of cancer present by its functional differentiation (mainly by *cytoplasmic* features Chaps. X–XVIII).

While atypical functional differentiating features may be used to arouse suspicion and to encourage more specimens and more searching for discriminating features, they should *not* be admitted and used as discriminators in identifying cancer.

In applied diagnostic cytopathology, it is wise to disregard functional differentiating features until the diagnosis of cancer has been made. Then, once the presence of cancer has been established, these features are evaluated to determine the functional differentiation of the cancer cells and tissue, i.e.: the 'type of cancer' or 'cell type'.

First, cancer is identified by evaluating malignant criteria. Then, *second,* type is determined by evaluating functional differentiating characteristics.

The cells of squamous cell carcinoma are malignant neoplastic cells attempting to differentiate in the ways of euplastic cells of stratified squamous epithelium. Some, with properly unmasked genomes apparently intact, succeed in differentiating in typical fashion (Chap. X). Others are atypical; at times, macabre.

Malignant neoplastic epithelium functionally differentiates virtually across the total spectrum, typical and atypical. As its atypicalities can be so extreme, it serves as a good *example* for *atypical differentiation.*

B. General Characteristics

Cells shed from well-differentiated squamous cell carcinoma tend to separate from each other and to shed *singly* more often than virtually any other cancer cell type except for melanoma, lymphoma, and leukemia

---

Figures XII-1 through XII-4: Atypical cytoplasmic thinning (tail formation) in nonmalignant squamous epithelial cells, showing varying degrees of mild to moderate proplasia from irritation, inflammation, and infection.

Fig. XII-1: Oral. An irritated squame is forming the single tail of a tadpole cell, with abnormal thinning of the cytoplasm into a tail which has an ecto-endoplasmic border extending down it, forming a Herxheimer's fibrillary spiral. The tail ends in a small but definite *rounded* bulb, evidencing that it is *not* a pulled-out whisp of degenerating cytoplasm. The other squames are typically thinned symmetrically about their nucleus. Note that the nucleus is only very slightly enlarged (mild karyomegaly).

Fig. XII-2, XII-4: Sputum. Irritated squames forming the two tails of spindle cells. The elongated nuclei are mildly hyperchromatic, with patterns of slight proplasia. Many are extremely karyopyknotic so that their exact identity cannot be determined.

Fig. XII-3: Vaginal smear. An irritated parabasal cell is binucleated (from proplasia) and is beginning to thin its cytoplasm to become an intermediate squame (Fig. XI-5), but is performing it atypically as a tail. It is evidencing another atypical functional differentiation, the formation of ecto-endoplasm. Genesis of a fibrillary spiral is noted with ridges in the ecto-endoplasmic border extending down into the tail. Nuclei are moderately proplastic.

Papanicolaou stain.

XII-1: × 500; XII-2: × 800; XII-3: × 1,320; XII-4: × 200.

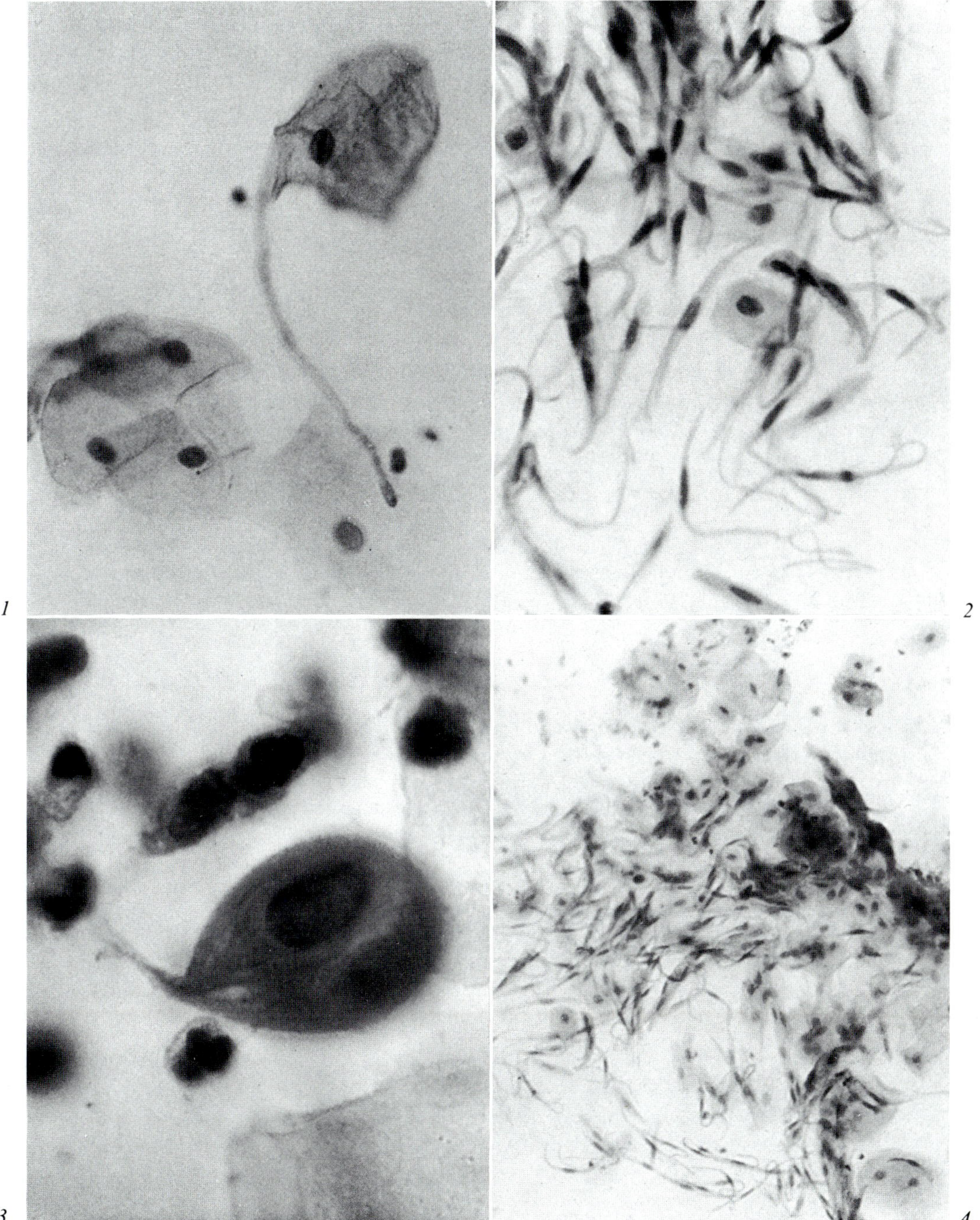

1
2
3
4

(Chaps. XV, XVII, XVIII). This feature also is found more in normal keratinizing stratified squamous epithelium (Chap. XI) than in other nonmalignant cell types (Chap. XIV) except for hematologic cells.

When *tissue fragments* do shed from squamous cell carcinoma, they are most often from the *less differentiated* tumors, or are in specimens obtained by *abrasive* techniques (e.g.: scrape, brush) or by *needle aspiration.*

Even though the cells have fallen off or have been torn loose from the stratified squamous epithelium, the cytoplasm of the exfoliated cells tends to be intact. This is especially so in the better differentiated squamous cell carcinomas having more sturdy cytoplasm containing more mature keratin proteins.

C. Nuclear Features

1. Placement

As with typical squamous maturation (Fig. XI-5), nuclei tend to seek the *center* of the cell in atypical squamous maturation (Figs. XII-1 through XII-4). Even with the very macabre cell alterations (i.e.: ecto-endoplasm (Fig. XII-7), pearl (Fig. XII-5), spindle (Figs. XII-2, XII-4, XII-11, XII-12),

---

Figures XII-5 through XII-9: Kertinized stratified squamous epithelial functional differentiation in cancer (squamous cell carcinoma).

Fig. XII-5: Pearl ('pearly body'). This pearl is composed of cancer cells (note malignant criteria) which are lying in a concentric arrangement on top of each other, in a truly stratified fashion. Note the intercellular bridges between some of the cells, especially near the bottom of the figure.

Fig. XII-6: Eleidin granule. This abnormal, brilliant orange, hyaline, keratinization particle lies in the cytoplasm below the nucleus. Note the clear zone about it and its somewhat irregular and sharp crystalline appearance.

Fig. XII-7: Ecto-endoplasm of atypical keratinization. The hyaline ectoplasm is separated from the textured endoplasm by a distinct ecto-endoplasmic border. Within the endoplasm sit multiple malignant nuclei. Multiple barely discriminable rings can be detected in the ectoplasm around 9-o'clock.

Figs. XII-8, XII-9: Intercellular bridges. Note the virtually *uniform periodicity* of the bridges across the relatively clear intercellular spaces; and the *parallel cell borders* of the adjacent cells, which are *not* scalloped (loop-the-loop) as they are in the presence of intercellular degenerative vacuoles.

Papanicolaou stain.

XII-5 through XII-8: × 2,000; XII-9: × 6,300.

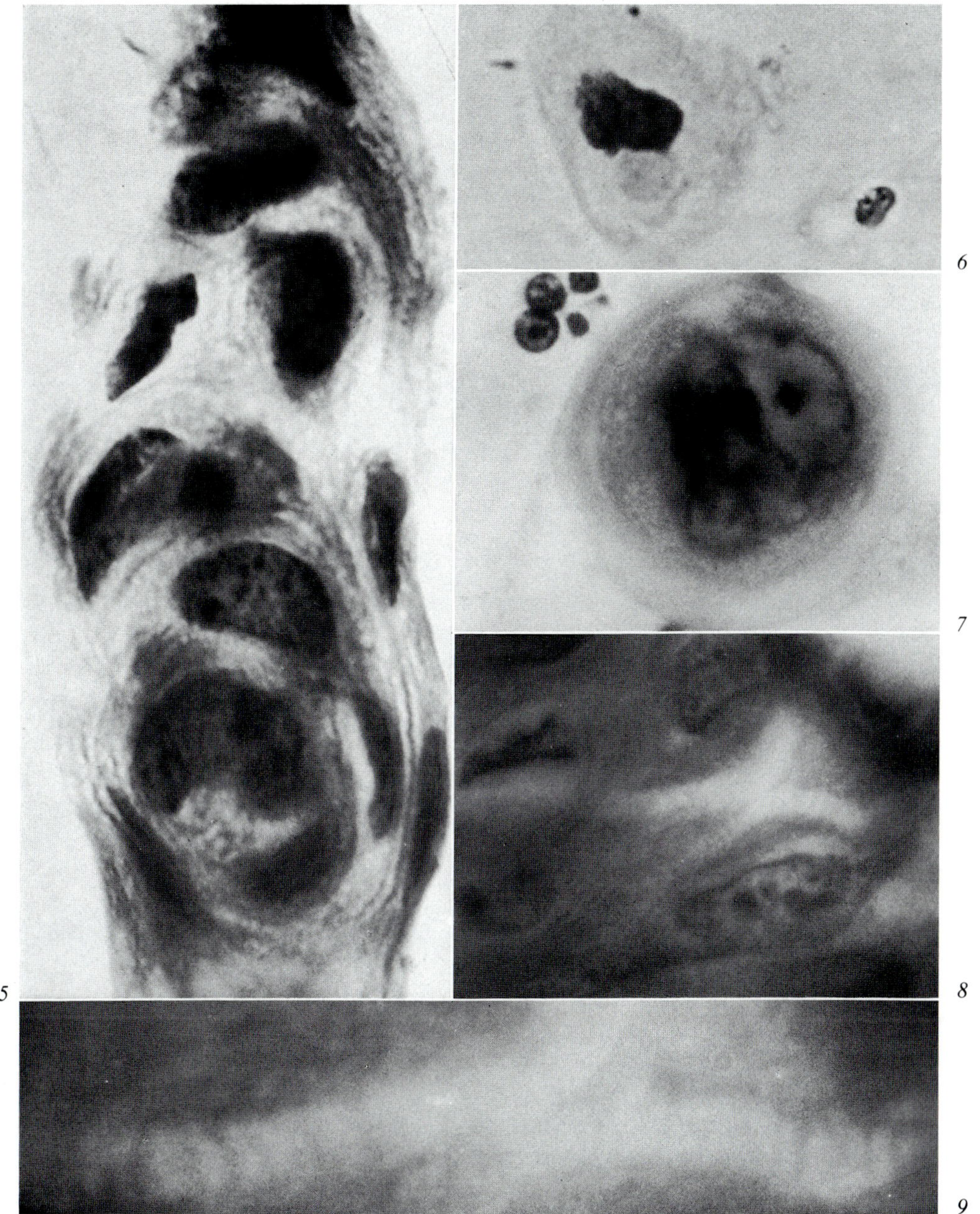

and tadpole (Figs. XII-1, XII-3, XII-10), the nucleus tends to be central to the main mass of cytoplasm. In the spindle and tadpole cells, the nucleus appears to be eccentric when viewed in profile while it is lying on the microscopic slide; in actuality, however, their symmetry is really about their *long* axis which orients parallel to the surface of the glass slide.

2. Chromatin Pattern

The thread-like and fine granular chromatin pattern, so typical of non-malignant cells with squamous maturation (Fig. XI-5), is at times retained. A 'ground-glass' nucleus with uniform fine granules uniformly dispersed throughout the parachromatin, reminiscent of proplasia (Fig. VIII-2), occurs frequently in squamous cell carcinoma (Fig. XII-13). It is *not* a good functional differentiation feature of squamous cell carcinoma, however, as it also frequently occurs in many other cell types of cancer (e.g.: adenocarcinoma, transitional cell carcinoma, sarcomas).

Thus, it is of virtually *no* value in discriminating and identifying squamous cell carcinoma. With the overwhelming disruption of chromatin pat-

---

Figures XII-10 through XII-12: Atypical cytoplasmic thinning (tail formation) of keratinizing stratified squamous epithelial functional differentiation in malignant cells (squamous cell carcinoma). Note the small granules of hemosiderin which decorate the fine fibrils of fibrin in the background indicating old chronic bleeding.

Fig. XII-10: Tadpole cell. The nucleus is pyknotic. It is extremely pointed and squeezed at one end, nearest to the cytoplasmic tail, by the tremendous forces of the cytoplasm's thinning. A thin rim of cytoplasm is all the nucleus has around its other end, so that it has been shoved for a short distance down the endoplasm of the tail by these pressures, and has thinned nearly to a point. The ectoplasm, endoplasm, and ecto-endoplasmic border are just barely discernible at the right of the figure. Note the fibrin and neutrophils of the cancer diathesis background.

Fig. XII-11: Spindle cell. Multiple nuclei, drawn out at both ends by the cytoplasmic forces of thinning, lie in the endoplasm. The ectoplasm extends over them and down the tails at both ends.

Fig. XII-12: Spindle cell with a monstrous bulbous enlargement of the tip one of its tails. The ecto-endoplasmic border extends out both tails, appearing as cytoplasmic fibrils. One end gradually widens (right upper bending to right lower). The other end (left) suddenly opens into a wide bulb containing copious endoplasm, as though it is awaiting a second nucleus. Leukocytes lie above the cell. There is a hyaline ectoplasmic rim about the bulb, which contains some 'ringing'. Note the red blood cells, neutrophils, and cellular debris of the cancer diathesis background.

Papanicolaou stain.

XII-10, XII-11: × 2,000; XII-12: × 1,320.

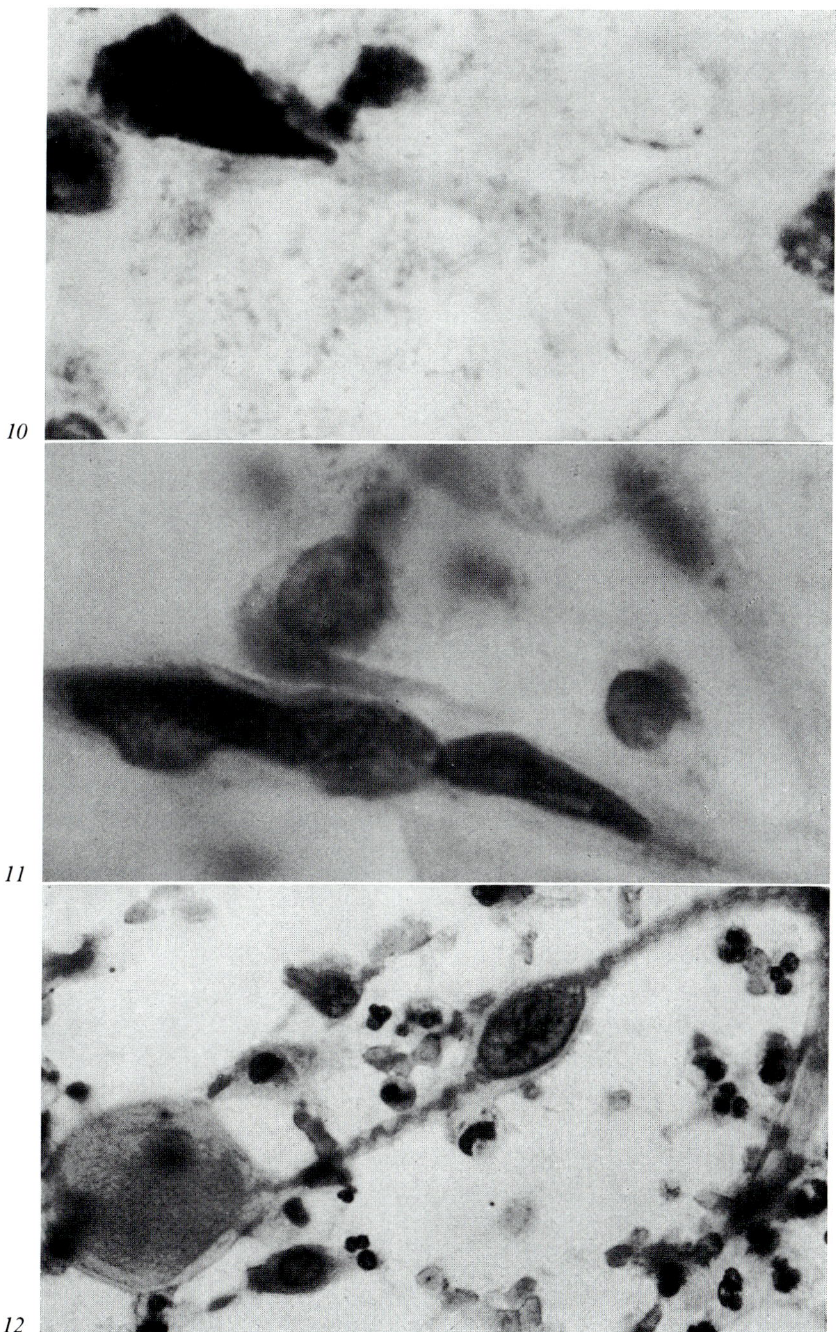

10

11

12

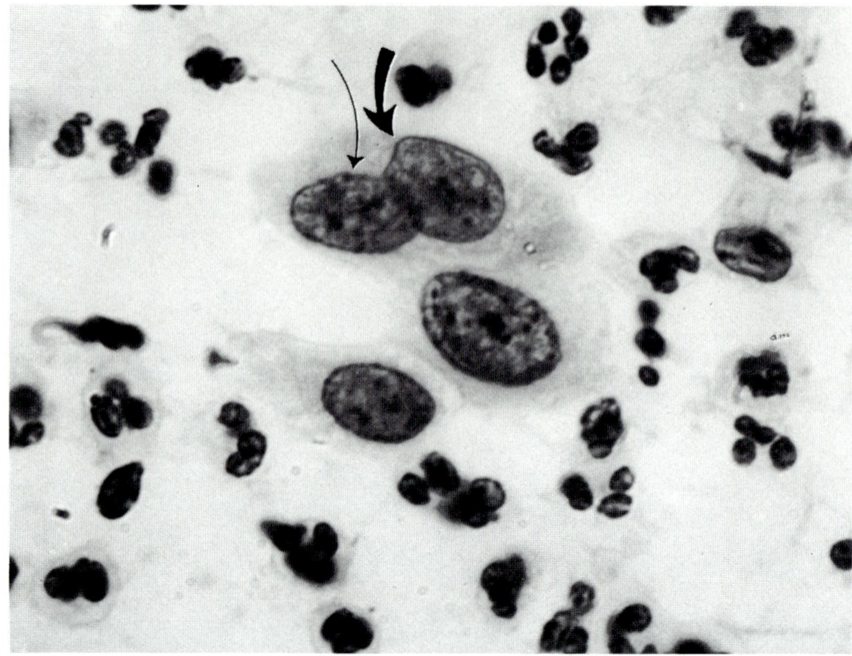

Fig. XII-13: Invasive squamous cell carcinoma. These four disarmingly benign-appearing cells have 'ground-glass' dysplasia-type nuclei (Chap. VIII), but with multiple and prominent *nucleoli.* The latter serves as a warning that these might not be from simple proplasia, which rarely has nucleoli (cf.: Fig. VIII-2, VIII-3); but, rather, could well be from invasive cancer. This mandates careful evaluation of other features which may indicate malignancy. Note the small but definite jag in the nuclear membrane (thin arrow) and the concavity on the end of the adjacent nucleus (thick arrow), each for *no* apparent reason. There are also areas of parachromatin clearing which are *une*venly dispersed throughout the nuclei.
Papanicolaou stain.
XII-13: × 1,250.

tern by malignant characteristics, and by karyopyknosis, however, this tendency is usually obscured except in poorly-differentiated portions of a moderately well-differentiated tumor.

3. Nucleolus

The prominent nucleolus and multiple nucleoli, which are associated with rapid cell growth, are usually less frequent in moderately well-differentiated and in well-differentiated squamous cell carcinoma than in the poorly differentiated tumors. As normal cells mature from parabasal to intermediate cell types in the maturation of normal keratinizing stratified squamous epithelium (Chap. XI), the nucleus assumes a more resting pattern of decreased protein production. With decreased demand for ribosomes, the nucleolar material decreases and, eventually, disappears from the nucleus.

Thus, nucleoli become inconspicuous in the better differentiated squamous cell carcinomas, as compared to the extremely common finding of prominent nucleoli in the better differentiated adenocarcinoma (i.e.: secretory). A dangerous *fallacy* has thus grown that nucleoli are good differentiating features of adenocarcinoma, and their absence is a valuable feature in favor of squamous cell carcinoma.

The fact of the matter is that this holds true *dependably only* in the better differentiated tumors – where there is more than enough, abundant *other* evidence for *reliable* functional differentiation in both tumors. *Conversely,* in those situations where the other reliable evidence (e.g.: secretion *vs.* keratinization) is *not* available (such as in the poorly-differentiated tumors, adeno- or squamous), this evidence regarding nucleoli actually *reverses* its role and provides *falsely* discriminative information. The *poorly-differentiated squamous cell carcinomas* frequently have just as *prominent nucleoli*, evidencing their increased biologic activity with high protein production (Figs. XII-13, XII-14). Likewise, prominent nucleoli occur in high-grade transitional cell carcinomas, large cell undifferentiated carcinoma (Fig. XVIII-1), sarcomas, and poorly-differentiated or high-grade *any*-type of cancer.

The presence of *prominent nucleoli,* therefore, is a *poor* and a *dangerous* feature of functional differentiation, and should *not* be used as a functional differentiating feature. The only dependable information it provides, is that of indicating *increased protein production.* It does not discriminate between synthesis for use *intra*cellularly (e.g.: growth, keratin synthesis, mucus production) (Figs. I-5; XII-13, XII-14) and *extra*cellularly (e.g.: mucus secretion, collagen formation) (Figs. I-1, I-2; XV-4).

4. Karyopyknosis

As euplastic squames differentiate from the intermediate squamous cell to the superficial squamous cell, the major characteristic is that of pyknosis of the nucleus. This, likewise, is a prominent feature as the malignant neoplastic cell undergoes maturation in a manner of keratinizing stratified squamous epithelium, to produce squamous cell carcinoma.

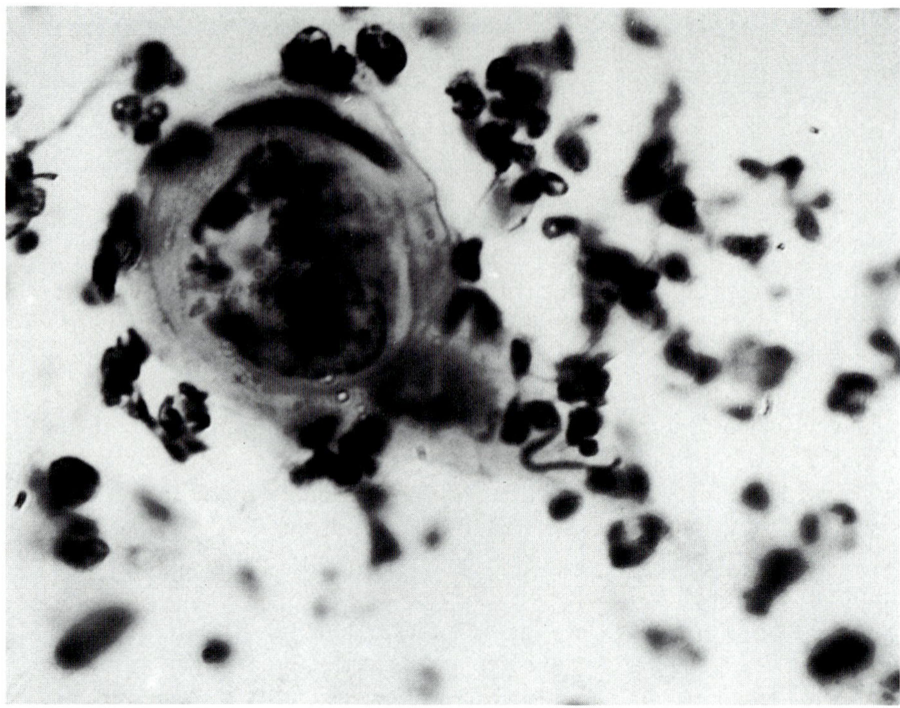

Fig. XII-14: An 'atypical' pearl. While the cell which extends underneath has a crescentic karyopyknotic nucleus (at 12 o'clock), the cell lying on top of it is undifferentiated and shows virtually no features of squamous differentiation. Note the latter's large, rounded and prominent *nucleolus* (it is actually brilliantly acidophilic) at 10-o'clock. This hallmark of increased protein production is frequently found in the *poorest* differentiated, most rapidly dividing cells of squamous cell carcinoma, and is *not* a good identifying or discriminating feature of secretion in adenocarcinoma.
Papanicolaou stain.
XII-14: × 1,250.

In this process of karyopyknosis (see Chap. XI), the nuclear contents become hyperchromatic due to loss of water, denaturation of proteins, and condensation of degenerating chromatinic material. This includes chromatin threads which were finely suspended in the parachromatin, so that the latter becomes abnormally cleared. The chromatin masses become rounded and smudged from protein denaturation with degeneration, so that pre-existing malignant criteria are obscured and lost to interpretation of biologic behavior (Fig. XII-11).

As more water is lost from the nucleus, the abnormal clearing of the parachromatin decreases in amount, the nucleus contracts or collapses so that its overall size diminishes, and the nuclear membrane wrinkles. The resulting pyknotic nucleus is a small mass of dark, amorphous, hematoxylinophilic material (Fig. XII-10).

As stages in this process of karyopyknosis proceed, the nuclear morphologic criteria become progressively lost. They eventually are so obscured that one cannot determine with any accuracy as to whether one is dealing with euplasia, retroplasia, proplasia, or malignant neoplasia. One of the latter two states existed before karyopyknosis obscured the picture, however, if the resulting pyknotic nucleus is *large* and dark – the so-called 'India ink nucleus'. One cannot discriminate between them, and one cannot presume cancer. These should *not* have the diagnosis of cancer based upon them, but should serve to identify a very disturbing finding whose true biologic significance should be determined by further careful studies.

The process of karyopyknosis is basically one of degeneration as a cell hypermatures. Identical nuclear retroplastic processes take place in other cell types and in degenerating cancers *other* than squamous cell carcinoma. Karyopyknosis is not specific for this cell type, therefore, but is highly suggestive.

D. Cytoplasmic Features

Most of the more valuable and highly specific information for determining squamous cell functional differentiation, is contained in structures of the cytoplasm. While their characteristics are numerous and frequently striking, two are of major identification value – the production of *keratins*, or keratinization, and extreme symmetrical *thinning* of the cytoplasm.

1. Keratinization

When a cancer cell differentiates toward keratinizing stratified squamous epithelium, many of the features of keratinization occur in the same typical way as they do in the euplastic cell (see Chap. XI). Frequently, however, malignant cells *keratinize atypically.*

This abnormal keratinization, or *dyskeratosis* (*dys-*: Gr. = ill, difficult, faulty, abnormal; *-keras*: Gr. = horn; *-osis*: Gr. = condition, state, process), is not to be confused with dyskaryosis (Chap. VIII, C). The former involves atypical keratinization in the *cytoplasm;* the latter involves atypical morphology of the *nucleus.*

Dyskeratosis occurs in two major forms:
- *typical* keratinizing changes appearing *out-of-time* with the rest of the surrounding tissue development, and
- *atypicalities* of the actual keratinizing tendencies themselves.

The morphologic features of the former have been discussed (Chap. XI). The latter is dealt with mainly in this part.

In *typical* keratinization, the various phases of the process (i.e.: hyalinization, color) occur: 1) throughout the *whole cytoplasm,* appearing relatively uniformly distributed throughout the cytoplasm from nuclear envelope to cell membrane; and 2) in a *predictable order,* from the less mature prekeratins in the lower immature cells to the more mature keratins in the overlying cells (Chap. XI).

In *atypical* keratinization, or dyskeratosis, the process can occur: 1) anywhere in the cytoplasm (usually most marked at the periphery, and appearing to move inwards); and 2) in any order (even different phases present in the same cell). The changes involved include *hyalinization, extreme color, keratinization particles, 'hard' cellular border, ectoplasm and endoplasm,* and *refractile ringing.*

a. Hyalinization. A highly glassy appearance characteristically occurs with extreme, mature keratinization. Typically, hyalinization (*hyalos:* Gr. = glassy), appears in the latter stages of keratinizing stratified squamous cell differentiation (i.e.: anucleate, horny). Atypically, it appears *earlier* in areas where it normally should not be, and in *abnormal forms.*

An *extreme* quality of hyalinization to the highly mature keratin, is characteristic. With the microscope light shining through it, it appears as the stained glass in the window of a church with the sun streaming through it from behind. This highly refractile appearance of glass is found in virtually any cytoplasmic hue of the spectrum (i.e.: red, orange, yellow, green, blue) or distribution (i.e.: total cytoplasm, keratohyaline clumps, eleidin granules, ectoplasm, cell border).

b. Extreme Color. Although an acidophilic cytoplasmic color is frequently referred to as a differentiating characteristic of keratinization, it discriminates *poorly* between keratin and acidophilic artifacts. These latter are frequently indistinguishable from keratinization, and are thus mistaken for it. They can occur in such situations as *inflammation, necrosis, drying before fixation,* use of *improper fixative* (e.g.: isopropyl alcohol), and *pH change* when fixation of the center of a clump of cells is not adequate. One tinctorial reaction which is extremely suggestive of keratinization, however, is the *brilliant* orange-red (i.e.: 'Halloween orange') which at times can be dazzling in dyskeratosis (Chap. XI). This, with hyalinization, is most suggestive of atypical keratinization.

c. Keratinization Particles. Keratinization particles (i.e.: keratohyaline, eleidin) can appear in the cytoplasm of malignant cells differentiating toward keratinizing stratified squamous epithelium. Particularly, *eleidin* granules can be of great help in identifying keratinization.

Inspissated mucus secretion can closely mimic and be mistaken for eleidin granules. While mucus is usually basophilic, it can be acidophilic and, even though rarely, hyaline and orangophilic. Its borders are more rounded and appear more amorphous than the somewhat crystalline-appearing eleidin.

In proplasia and in malignant neoplasia, eleidin granules can be much larger than those occurring in euplastic cells. They may be suggestive of crystalline structure by their glassy appearance, and by their angled and markedly irregular borders. At times, a small cytoplasmic clearing or vacuole occurs around an eleidin particle (Fig. XII-6). It may be markedly *orangophilic* and strikingly *hyaline.*

d. 'Hard' Well-Defined Cellular Border. Cells which are keratinizing have a distinct, crisply defined cellular border, giving it the appearance of an extremely thin, hard, glassy substance (Fig. XII-15). This feature characterizes normal intermediate and superficial cells (Figs. XI-3, XI-5), and occurs even in the early stages of keratinization in disturbed cells (Figs. XII-1, XII-3) and in cancer cells (Fig. XII-6), where this tendency may be the only available evidence of functional differentiation toward squamous. This extremely clear-cut cellular border, without other evidence of keratinization in the rest of the cell, appears to bespeak a beginning of the keratinization process in the outer portions of the cell, which appears to be more advanced (i.e.: mature) in this regard than the rest of the cytoplasm.

e. Ectoplasm and Endoplasm. In this atypicality, cytoplasmic keratinization appears to proceed from the 'hard', well-defined outer cellular border, just discussed, inward toward the nucleus. In the process, the cytoplasm becomes *biphasic* with two distinctly different types of cytoplasm: this hyaline outer *ectoplasm,* and the remaining textured *endoplasm* (Figs. XII-7, XII-12, XII-16, XII-17). The interface between these two phases, is sharply and distinctly delineated as the *ecto-endoplasmic border.*

The *ectoplasm,* extending from the cellular border to the ecto-endoplasmic border, is more *hyaline* than the endoplasm (Figs. XII-7, XII-12, XII-16, XII-17). At times it is *extremely* glassy; however, it can be of any cytoplasmic color (i.e.: blue, green, yellow, orange, red), but in extreme keratinization it usually is strikingly orangophilic.

The *endoplasm,* extending within the ecto-endoplasmic border, is more highly textured. All cellular organelles, including the nucleus, are located within the endoplasm (Figs. XII-7, XII-12, XII-16, XII-17).

The nucleus is never found in the ectoplasm. It lies within the endoplasm, confined by the ecto-endoplasmic border – a prisoner of its own production. In tissue culture, the nucleus is observed to spin and turn within the endoplasm, in active metabolic motion.

The ecto-endoplasmic border, as a distinct interface, is prominent when viewed in profile. The ectoplasm extends completely around the cell as a shell, but is best studied microscopically at its lateral margins when the microscopic plane of focus lies perpendicular to the ecto-endoplasmic border and to the cell membrane. By fine focusing up and down through it in this position, one can best delineate the structure (Figs. XII-7, XII-12, XII-16, XII-17).

The ecto-endoplasmic border frequently is thrown into linear folds, creases, or ridges which may appear as lines or fibers (Figs. XII-16, XII-17). When the cytoplasm is abnormally thinned and extending out in unipolar (tadpole) or bipolar (spindle) fashion, the wrinkled ecto-endoplasmic border extends down the thinned tail, with its exterior ectoplasm encasing its endoplasm (Figs. XII-10 through XII-12). These creases in the ecto-endoplasmic border give the appearance of fine cytoplasmic fibrils extending down the tail, the so-called 'keratinizing fibrils', fibrillary apparatus, or 'Herxheimer's spiral' (Fig. XII-12). The terminal portion of this atypically thinned cytoplasm frequently forms a bulb or endoplasm surrounded by ectoplasm (Figs. XII-1, XII-15). At times its size is appreciable, as if the cell were preparing to accept another nucleus in this bulb at the end of its tail (Fig. XII-12).

Fig. XII-15: Tadpole cell with pearls. Keratinizing stratified squamous epithelial functional differentiation in malignant cells from squamous cell carcinoma. A diagnostic true tissue fragment (DTTF) of cancer cells lying in stratification forming two 'pearls', one at the 'head' end (right) and one halfway down the tail (left). The cell membrane has the 'hard' appearance of an extremely thin glass shell around the head of the tadpole. It then thickens into a thin, clear, glassy ectoplasm, around 3 o'clock and again visible along the midsection of the tail. The extreme end of the tail flares out into a bulb.

Papanicolaou stain.

XII-15: × 2,000.

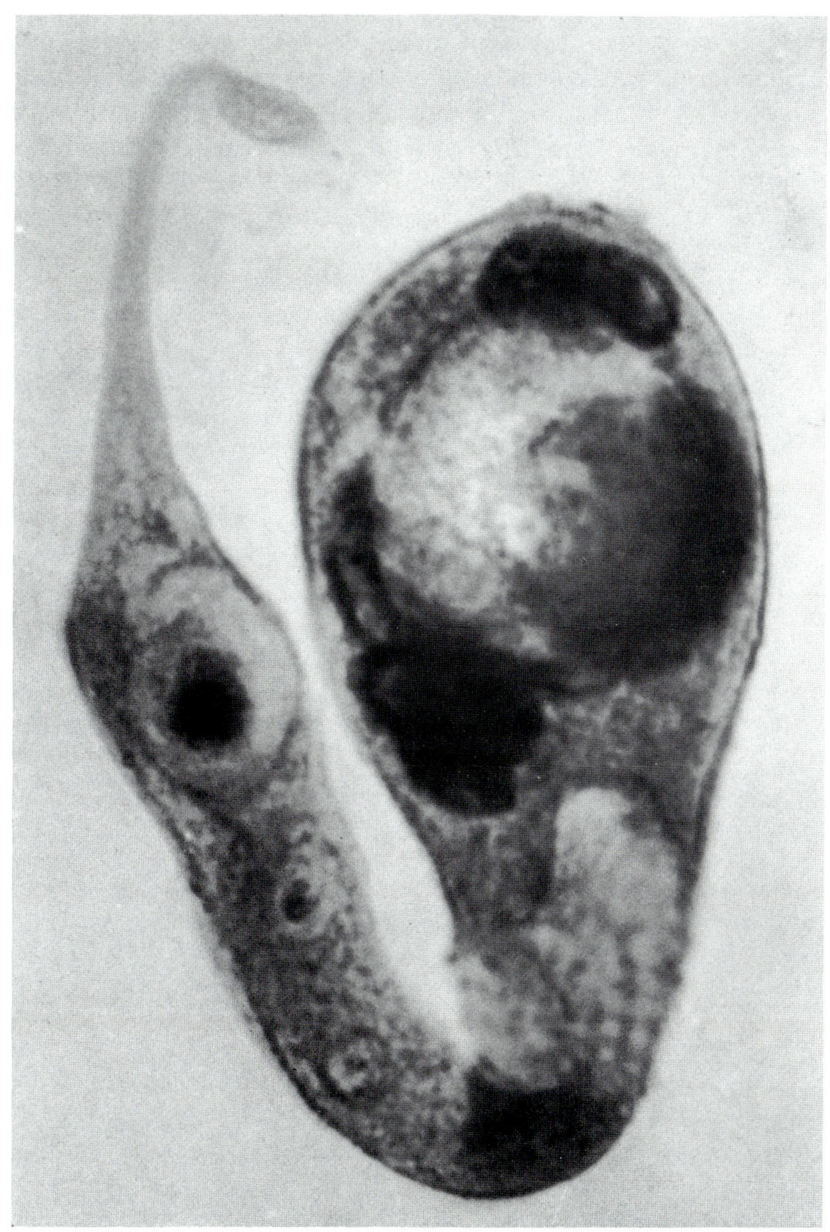

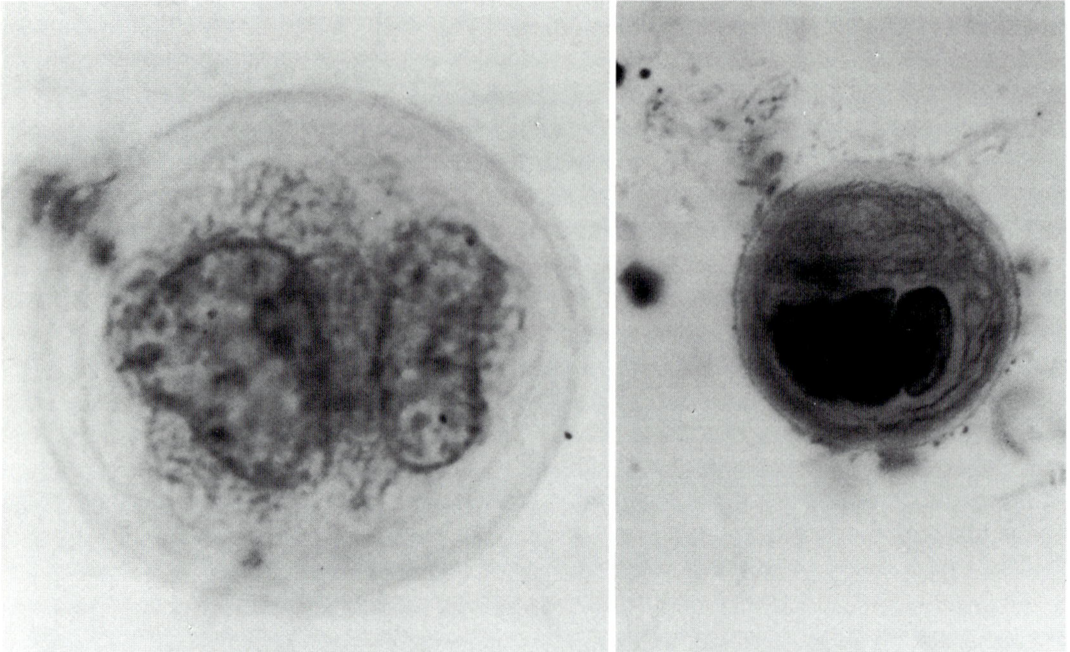

Fig. XII-16, XII-17: Ectoplasm and endoplasm of atypical keratinization (squamous cell carcinoma). The cell membranes are extremely well defined, crisp, and appear 'hard'. The glassy hyaline *ectoplasm* extends from the cell membrane inward to the well-defined ecto-endoplasmic border. The ectoplasm contains numerous thin and delicate concentric shells of 'ringing'. From the ecto-endoplasmic border, inward, is the reticulated, vacuolated, highly textured *endoplasm* which contains all of the cell's organelles including the multiple nuclei with nucleoli, mitochondria, and endoplasmic reticulum. The *ecto-endoplasmic border* is the highly refractile interface between these two phases of cytoplasm. The background contains old hemorrhage and other elements of a cancer 'diathesis'.
Papanicolaou stain.
XII-16: × 2,000; XII-17: × 1,500.

f. Refractile Ringing. Refractile rings usually are detectable within the hyaline ectoplasm, even though they are frequently extremely thin and pale, and only discerned with great difficulty. These appear to represent the interfaces of successive shells of cytoplasmic gelation, or layers of hyaline, as if successive zones of hyalinizing gelation of the cytoplasm occurred in this atypical process of keratinization. As these concentric, glassy shells consist of virtually identical material, their interfaces can only be seen when they lie perpendicular to the plane of focus. By focusing up and down through the cell and carefully examining its lateral equator, where the shells are parallel with the rays of light, their interfaces can be observed to appear as refractile rings in the ectoplasm (Figs. XII-7, XII-12, XII-16, XII-17).

These refractile rings are not to be confused with either vacuoles or amorphous granules which may occur in a somewhat paler outside zone of cytoplasm of cells which are abnormal secretory cells, macrophages, or cells which are degenerating in watery fluid (e.g.: pleural fluid) (Fig. VII-7). True refractile ringing of keratinization occurs in a hyaline ectoplasm of a keratinizing cell, and lies parallel to both the cell membrane and the ecto-endoplasmic border.

2. Thinning of the Cytoplasm

In *typical thinning,* the euplastic cell of keratinizing stratified squamous epithelium functionally differentiates by squamifying, thinning uniformly in all directions in *one plane* to a wafer thickness (Fig. XII-1).

Atypical thinning, however, occurs along *one axis* into an elongated, bizarre, *spindloid* structure. Thus, in reactive epithelium 'tadpole' cells form with the nucleus at one end and a cytoplasmic tail actively extending out the *other* end (Figs. XII-1, XII-3). Cytoplasmic thinning from *both* sides of the nucleus (bipolar) along the same axis, has been referred to as a 'fiber' cell [202, 204]; however, this term may be easily confused with the true fibrocyte, with which it has no relationship, so that it is better referred to as a *'spindle' cell* (Figs. XII-2, XII-4).

Likewise a malignant squamous cell, as it differentiates, can thin uniformly in the typical fashion into a squamous plate (Fig. XII-6). It can, however, differentiate atypically by thinning into a 'tadpole' (Figs. XII-10, XII-15) or a 'spindle cell' (Figs. XII-11, XII-12).

This spindle thinning can be very striking in appearance. As the cytoplasm thins, the nucleus may be drawn out along the lateral axis of thinning into the endoplasm, and become extremely pointed (Figs. XII-2, XII-10, XII-11). The nuclei of 'spindle' cells tend to be pointed at both ends (Figs. XII-2, XII-11, XII-12). On the other hand, nuclei of 'tadpole' cells tend to be rounded at the 'head' end and pointed toward the 'tail' (Fig. XII-10); or they may be completely round and located near the head end, with adequate surrounding cytoplasm to buffer it from the constrictive thinning forces of the tail (Fig. XII-15).

Nuclear thinning requires a restricted amount of cytoplasm in the region of the nucleus, and is most marked when the cell also has differentiated ectoplasm and endoplasm (Figs. XII-10, XII-11). At the start of the tail, the ecto-endoplasmic border narrows down virtually into a pointed funnel. Into this funnel if there is no excess cytoplasm into which the nucleus can escape, the nucleus is jammed and extends for a distance down the tail. Thus, thinning is neither a passive process from degeneration, nor one of external mechanical distortion. It is an *active,* forceful functional differentiation of the cell toward a thinning, keratinizing stratified squamous epithelial cell.

E. Tissue Fragments

While cancers, which are differentiating toward keratinizing stratified squamous epithelium, *tend* to shed cells singly, they also shed some cells in diagnostic true tissue fragments (Table VI-1). In addition to the above *individual* cell features of functional differentiation, two features of these *diagnostic true tissue fragments* characterize them as being of squamous cell maturation. Both features have to do with true *stratification* of the cells in the tissue: *pearls* ('pearly bodies') and *intercellular bridges.*

1. Pearls ('Pearly Bodies')

When keratinizing stratified squamous epithelium is thrown into folds (e.g.: irritation, acanthosis, neoplasia), some deep, concentric pits are formed on the surface. When shed, the surface cells of these pits retain their concentric relationships and their stratification upon one another, and are recognized as *'pearly bodies'* or *pearls,* either benign (Fig. XI-6) or malignant (Figs. XII-5, XII-14, XII-15). This form of functional differentiation is frequent in well-differentiated squamous cell carcinoma, and is not difficult to recognize when there are multiple cells involved in the pearl structure (Figs. XII-5, XII-15).

Pearl formation was first named 'cannibalism' [202]. This is an easily misunterstood term as, when only two cells are involved, it can be misinterpreted as cytophagocytosis, or true cannibalism (Fig. XII-14). Careful focusing up and down under high power, however, delineates that *one* cell lies *on top of another,* as a cup in a saucer; or tends to wrap itself around another, as a catcher's mitt about a baseball.

Discriminating true cellular phagocytosis from a pearl formation, is not merely of academic interest. It can be of great practical, diagnostic importance.

Some *undifferentiated* cancer cells actually do engulf other cells (cyto-phagocytosis), as has been observed in tissue culture. While this cannibalistic behavior is not rare in undifferentiated or poorly-differentiated cancers of *any* cell type, it is *not* observed in well-differentiated squamous cell carcinoma, with which 'pearl' formation is diagnostically associated and of which this is a characteristic and valuable feature of functional differentiation.

At times, an active histiocyte will engulf a malignant cell. In this close association, and perhaps under an examiner's enthusiasm, the active histiocyte may be interpreted erroneously to be malignant, and 'cannibalism' is misidentified.

There remains the possibility of an abnormal mitotic division, with cytoplasmic separation of a cell completely within its sister cell. This is extremely rare if, in fact, it even exists.

Thus by determining that one cell lies *on top of* another cell within a cytoplasmic concavity or pit *in true stratification,* the true process of functional differentiation maturation toward stratified squamous differentiation is identified. This is a valuable feature of squamous cell carcinoma.

However, after carefully focusing up and down with highest resolution (i.e.: 100 X oil immersion objective with Koehler illumination, Table IV-1), if one cannot rule out a cancer cell actually *within* another cancer cell (i.e.: cytophagocytosis), the structure could be from undifferentiated cancer or poorly-differentiated cancer of *any* cell type. In this situation, it should *not* be considered a discriminating feature of a squamous cell carcinoma.

Again, a pearl is *not* a malignant criterion. It is a *functional differentiation feature* of keratinizing stratified squamous epithelium, either benign (e.g.: squamous atypia) or malignant (e.g.: squamous cell carcinoma).

2. Intercellular Bridges

Intercellular bridges occur in the Malpighian layer of keratinizing stratified squamous epithelium between the true basal cells, parabasal cells, and the lower intermediate cells (Fig. XVI-9). They do *not* interconnect the flowing cytoplasm between adjacent cells, in the older 'intercellular bridge' concept.

Intercellular bridges represent finger-like ridges in the cell membranes of the adjacent cells, which interdigitate and attach to each other's lateral surface (Fig. XII-18). The ridges lie closely together with firm cell-to-cell attachments of the cell membranes by adhesive material (e.g.: mucopolysaccharide) in the intercellular cleft between adjacent cells, and by desmosomes.

After a pair or so of ridges is so attached, a pair lie unattached, creating an intercellular space (Fig. XII-18). These attached 'fingers', alternating with the intercellular spaces, are very uniformly arranged, giving the uniform periodicity under light microscopy which is so characteristic of the intercellular bridges (Figs. XII-8, XII-9, XII-19).

Thus the cytoplasm of adjacent cells does not flow from cell-to-cell as the old histologists speculated, as each ridge is covered with its own intact cell membrane. The intercellular fluid lying between cells, however, is able to circulate around the cells, between the intercellular bridges, for nutrition and removal of metabolites.

While intercellular bridges occur frequently in well-differentiated squamous cell carcinoma, rarely are they definitely identifiable as a differentiating feature in exfoliated cellular material. When they *do* occur and are *characteristic,* however, they can be most helpful and discriminating to identify the cells as maturing toward keratinizing stratified squamous epithelium.

Figures XII-18, XII-19: Intercellular bridges in keratinizing stratified squamous epithelial functional differentiation.

Fig. XII-18: Squamous metaplasia, bronchial biopsy. Finger-like ridges in the cell membrane of the two adjacent cells, extend across the intercellular space (thick arrows). They are adherent to one or two similar ridges of the neighboring cells, and then have a wide expanse of intercellular space before two or three more ridges extend and are adherent. This is repeated in a fairly uniform fashion, producing the periodicity noted by light microscopy (Fig. XII-19). Occasional desmosomes (d) increase the adherence of the finger-like ridges. Each desmosome represents a button-like structure consisting of two dense plaques, one on each of the opposing cell surfaces, with tonofibrils extending from each plaque and fanning out into the cytoplasm of its respective cell.

Fig. XII-19: Squamous cell carcinoma, oral smear. Prominent intercellular bridges extend between adjacent cells. Note the fairly *uniform periodicity* of the bridges, their *uniform lengths,* and the relatively *parallel cell borders* of opposing adjacent cells. This is in contrast to vacuoles of degeneration which, when they are present between two cells, may dangerously resemble intercellular bridges; however, degenerative vacuoles vary in size, so that their intervacuolar borders bridging between the cells do not have a uniform periodicity, and the cellular borders along the sides of the vacuoles have a scalloped, loop-the-loop appearance rather than lying parallel to each other, as do the cell borders of true intercellular bridges.

XII-18: Osmium/methacrylate; XII-19: Papanicolaou stain.

XII-18: × 25,000; XII-19: × 1,500.

XII-18: (Frost, Erozan, Donovan; unpublished electron micrograph.)

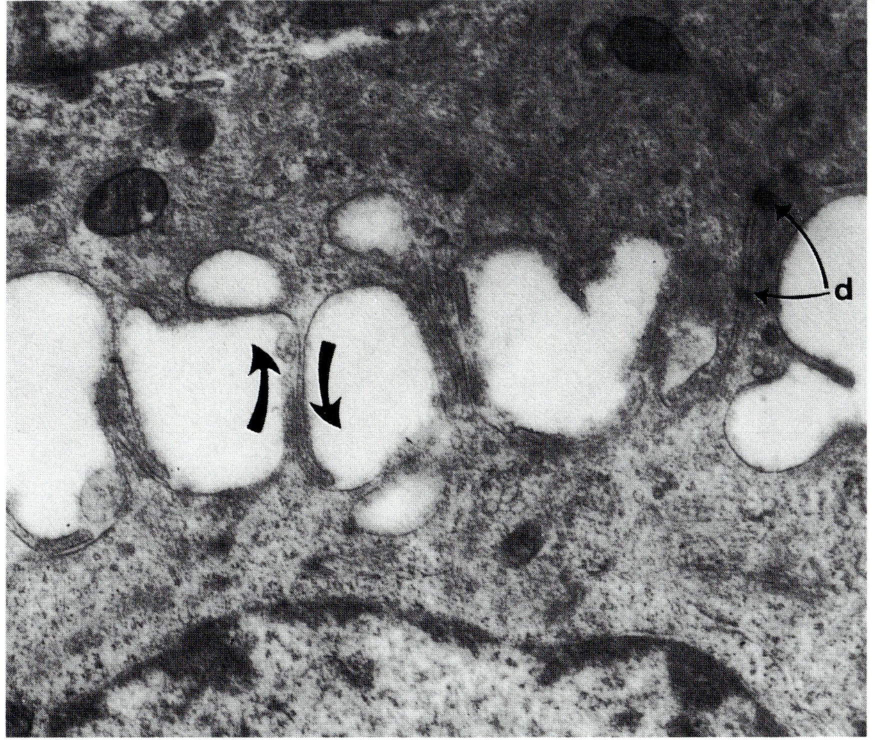

18

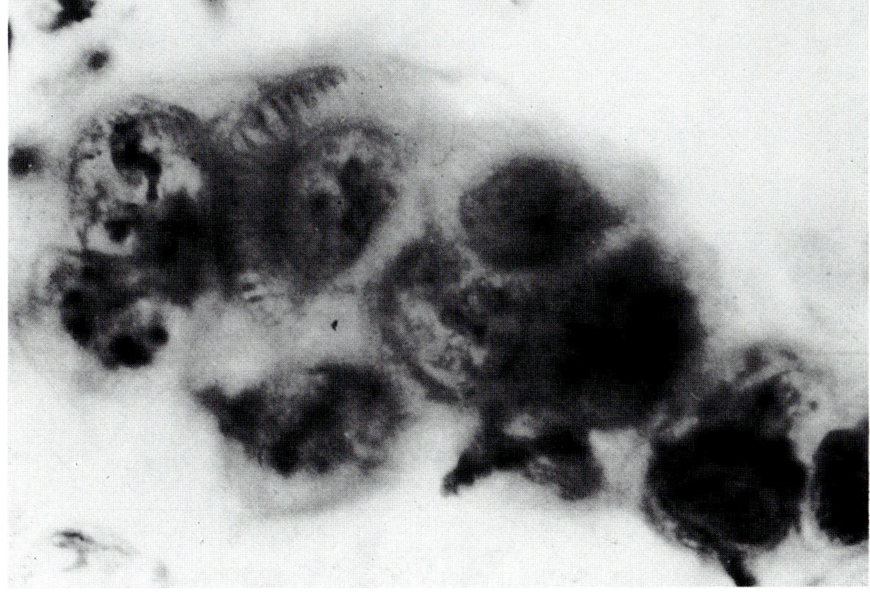

19

Characteristic intercellular bridges can be definitely identified by careful, methodical examination. Those which are distorted usually cannot be definitely identified, however, as their structures are mimicked by other situations. There is a *uniformity* and a *periodicity* about intercellular bridges, which is lacking in structures mimicking them.

The cell membranes at the base, or trough, of the finger-like folds (Fig. XII-18) line up and are visible in light microscopy as the apparent 'cell membranes' from which the 'bridges' extend (Fig. XII-19). The *parallel* relationship of these *cell membranes* of the two adjacent cells, is strikingly retained over great distances (Figs. XII-8, XII-9, XII-19). This gives the bridges a uniform length.

The *periodicity between the bridges* appears strikingly uniform (Figs. XII-8, XII-9, XII-19). The impression can be likened to that of tracks made by a track-laying tank, or caterpillar tractor (Fig. XII-19).

The most frequently encountered change which dangerously mimics these features diagnostically is small, intercellular vacuoles of degeneration occurring between the cells (Fig. VII-7). The cells, themselves, do not have to exhibit other degenerative features in order to have early intercellular vacuoles of degeneration. This occurs with all types of cells, benign or malignant (i.e.: adenocarcinoma, undifferentiated), and can be a source of *false* identification as keratinizing stratified squamous epithelial maturation if one is not extremely cautious in the identification of intercellular bridges.

Two helpful, diagnostic differences occur between the *intercellular vacuoles* of *degeneration* and the *intercellular bridges* of *squamous maturation:*

1) *Vacuoles of degeneration* vary markedly in size, so that the apparent bridges of cytoplasm are *not uniformly spaced* between one another. True *intercellular bridges* appear to be of almost uniform periodicity by light microscopy (Figs. XII-8, XII-9, XII-19);

2) Vacuoles of degeneration have rounded sides, which cause concavities in the cell membranes. This gives the adjacent cell borders a looping, or a loop-the-loop appearance, rather than the straight parallel cellular borders flanking true intercellular bridges (Figs. XII-8, XII-9, XII-19).

XIII. In situ Squamous Carcinoma, Invasive Squamous Cell Carcinoma, and Developing Cancer

A. The Cells of in situ Squamous Carcinoma

The features of cells shed from *in situ* (intraepithelial, preinvasive) squamous carcinoma, have been best characterized in the cervix uteri. This site has provided the greatest material for study and follow-up. Increasing information from other sites (i.e.: lung, esophagus, urinary bladder, oral cavity, skin) gives evidence that many of these features may be generally applicable.

There are certain variations site-to-site which occur in *in situ* squamous carcinoma, however. The most critical of these involves the degree of keratinization.

The *least* keratinization occurs with *cervical in situ* carcinoma, *more* with *in situ* carcinoma of the *lung* and *esophagus,* higher average levels of maturation are reached with *in situ* of the *mouth* and larynx and the highest with the *skin.* This must be more completely characterized, but most of the features from cervical *in situ* carcinoma hold true generally in these other sites, also.

In uterine cervical *in situ* carcinoma, the most characteristic and discriminable cells which are shed show little or no cytoplasmic maturation toward squamous. Their nuclear pattern is that of extremely severe proplasia or dyskaryosis, rather than having the macabre malignant criteria characteristic of invasive cancer. The nucleo-cytoplasmic ratio is extremely high. The most characteristic and discriminable cells from *in situ* carcinoma are strikingly similar, however, without the marked pleomorphism of the cells which is present in invasive cancer.

1. Nucleus
a. Shape and Orientation within the Cell. The nucleus is disarmingly *rounded* and *symmetrical.* Its membrane is wavy and can be molded by adjacent nuclei.

The nucleus is centrally located within the cell. The cytoplasm, in spite of its scant amount, is uniformly and concentrically distributed about the nucleus (Fig. XIII-2).

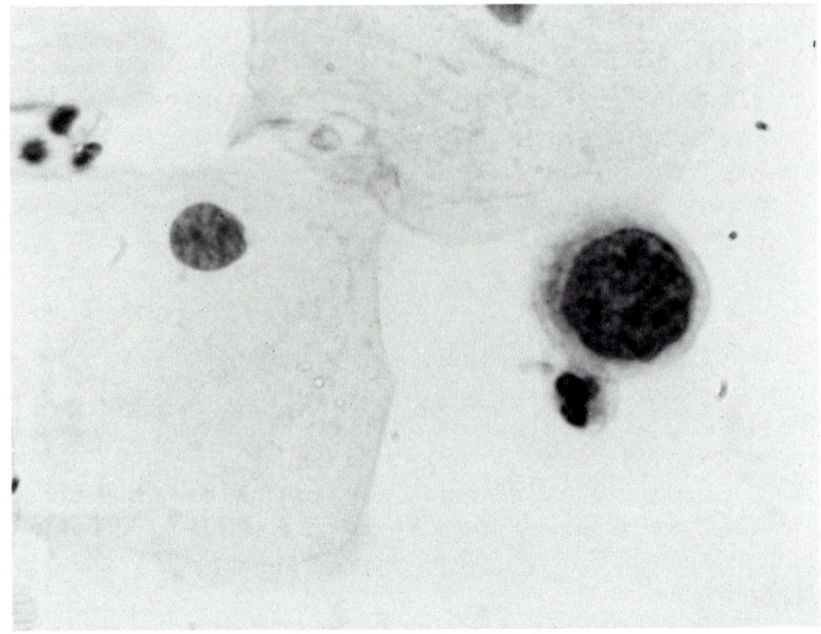

Fig. XIII-1: In situ carcinoma, cervix uteri. Fast (vagino-pancervical) smear [99]. The nuclear features, of the cell on the right, are those of a very severe proplasia (i.e.: marked dyskaryosis) (Chap. VIII). The nucleus is large (i.e.: 4–5 times the entire cell of the neutrophil or of the intermediate cell nucleus, left). There is extreme *hyperchromasia* in a well-preserved cell, evidencing increased DNA and its associated proteins. There are *coarse granules* in the chromatinic net, which are *uniformly* dispersed throughout *cleared* parachromatin. The chromatinic rim is uniformly *thickened,* and there is severe *undulation* of the nuclear membrane. There is an extremely *high* nucleo-cytoplasmic ratio, with the cytoplasmic membrane *intact* around a *scanty,* thin, well-preserved rim of cytoplasm. There are no hallmarks suggestive of invasive cancer (e.g.: no nucleoli, no irregular chromatin clumps larger than coarse granules, no irregularly-dispersed or irregularly-sized areas of parachromatin clearing).

Papanicolaou stain.

XIII-1: × 960.

Fig. XIII-2: In situ carcinoma, bronchus. Chromatin is densely condensed in rounded granules. These are uniformly distributed throughout cleared parachromatin as the chromatinic net, and against the inner nuclear membrane and lamina as a granular but uniformly-thickened chromatinic rim. While the parachromatin is uniformly cleared, there still remains extremely fine chromatin threads regularly dispersed throughout it. The nuclear membrane is thrown into gentle waves with numerous pores throughout.

Osmium/methacrylate.

XIII-2: × 29,600.

(Frost, Erozan, Donovan; unpublished electron micrograph.)

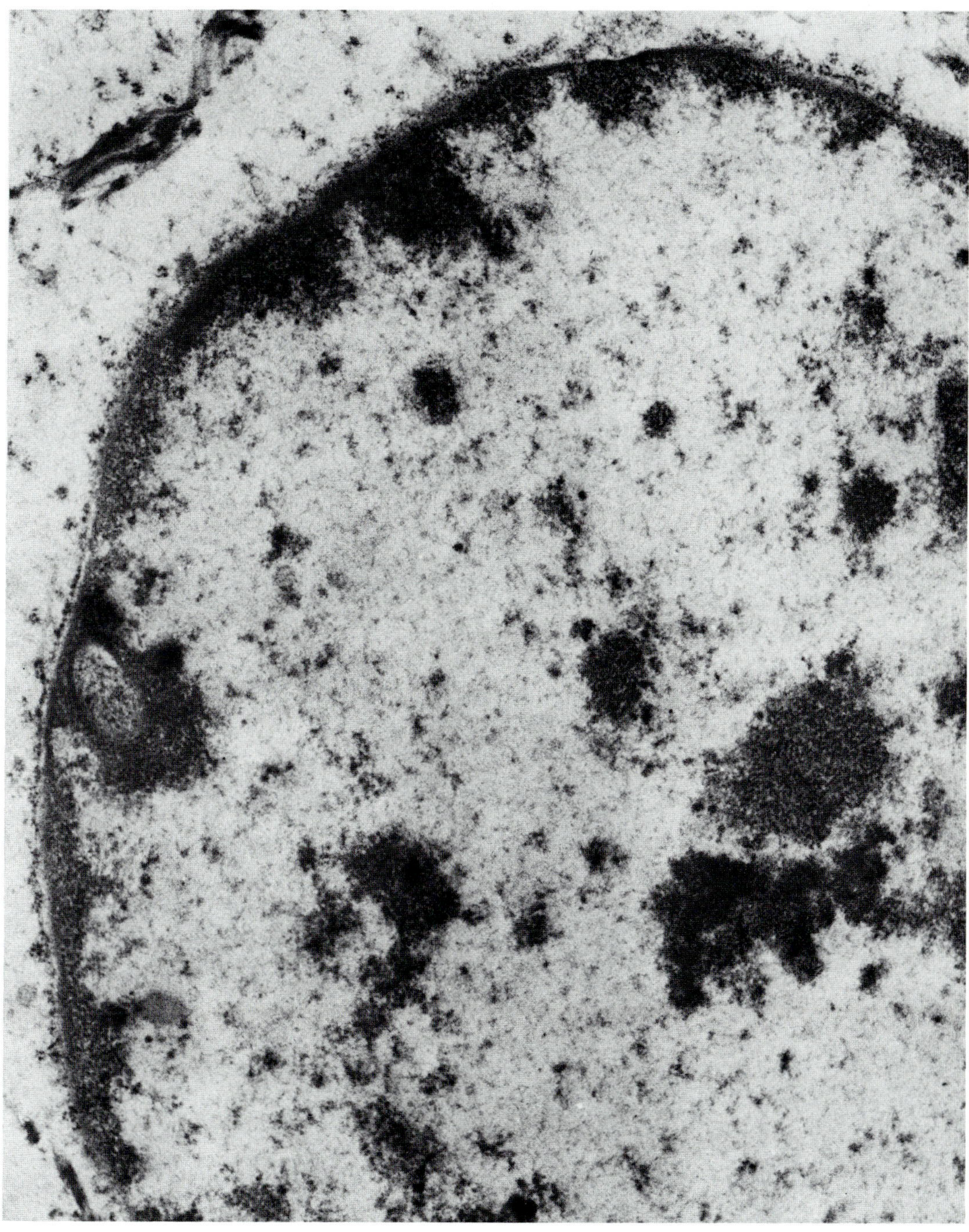

2

b. Hematoxylinophilic chromasia. The nuclei of *in situ* carcinoma are hyperchromatic. At times, this is only slightly increased over euplastic intermediate cell nuclei, but their increased size bespeaks increased DNA and associated proteins. Usually they are very hyperchromatic and, occasionally, can be extremely hematoxylinophilic.

c. Nuclear membrane shape. While the nucleus is basically round or oval, its nuclear membrane is thrown into waves, from gentle ripples to extreme undulation (Figs. XIII-1, XIII-2).

It does not have the macabre and grotesque nuclear membrane irregularities, however, which classically occur in invasive cancer (Chap. IX). Nor is this the wrinkling of shrinkage found in degeneration (Chap. VII). It is the undulation of increased general activity (Chap. VIII), but is frequently carried to extreme.

d. Chromatinic Rim. The *chromatinic rim* is granular, but it is essentially of *uniform thickness* with*in* each cell. While the variation in thickness of the chromatinic rim may be great from cell-to-cell, its thickness in a given cell is very uniform. This is in striking contrast to the extreme variation in thickness which can be encountered along the same chromatinic rim of cells from invasive cancer.

e. Chromatinic Net. The pattern of the *chromatinic net* is essentially *granular.* While there are countless fine granules and many moderate-sized granules, coarse granules may be few but they constitute characteristic hallmarks. The granules are *uniformly distributed* throughout the parachromatin.

The pattern is basically that of *extreme* proplasia. In those *in situ* carcinomas which appear most active, the extreme granularity of the chromatinic rim and of the chromatinic net resemble the granularity of the chromatinic material of cells in *impending* or *early prophase* [226], frequently with an abnormally indistinct or 'disappearing nuclear membrane' – the lamina, inner nuclear membrane, and marginating chromatin disassociate, so that the chromatinic rim disappears. This leaves the granular chromatin to organize into chromosomes in preparation for the upcoming division.

Not present, are those features which characterize invasive cancer and which are lacking in the classical *in situ* carcinoma (e.g.:irregularly distributed large spiculated and angulated chromatin clumps, nonuniformly-sized areas of abnormally cleared parachromatin, nucleoli). These changes do appear, however, as the tumor develops the capacity for microinvasion (Fig. XIII-3).

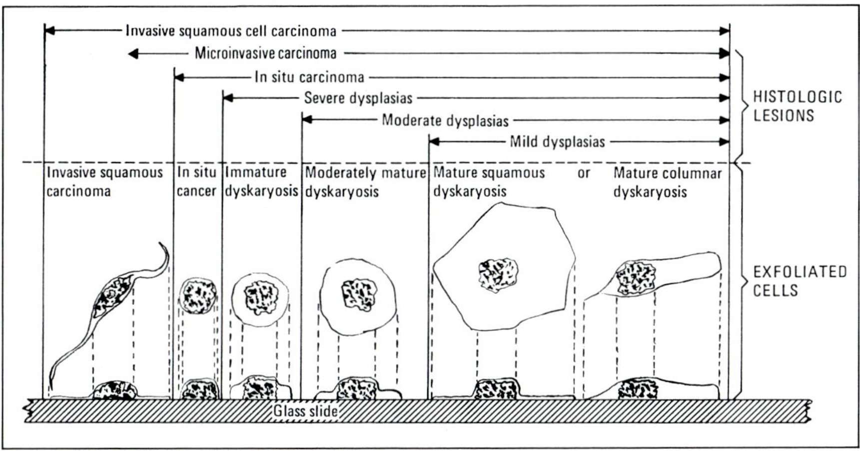

Fig. XIII-3: see text.

f. Parachromatin. The *parachromatin* is extremely *cleared,* with crisply defined chromatin/parachromatin interfaces, reflecting increased general activity in a well-preserved cell. The degree of clearing, while extreme, is *consistent* throughout a given nucleus. Furthermore, the regions of cleared parachromatin are *equally sized* in *area,* and are *uniformly distributed* throughout the granular chromatinic net.

g. Nucleolus. In situ carcinoma is characterized by the virtual absence or extreme paucity of nucleoli. They are actually either *absent* or *inconspicuous* in over 95 percent of the cells from *in situ* carcinoma [198]. As invasion develops, nucleoli begin to appear. As the tumor's invasive capabilities increase, nucleoli become prominent and more numerous per cell, and the proportionate number of cells bearing them increases.

h. Nuclear Size. The nucleus of *in situ* carcinoma is larger than its normal euplastic counterpart (e.g.: intermediate cell nucleus). This increase in size is in the range of *only* four to six times. Increase over six times usually is associated with degeneration, invasive cancer, chemotherapy or other damaging metabolic effects on the cell (e.g.: radiation, folate deficiency).

This size increase is accompanied by *hyperchromasia,* and not by pallor, as with size increase due to degeneration. While cells from *in situ* carcinoma can also be degenerated, chromatin blurring, nuclear membrane wrinkling and other features of retroplasia are *not* a part of the processes of *in situ* carcinoma, nor are they features thereof.

2. Cytoplasm

The discriminable cell of *in situ* carcinoma frequently has an extremely *scanty cytoplasm* symmetrically arranged about its nucleus in a thin, tight rim (Figs. XIII-1, XIII-3). At times the cytoplasm is so attenuated that high resolution (100 × oil objective with Koehler illumination, Table IV-1) may be needed to adequately evaluate the cell membrane. This procedure is of diagnostic importance when an intact cell membrane is needed to give evidence that the cell is *well preserved* and that the cytoplasm is, indeed, *scanty* rather than partially missing into the background due to degeneration (Fig. VII-1). If the membrane *is* found to be defective, on the other hand, it may indicate a *degenerated* cell of one of many types, benign or malignant, and a diagnosis should *not* be based upon an apparent high nucleo-cytoplasmic ratio.

Cytoplasmic functional differentiation of discriminable cells shed from *in situ* carcinoma, is usually *absent* or *minimal* in cancer of the cervix uteri. Here a prominent, crisp, clearly defined cellular border ('hard'-appearing cell border, as in squamous differentiation, Chap. XI) may be the only differentiating feature (see Chap. XII).

In other sites (i.e.: bronchus, mouth, esophagus) and in the few percent of keratinizing carcinomas of the cervix, more cytoplasmic features of functional differentiation occur in the *in situ* stage [70, 103]. Even in these more highly keratinizing sites, however, more capacity to invade is accompanied by more functional differentiation unless the cancer develops into a pleomorphic, poorly-differentiated tumor.

Figures XIII-4 through XIII-7: Moderate atypia of the cervix uteri [162].

Fig. XIII-4: Many moderately mature and mature dyskaryotic cells have been shed into the cervical mucus. Fast (vagino-pancervical) smear.

Fig. XIII-5: In addition to being proplastic, the lower dyskaryotic nucleus (cytoplasm is out of focus) is also retroplastic, with smudging along the inner borders of the chromatinic rim and of the chromatin/parachromatin interfaces. Above is a moderately mature dyskaryotic cell, with moderately abundant and moderately thinned cytoplasm (see Fig. XIII-3). Same specimen as Figure XIII-4.

Fig. XIII-6: The cells in the lower and middle layers are 'active', or proplastic with rubbery, rounded cytoplasm. There is a marked morphologic change in the cells as they mature toward the lumen. The nuclei shrink and become pyknotic, yet they are larger than a normal superficial cell nucleus. There is good orientation, stratification and squamification, with more mature keratinization on the surface. Directed cervical biopsy of the same patient.

Fig. XIII-7: There is moderately abundant cytoplasm in these mildly proplastic ('active') cells. Same specimen as Figure XIII-6.

XIII-4, XIII-5: Papanicolaou stain; XIII-6, XIII-7: Hematoxylin and eosin stain. XIII-4, XIII-6: × 200; XIII-5, XIII-7: × 2,000.

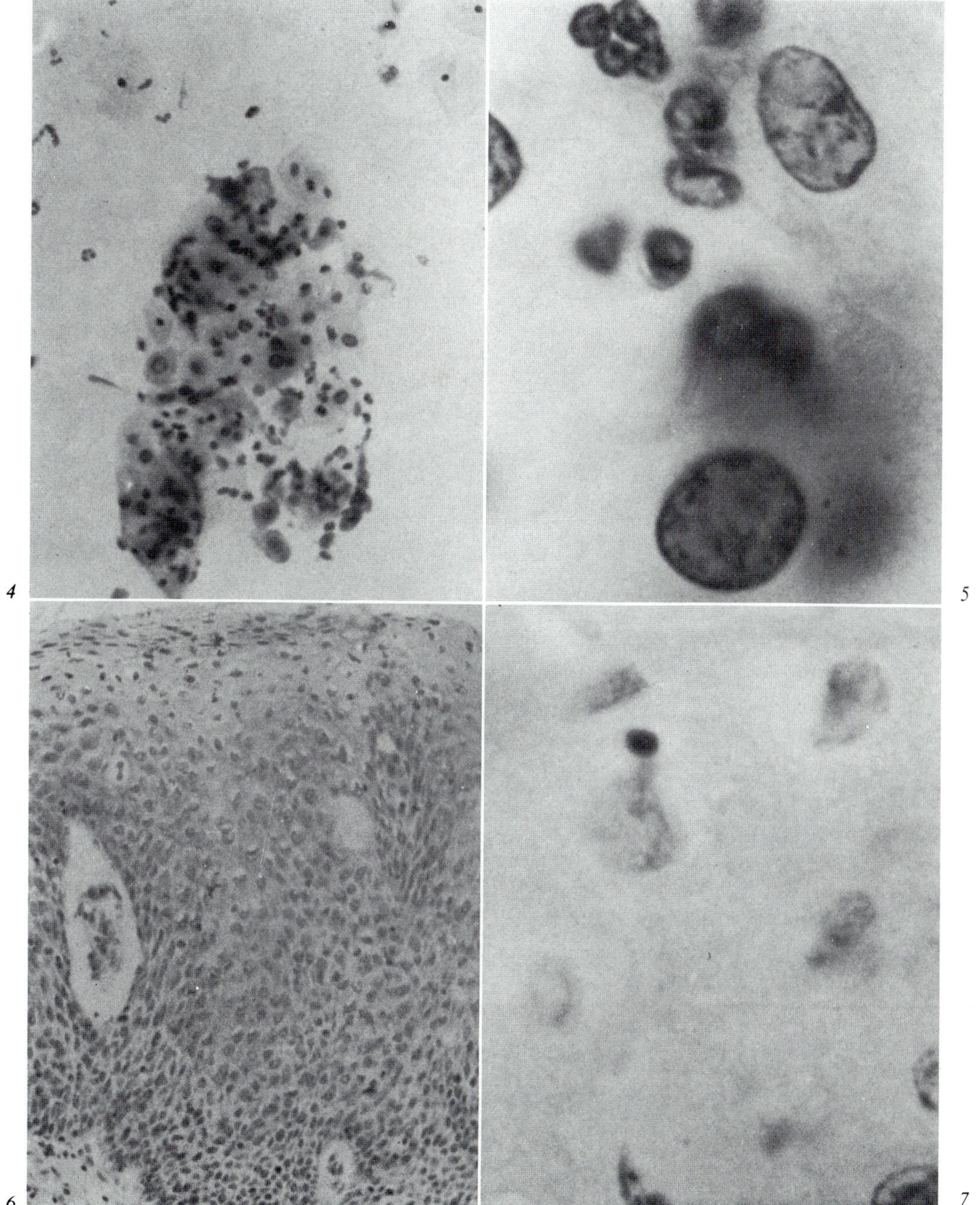

In *in situ* carcinoma of the cervix uteri, however, these features of cytoplasmic differentiation (i.e.: ecto-endoplasm, ringing, tail formation, keratinization) (Chap. XII) are either from the infrequent keratinizing *in situ* carcinomas or are, frequently, associated with 'early', developing invasive carcinoma. When these features are present, they indicate need for serious consideration of adequate repeat specimens for careful evaluation and for adequate further studies as the clinical situation warrants to locate a small focus of invasion.

B. Apparent Paradox of the Discriminable in situ Cell

The cytoplasm of the discriminable cell shed from most *in situ* carcinomas of the cervix uteri, does *not* appreciably differentiate (Figs. XIII-1, XIII-3). At this stage in its changing biologic behavior, the cancer cell's functional differentiation appears at its lowest ebb.

The cytoplasm of cells on either 'side' of *in situ* carcinoma in the biologic spectrum (Fig. XIII-3), *does* functionally differentiate. On the *one* hand (i.e.: toward the right in Fig. XIII-3), proplastic and retroplastic epithelia (i.e.: atypias, atypical metaplasias, dysplasias, intraepithelial neoplasias) as well as euplastic epithelia have more cytoplasm and differentiate mainly *typically* (Figs. XIII-4 through XIII-7) (Chaps. VI, VII, VIII, XI,

---

Figures XIII-8 through XIII-11: In situ carcinoma of the cervix uteri with surface and gland neck involvement [162].

Fig. XIII-8: There are moderate numbers of individual dyskaryotic and discriminable *in situ* carcinoma cells scattered throughout in a serous background. Fast (vagino-pancervical) smear.

Fig. XIII-9: This large severely dyskaryotic nucleus has hyperchromasia, uniformly dispersed granular chromatin and an undulating membrane. This is a discriminable *in situ* cell. Same specimen as Figure XIII-8.

Fig. XIII-10: The nuclei remain essentially unchanged from basement membrane to luminal surface, being hyperchromatic and dyskaryotic. There is virtually no stratification or squamification as the cells approach the lumen. Directed cervical biopsy of the same patient.

Fig. XIII-11: The hyperchromatic dyskaryotic nuclei have scanty cytoplasm and show virtually no differentiation as they reach the surface (upper right) preparatory to exfoliation. Note the granular chromatin and nuclear membrane undulation, as with the exfoliated cells in Figure XIII-9. Same specimen as Figure XIII-10.

XIII-8, XIII-9: Papanicolaou stain; XIII-10, XIII-11: Hematoxylin and eosin stain.

XIII-8, XIII-10: $\times$ 200; XIII-9, XIII-11: $\times$ 2,000.

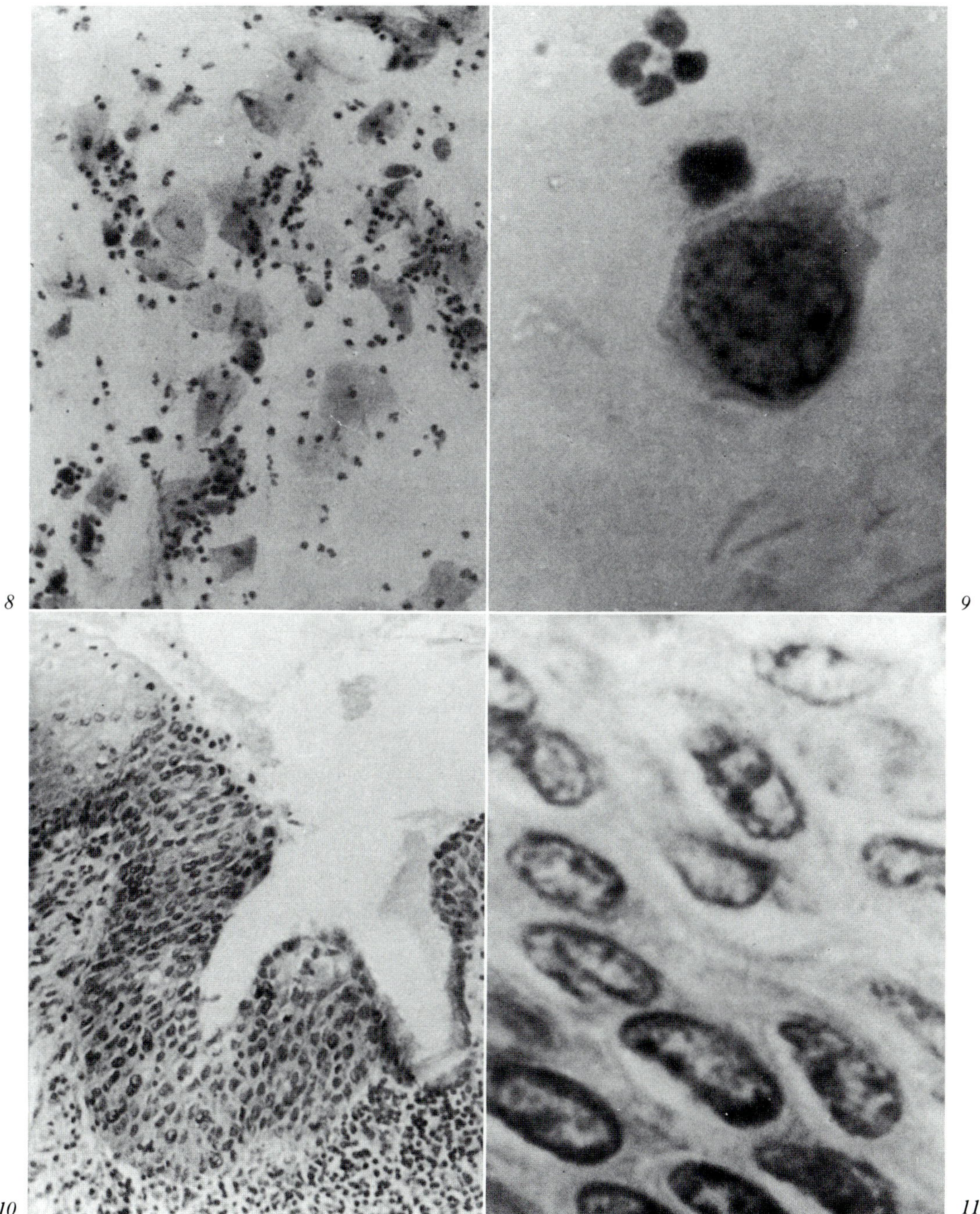

8

9

10

11

XVI). *In situ* carcinoma, then, has scanty cytoplasm with apparent inability to differentiate (Figs. XIII-8 through XIII-15). On the *other* hand, invasive squamous cell carcinoma now, again, has more cytoplasm than *in situ* carcinoma and functionally differentiates, but mainly *atypically* (Figs. XIII-16 through XIII-19).

It is as though *in situ* carcinoma is at its lowest level of functional differentiation, in this phase of its neoplastic progression toward development of fully invasive malignant neoplasia. It appears to be most defensive, in a hostile environment, and concerned principally with its own survival.

C. Invasion: Morphologic Hallmarks of a Biologic Behavioral Change

When *in situ* squamous carcinoma of the cervix uteri develops the ability to invade its host, it progresses in its neoplastic development to the stage of malignant neoplasia, e.g.: invasive squamous cell carcinoma. From this point in its development, morphologic hallmarks usually appear within the cancer cells attesting to this change in biologic activity (Fig. XIII-3) [94, 162, 194].

The rounded dyskaryotic *nucleus,* which is typical of *in situ* carcinoma (Figs. XIII-1, XIII-9, XIII-12, XIII-13) becomes elongated, angular, or

---

Figures XIII-12 through XIII-15: In situ carcinoma of the cervix uteri with extensive deep gland involvement.

Figs. XIII-12, XIII-13: The chromatin pattern basically is coarsely granular. The nuclei are rounded, with severely undulated membranes. There is marked hyperchromasia with extreme parachromatin clearing but uniform distribution throughout the chromatinic net. The inner surface of the chromatinic rim is as sharply defined as is its outer surface, evidencing that any blurring is due to the excessive magnification and not to degeneration. The nucleo-cytoplasmic ratio is extremely high in this discriminable *in situ* carcinoma cell. There is moderate molding of the nuclei in the tissue fragment of Figure XIII-12, and intercellular borders are faint and difficult to discern, a so-called 'pseudosyncytium'. Fast (vagino-pancervical) smear.

Figs. XIII-14, XIII-15: In situ carcinoma with extensive gland involvement. The large, dyskaryotic nuclei are virtually unchanged from basement membrane to lumen with no attempt at maturation. Directed cervical biopsy of the same patient.

XIII-12, XIII-13: Papanicolaou stain; XIII-14, XIII-15: Hematoxylin and eosin stain.

XIII-12: × 800; XIII-13: × 3,200; XIII-14: × 40; XIII-15: × 400.

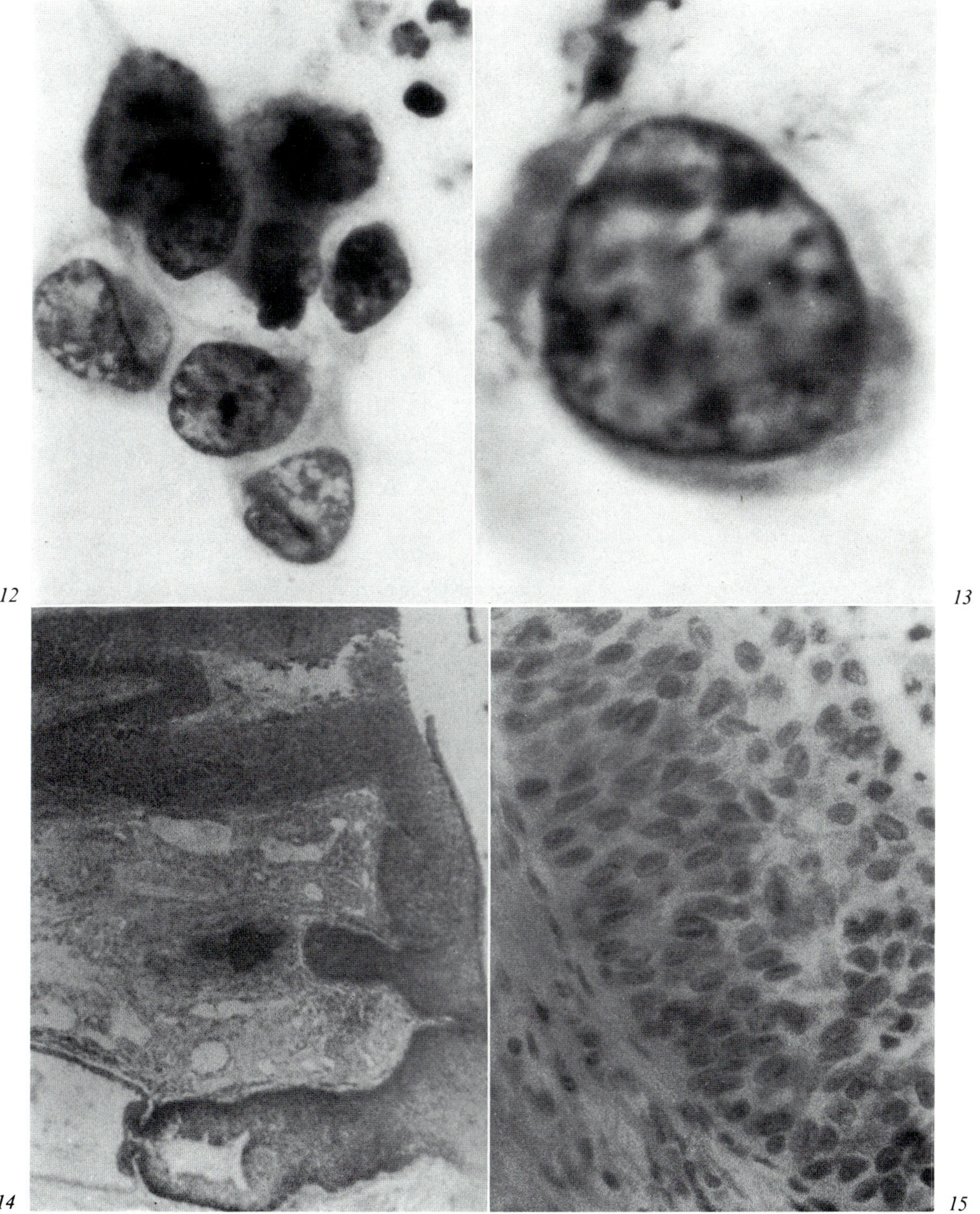

otherwise grotesque (Figs. XII-5, XII-10 through XII-12, XII-14, XII-15, XII-17; XIII-16, XIII-17). The chromatinic rim, which is so uniformly thick in *in situ* carcinoma (Figs. XIII-9, XIII-13) becomes markedly varied in thickness throughout the same nucleus with invasion (Figs. IX-1, IX-6; XII-7). Uniformly dispersed granular chromatin (Figs. XIII-1, XIII-9, XIII-13), becomes irregular masses (Figs. IX-1, IX-8) which are unevenly dispersed between anomalously large and nonuniform areas of abnormally cleared parachromatin. Nucleoli go from absent or inconspicuous in *in situ* (Figs. XIII-1, XIII-9, XIII-13), to become prominent and irregular as the cell develops ability to invade (Fig. IX-8, XIII-17).

Even within the *in situ* carcinomas, there can be noted a progression of lesions of higher and higher potential for biologic activity (i.e.: degrees of gland involvement). As the *in situ* carcinoma is confined to the surface, as it then involves the necks of the glands (Figs. XIII-10, XIII-11), as it extends

---

Figures XIII-16 through XIII-19: Invasive squamous carcinoma of the cervix uteri [162].

Fig. XIII-16: Squamous dyskaryotic cells, round discriminable *in situ* cells, and macabre cells characteristic of invasive squamous cell carcinoma are all distributed throughout this inflammatory smear (see Fig. XIII-3). They occur both singly and in small tissue fragments. The background is somewhat more serous and necrotic than Figures XIII-4 or XIII-8, but this is *not* a *reliable* criterion for degree of neoplastic progression. Fast (vagino-pancervical) smear.

Fig. XIII-17: The nucleus is hyperchromatic and elongated along the major direction of the cytoplasmic thinning. There are moderate-sized and coarse irregular chromatin clumps. There are some irregularly-placed areas of extreme parachromatin clearing and a prominent nucleolus near the center of the nucleus, both features which help to discriminate this invasive carcinoma from the usual squamous atypia of inflammation and repair (Figs. XII-1 through XII-4) and of dysplasia (Figs. XIII-4, XIII-5). The cytoplasm is thinning bilaterally to form two tails of a spindle cell, with an ectoplasm barely visible extending up and down the tail. Some longitudinal foldings of the ecto-endoplasmic border ('fibrillary apparatus') extend down the tail. Same specimen as Figure XIII-16.

Fig. XIII-18: There is a cellular pleomorphism, with karyopyknosis scattered in the deeper layers of this malignant epithelium, as well as near the surface. Cervical biopsy.

Fig. XIII-19: Cells are thinning parallel to the luminal surface (upper). Nuclei are pleomorphic in size, shape, chromasia and pattern. There are moderate-sized and coarse irregular chromatin clumps (lower mid field) and abnormal parachromatin clearing (lower) with nucleoli. Same specimen as Figure XIII-18.

XIII-16, XIII-17: Papanicolaou stain; XIII-18, XIII-19: Hematoxylin and eosin stain.

XIII-16, XIII-18: × 200; XIII-17, XIII-19: × 2,000.

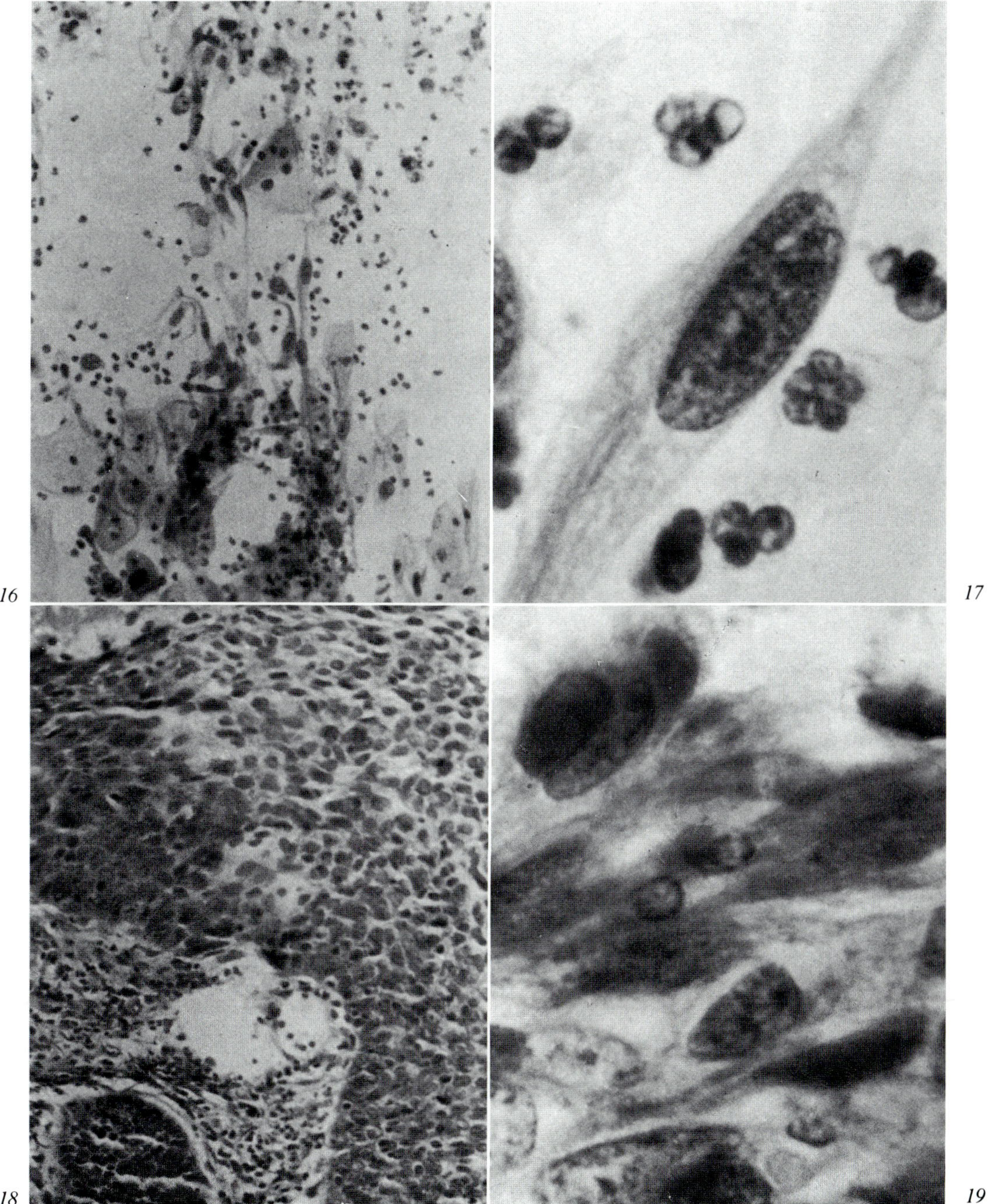

into the glands and, finally, as it involves deeply into the entire depths of the glands (Figs. XIII-14, XIII-15), corresponding nuclear changes occur in the cells of the lesion examined histologically and of the exfoliated cells. Hematoxylinophilic hyperchromasia increases, chromatin granules become more coarse and numerous, parachromatin clears more to become as strikingly clear as the background, the chromatinic rim thickens, nuclear membrane undulation becomes more severe, and the cytoplasm becomes even scantier (Figs. XIII-1, XIII-9, XIII-12, XIII-13).

As a carcinoma becomes invasive, its cytoplasm manifests additional hallmarks of this biologic change. In tissue stained with hematoxylin and eosin, a marked eosinophilia occurs in abundant cytoplasm of cells down in the regions of the microinvasion. Generally believed that this represents keratin of atypical maturation down in the depths of the lesion (dyskeratosis), there is evidence that this might represent intermediate filaments of motility (i.e.: actin, myosin) [214] of a tumor which is beginning to acquire the ability to invade its host.

The cells exfoliated from the *surface* of a lesion which is becoming invasive in its *depths,* show many features of atypical maturation. These include beginning cytoplasmic tailing (Figs. XII-10 through XII-12, XII-15; XIII-16, XIII-17), ecto-endoplasm formation (Figs. XII-7, XII-12, XII-15 through XII-17, XIII-17), ringing (Figs. XII-7, XII-16, XII-17), extreme hyalinization and pearl formation (Figs. XII-5, XII-14, XII-15).

This same neoplastic progression occurs in many other sites (e.g.: lung, esophagus, mouth, pharynx, bladder). These changes in the lung have been described and illustrated in full color [103], ranging from the atypias through increasingly severe *in situ* involvement of the submucosal bronchial glands, and increasing degrees of invasion and tumor aggressiveness, as encountered, studied, and followed up in the Johns Hopkins Early Lung Cancer Program [101–103].

D. The Spectrum of Differentiation in Squamous Cell Carcinoma

1. The Same Lesion Sheds Many Types of Cells

The total spectrum of cells (Fig. XIII-3) can be shed from invasive squamous cell carcinoma. This includes not only those which are morphologically discriminable of invasive squamous cell carcinoma (Figs. IX-1 through IX-8; XII-5, XII-7, XII-10 through XII-12, XII-14 through XII-17; XIII-17), but also those discriminable of *in situ* carcinoma (Figs. XIII-1,

XIII-9, XIII-12, XIII-13), of the dysplasias (Figs. VIII-2, XIII-4, XIII-5) (Chaps. VIII, XVI), and of normalcy (Figs. XI-5, XI-6) (Chap. XI). Thus, cells discriminable of *in situ* carcinoma (Figs. XIII-1, XIII-9, XIII-12, XIII-13), discriminate it from lesser lesions (e.g.: dysplasias, atypias) but *not* from invasive squamous cell carcinoma.

In situ carcinoma in like manner, sheds cells of this same spectrum (Fig. XIII-3), *except* for those discriminable of invasive squamous cell carcinoma. Likewise, dysplasias shed cells of this same spectrum, *except* for those cells which are discriminable of invasive and *in situ* carcinoma (Fig. XIII-3).

2. Multiple Neighboring Lesions Are Frequently Present

An additional situation to be considered is that *in situ* cancer is frequently found adjacent to invasive squamous cell carcinoma, adjacent to dysplasia, or adjacent to normal epithelium. These each shed their own characteristic spectrum of cells (Fig. XIII-3).

Thus, if one has invasive squamous cell carcinoma with the total spectrum of cells mentioned above present in a cellular specimen, one cannot be certain whether there are separate lesser lesions or whether the major lesion itself is shedding all of the cell types. The presence of great percentages of cells discriminable as shed from lesser lesions, speaks in favor of the lesser lesions being present; however, this is slim evidence, indeed, as frequently the major lesion sheds only a small percentage of *discriminable* cells (Figs. XIII-4 through XIII-19).

Papanicolaou recognized and described two types of cells associated with invasive squamous cell carcinoma, and dubbed them 'tadpole' and 'fibercell' (spindle cells) (Chap. XII) [202]. Graham then recognized a third cell type associated with invasive squamous cell carcinoma, and simply named it the 'third-type' cell [126]. Grahams 'third-type' cell encompassed the immature dyskaryotic cell (Fig. XIII-3) and the discriminable *in situ* cell. As the immature dyskaryotic cell can be shed from severe dysplasia as well as from *in situ* carcinoma and invasive squamous cell carcinoma, it does not discriminate between them.

The ability to discriminate between the presence or absence of a more severe lesion, is *solely* based upon the *presence* or *absence* of those cells which are *characteristic* of the *more* severe lesion. Thus, as in tissue biopsies, the *accuracy* of diagnosing the most severe lesion which is present, depends directly upon the *adequacy* of the sample, its thorough screening, and a proper *interpretation* thereof.

E. Degree of Differentiation of Squamous Cell Carcinoma

The pattern which a squamous cell carcinoma takes in tissue and its degree of differentiation, are thus closely mirrored in the cells it exfoliates.

In Broder's classification [48], well-differentiated and hyperkeratinizing Grade I cancer sheds mainly hyperkeratinizing cells with many bizarre, elongated tadpole and spindle types with pyknotic nuclei (e.g.: Figs. IX-2, IX-5; XII-5 through XII-12, XII-14 through XII-19; XIII-17). In the lung, this well-keratinizing squamous cell carcinoma frequently cavitates. It is *the* cancer of the lung which most frequently presents clinically as a cavitating lesion. In contrast, the poorly-differentiated cancer of Broder's IV exfoliates mainly undifferentiated cells (e.g.: Figs. IX-1, IX-3, IX-4, IX-6, IX-8; XVIII-1, XVIII-5, XVIII-6).

In Wentz and Reagan's valuable classification of squamous cell carcinoma of the cervix uteri [297], three major groups are recognized: large cell keratinized, large cell nonkeratinized, and small cell (nonkeratinized). The first sheds highly keratinizing, karyopyknotic, and otherwise well-differentiated squamous cancer cells (e.g.: Figs. IX-2, IX-5; XII-5 through XII-12, XII-14 through XII-19; XIII-17).

The large cell nonkeratinizing squamous cell carcinoma of the cervix uteri sheds mainly pleomorphic large cell undifferentiated type cells (e.g.: Figs. IX-1, IX-7, IX-8; XII-13; XVIII-1). At times, minimal keratinization and other squamous maturation tendencies occur.

The small cell nonkeratinized squamous cell carcinoma of the cervix uteri sheds the monotonous small cell undifferentiated type of cancer cells (Figs. XVIII-2 through XVIII-4). They are uniformly small and nonkeratinizing, or extremely minimally keratinizing, and bear the poorest prognosis of the three groups. In the cervix uteri, the large cell nonkeratinizing group bears the best prognosis.

XIV. Typical Functional Differentiation of Columnar Epithelium

A. General Considerations in Columnar Epithelial Functional Differentiation

Tissues and cells in the euplastic state which are differentiating functionally as columnar epithelium, are characterized by numerous key morphologic features. These include *columnar* shape of the most mature (luminal) cells, orientation *perpendicular* to the luminal surface and to the basement membrane, *vesicular* metabolic nucleus, *granular* interphase chromatin pattern, *nucleolar* prominence, *basilar orientation* of the nucleus, abundant *thick* cytoplasm near the luminal surface, cellular *secretion* (e.g.: granules, vacuoles, mucoid border), *luminal* structures (e.g.: cilia, brush border, microvilli), and *organoid* tissue formations (e.g.: glands, polyps).

B. The Nucleus of Columnar Epithelial Differentiation

The nuclear morphology of the euplastic columnar cell is fairly characteristic of that cell type. Its changes mainly have to do with the functional state within that cell type.

1. Shape and Orientation

The nucleus of a columnar cell tends to be round or oval. It is frequently indented by secretion or by adjacent nuclei, either of a multinucleated cell or of neighboring cells in a true tissue fragment (Fig. XIV-1).

In nearly all forms and degrees of development, the nucleus is oriented away from the center of the cell toward the basilar, or tail-end of the cell. The degree of this orientation and the extent of its movement away from the center of the cell, depend in great measure upon the degree of maturation and the cell type of subdifferentiation within the class of columnar cells (e.g.: slight off-center nuclear placement in the 'exhausted' endometrial cell at the end of menstruation *vs.* the extreme polarity of the nucleus in columnar cells of the colon).

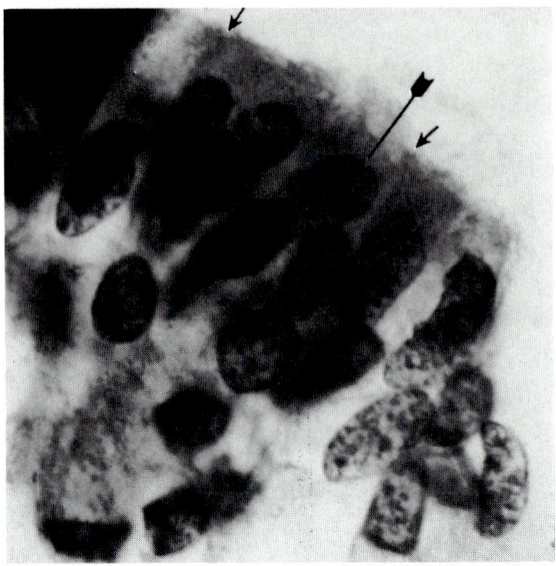

Fig. XIV-1: Human columnar cells shed in a diagnostic true tissue fragment (DTTF).
The nuclei are round or oval and appear vesicular, as if their inner fluid was under pressure. Chromatin is granular, mainly fine, with nucleoli (1 or 2 per nucleus). The luminal border has *terminal bars* discernible between cells, appearing as a small dot (arrows), representing the tight junctional complex viewed on end (cf. Figs. XIV-6, XIV-7). It is actually a zonula, or ring-like structure encircling each cell at its lumen, sealing off the intercellular space from the luminal contents. It frequently is most helpful in identifying a luminal border on a DTTF (Table VI-1). Note the nuclear indentation on the end of the nucleus facing the lumen (arrow shaft), due to cellular activity (e.g.: centriole, Golgi complex).
Papanicolaou stain.
XIV-1: × 1,200.

2. Nuclear Membrane Shape

The nucleus of a columnar cell tends to be *round* or *oval*. This is determined by the shape of the inner nuclear membrane which, while too thin to be revealed by light microscopy, can be accurately determined by the shape of the profile of the outer surface of the chromatinic rim (Figs. VI-1 through VI-3; XIV-1).

A nucleus is frequently pressed out of round by an object (e.g.: an adjacent nucleus) or a process (e.g.: secretion). Even though this is occurring, it usually maintains a *vesicular* appearance (Fig. XIV-1), as if it were a vesicle holding fluid contents which are under pressure, with globular distention of its limiting membrane (e.g.: a vehicle of chicken pox or herpes).

The nucleus is usually *single* in an euplastic columnar cell, which is at unadulterated baseline biologic activity, and of *uniform size* from cell to cell in a diagnostic true tissue fragment (DTTF; Table VI-1; Fig. XIV-1). However, columnar cell nuclei are particularly susceptible to becoming *multiple* and *enlarged* under even *slight* stress or increased activity, much more so than stratified squamous epithelial cells.

When nuclei do enlarge or become multiple, though, they retain uniformity and predictability to the other nuclei of the same cell and in the same tissue fragment. This is a characteristic which is in striking contrast to the very useful feature of nonpredictability among nuclei in malignant neoplasia (Chap. IX).

3. Chromatinic Rim

The chromatinic rim is of uniform thickness in a euplastic columnar cell (Fig. XIV-1). It is usually delicate; but when it does thicken, it does so moderately and uniformly throughout a nucleus, and predictably the same from nucleus to nucleus in a multinucleated cell or in a diagnostic true tissue fragment (DTTF).

4. Chromatinic Net

The chromatinic material of the interphase nucleus of the columnar cell, within the chromatinic net, usually is arranged in a predominantly finely *granular* pattern (frequently referred to as a 'ground glass' pattern). Occasional thin threads may be interwoven between the granules (Fig. XIV-1), but they are not the prominent feature which they are in the nucleus of a cell shed from euplastic keratinizing stratified squamous epithelium (Fig. XI-5). At times the chromatin granules can increase in size and number to be moderately prominent, and still be considered euplastic.

5. Parachromatin

The parachromatin of a euplastic columnar cell has the lightly hematoxylinophilic chromasia associated with euplasia (Chap. VI). The chromatinic net is uniformly dispersed throughout the parachromatin.

6. Nucleolus

Nucleoli are usually present in columnar cells, and are frequently prominent, multiple and large (Fig. XIV-1). Their number and size vary with cellular activity, increasing when the cell is manufacturing protein for increase of cellular substance (e.g.: proliferation, reproduction, repair) or secretion (e.g.: mucus, enzymes, hormones).

C. The Cytoplasm of Columnar Epithelial Differentiation

1. Shape and Orientation

As columnar cells mature and line the surface of the lumen, they elongate into a columnar shape with the cell's long axis *perpendicular* to the lumen and, thus, to the basement membrane (Fig. XIV-1). The major exceptions to this include when, for function, columnar cells do not reach the surface (e.g.: subluminal, germinal, basal, reserve), or when they shorten toward a simple low columnar or cuboidal shape (e.g.: bronchiolar cell, renal tubular cell) (Figs. XIV-2, XIV-3), or markedly attenuate to thinly cover a surface and facilitate transfer across a membrane (e.g.: simple squamous epithelial cells lining the pulmonary alveolus or the renal glomerulus). As an overwhelming general rule, however, the most mature cell lying on the luminal surface of epithelium functionally differentiating to columnar, is oriented perpendicularly both to the luminal border and to the basement membrane.

This orientation is usually retained with exfoliation if the cells lie in profile. In that situation, a small whisp of cytoplasm represents the tail which extended toward, and rested upon the basement membrane (Figs. XIV-2, XIV-3). The more abundant cytoplasm on the luminal end, often ending in specialized structures (e.g.: cilia, microvilli), may at times still retain, with its neighboring cells, an epithelial luminal border (Fig. XIV-1). If the cytoplasm is so scant that the tail is drawn up, the cell is frequently difficult to tell with certainty from a histiocyte (Fig. IV-1) or even a small parabasal cell.

2. Intracytoplasmic Structures

Relatively few of the many structures present within the cytoplasm can be specifically identified by *light* microscopy using *routine* diagnostic stains. Because of the numerous cytoplasmic structures present, the cytoplasm of columnar cells usually appears foamy or flocculent (Figs. V-2, V-3). These include membrane-bound organelles (e.g.: endoplasmic reticulum, Golgi complex, mitochondria, lysosomes, microbodies), inclusions (e.g.: secretion, pigment, glycogen, lipid) and others (e.g.: centrioles, microtubules, filaments) (Figs. XIV-1 through XIV-4). Special stains [211] and immunodiagnostic techniques [132] help in identifying these structures, as well as the many enzymes and other substances present in the cytoplasm [54].

The cytoplasm becomes more deeply basophilic as its RNA increases. Ribosomes increase in the cytoplasm, both membrane bound (i.e.: rough

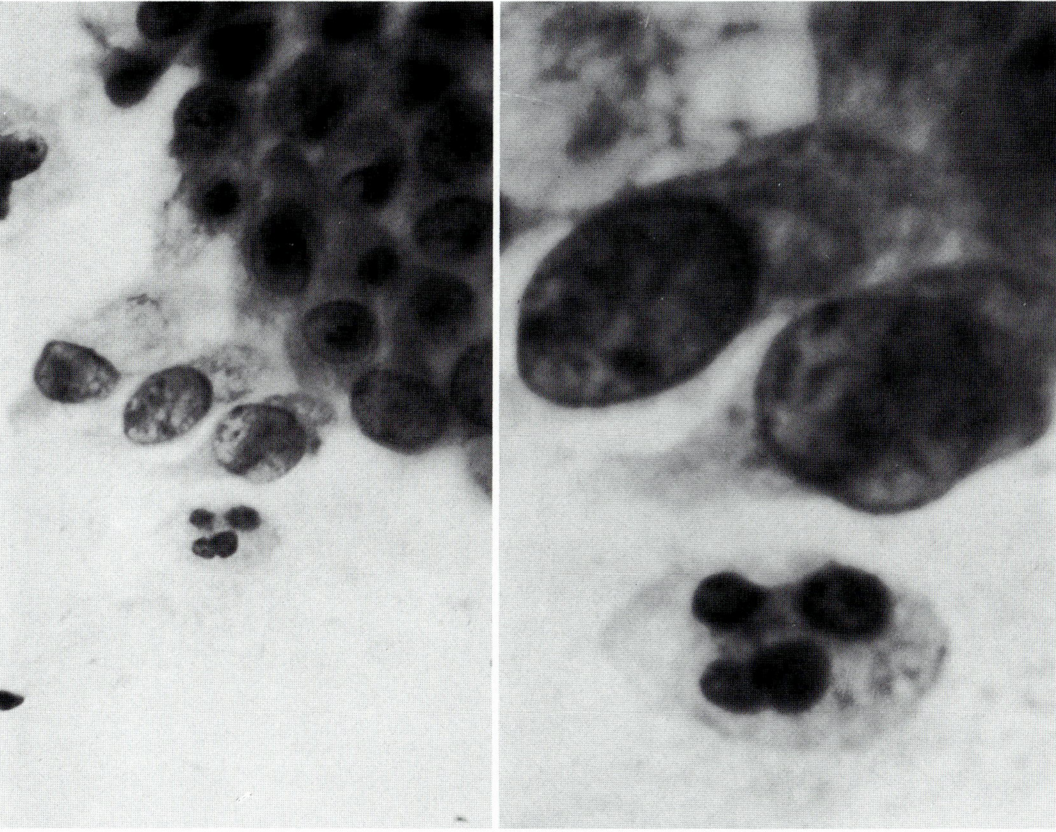

2   3

*Figures XIV-2, XIV-3: Short human columnar cells shed in a single cell sheet (i.e.:
simple columnar).*

Fig. XIV-2: The nuclei are round and vesicular with fine granules uniformly dispersed
throughout a pale hematoxylinophilic parachromatin. Each nucleus has one to three
nucleoli. The sheet of cells in the upper right are *en face,* in a somewhat distorted honeycomb arrangement, and some intercellular windows are forming at top center where they
are falling apart. Their exact cell type cannot be determined in the center of the *en face*
sheet. By moving to the edge of this diagnostic true tissue fragment (DTTF) where three
cells have rotated to lie parallel with the surface of the slide, their profile reveals their
identity.

Fig. XIV-3: The thin tail and the broad, abundant cytoplasm identify the former as
the basement membrane end of each cell, and the latter as the luminal end. Note the
well-preserved neutrophil, and that this *entire cell* (both the nucleus and cytoplasm of the
intact neutrophil) is virtually the same size as the *nucleus* of the columnar cell.

Papanicolaou stain.

XIV-2: × 1,200; XIV-3: × 3,250.

endoplasmic reticulum) and free, with increased protein production. The cytoplasm becomes more deeply basophilic (e.g.: green, blue) in proportion to the RNA-associated protein of the ribosomes.

The position of the *centrioles* can sometimes be recognized in the light microscope by their region of activity, as a small, prominent, extremely clear, rounded area of the cytoplasm on the side of the nucleus nearest the center of the cell. This region is called the centrosome, cytocentrum, or cell center. The rim of the clearing is not sharply defined, as it is with membrane-bound organelles, and appears fuzzy or hazy (Figs. XIV-2, XIV-3).

In this same general vicinity of the cytoplasm, the *Golgi* complex [75] and accumulations of the endoplasmic reticulum usually occur (Fig. V-3). When discernible in the light microscope, they appear as larger, less distinctly definable areas of clearing in the cytoplasm (Figs. XIV-1, XIV-3).

Great cellular activity takes place in this general region, next to the nucleus and either toward the center of the cell or on its luminal side. Its activity, and presumably the activity of that area of the nucleus, frequently causes flattening or indenting of the adjacent side of the nucleus, to form the so-called 'reniform', or kidney-shaped nucleus. As this resembles a house extending around a garden, the early Germanic microscopists referred to this portion of the cytoplasm as the hof (*hof:* Ger. = garden, courtyard) (Fig. V-7).

Mitochondria (Figs. V-3; XIV-4) appear as small, minute, cleared areas which are widely scattered throughout the cytoplasm. They, the endoplas-

---

Fig. XIV-4: Luminal end of bronchial columnar cells, human bronchial epithelium, transbronchoscopic biopsy (see survey electron micrograph, Fig. V-3). Numerous elongated, lengthwise-oriented *mitochondria* (m) are distributed throughout the luminal end of this pale ciliated cell, so that they lie near to the cilia which need their energy for ciliary motion [163, 164]. They impart to the cytoplasm a foamy appearance in light microscopy. Their double-walled cristae project into their cavities as transverse, shelf-like, partial septa. The free luminal surface of the cell has numerous slender, finger-like projections, or *microvilli* (mv). They have pale granular cytoplasm and are branching, and greatly increase the surface area of this cell. There are also numerous cilia (c) with their nine outer longitudinal fibrils and two inner central fibrils (the axial filament complex, or axoneme), covered by the cell membrane. As they enter the main body of the cell, they lose their cell membrane and terminate, each in a *basal corpuscle* (bc) which tails off as the striated *rootlet*.
Osmium/methacrylate.
XIV-4: × 17,300.
(Frost, Erozan, Donovan; unpublished electron micrograph.)

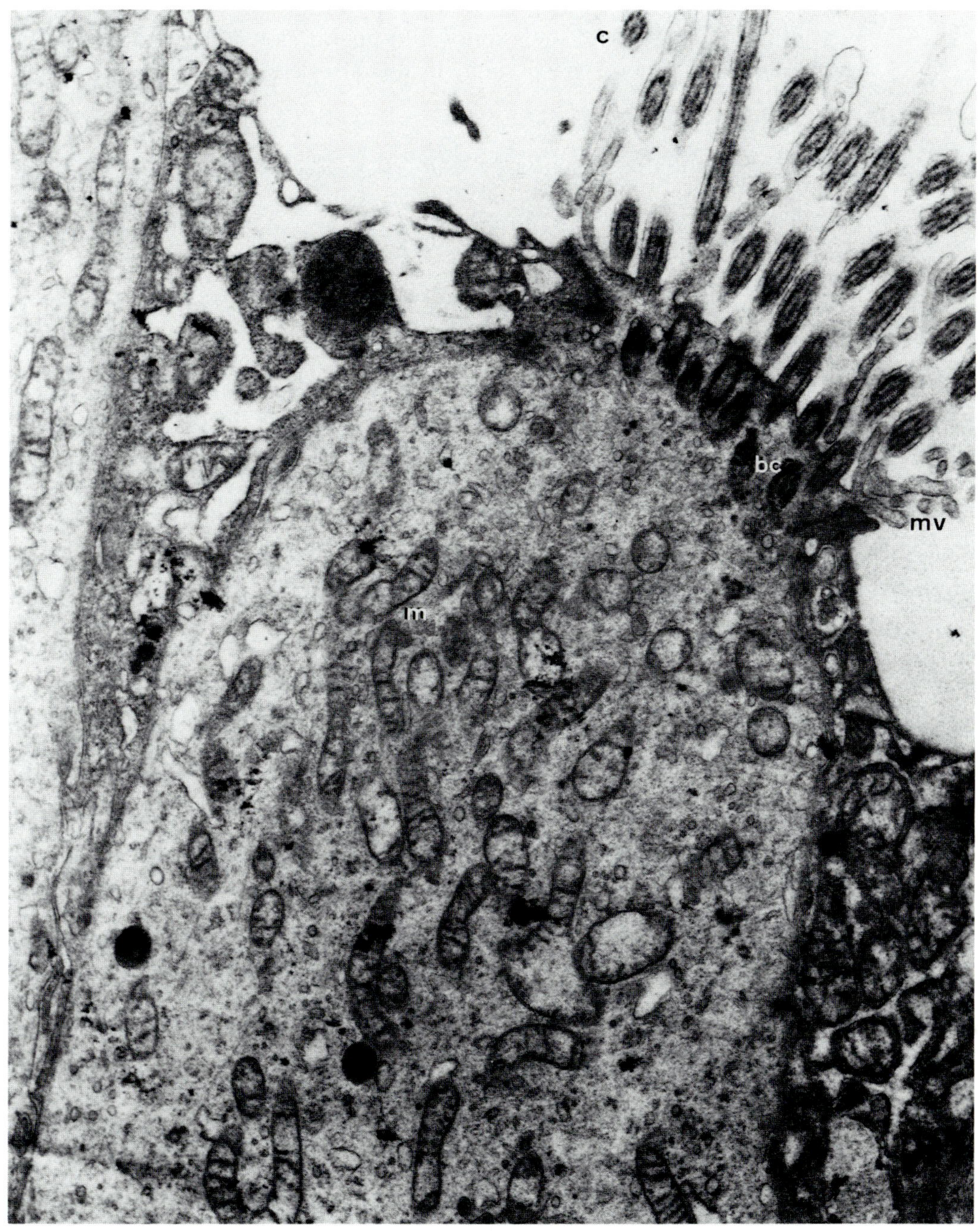

c
bc
m
mv

mic reticulum, secretion, and degeneration are probably the most frequent causes of imparting to the cytoplasm a *foamy* appearance. They cannot be distinguished from the other objects by the light microscope.

3. Secretion

Secretion occurs in many forms within the cell, depending upon the product and its stage of development. At times it appears as small granules, usually proteinaceous, which are either acidophilic, basophilic, or amphophilic. Occasionally, amorphous secretion of larger size is encountered in irritation and inflammation, which is usually inspissated protein (e.g.: mucus) appearing homogeneous with an occasional small rim of cytoplasmic clearing about it.

The most frequent evidence of secretion is the presence of cytoplasmic vacuoles, multiple or single. They are usually clearer than the surrounding cytoplasm, and occur mainly in the region above the nucleus within the large mass of cytoplasm at the luminal end of the cell (Fig. XIV-3). During the early stages of secretion in some cells (e.g.: endometrium), it may appear subnuclear.

Multiple vacuolation is the usual form of vacuolation of secretion encountered in exfoliated cells (Fig. XIV-3, XIV-5). However, this feature discriminates secretion poorly from the multiple vacuolation of either *cell degeneration* (Fig. VII-7) or *phagocytosis* (Figs. IV-1, IV-3).

The most valuable type of vacuolation to help identify *secretion* is the *large, single, hyperdistended* 'secretory-type' vacuole (signet ring cell). This

---

Fig. XIV-5: Columnar cells, human bronchial epithelium, transbronchoscopic biopsy. Two mucus secretory cells (i.e.: goblet cells) and one ciliated cell lying side by side forming an epithelial luminal border. The tight junctions between the cells are viewed on end. They are just large enough to be barely discernible by light microscopy as a dot, the terminal bar (Fig. XIV-1). Note the indentations in the surface of the nucleus facing the lumen, where there is increased activity. The semiseptated vacuoles in the cytoplasm of the two goblet cells, contain pale staining mucus. Note the uniform periodicity as one goes from cilia to cilia, particularly of their rootlets and where they emerge from the cell before they bend and wave about.
Osmium/methacrylate.
XIV-5: × 9,750.
(Frost, Erozan, Donovan; unpublished electron micrograph.)

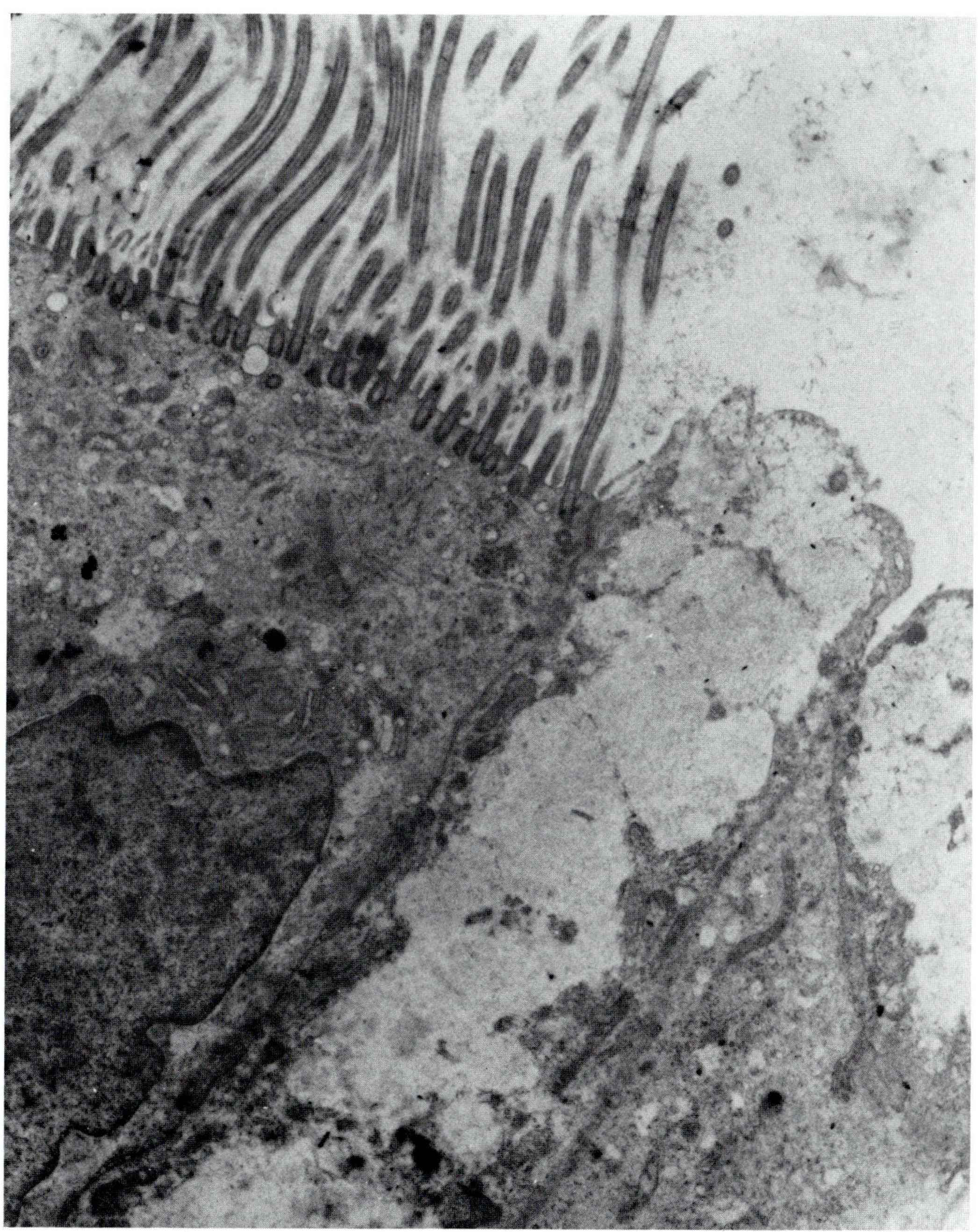

usually occurs as atypical functional differentiation (Fig. XV-2). It consists of one vacuole per cell, with a thin attenuated cell membrane about it (frequently unresolvable from the vacuole wall by light microscopy). The nucleus, usually pushed to one side of the cell, is markedly molded about the vacuole and may have one or more nucleoli.

These secretory vacuoles, however, can be mimicked by those arising in cells degenerating in watery fluids (e.g.: cell culture, body fluids) which can form single, large, pouched-out vacuoles (Fig. VII-7). But in nonwatery material (e.g.: mucus of the female genital tract, respiratory tract, alimentary tract; needle aspiration other than watery fluid), it is a discriminating finding in favor of secretion.

Large, hyperdistended single vacuoles resembling secretory-type vacuoles also occur in *fat* storage. Here the nucleus is small, however, has no nucleolus (i.e.: storing at low activity, not secreting at high activity), and is molded very little by the vacuole.

In *embryonal fat* and in *brown fat* of hibernation, there is more cytoplasm uniformly distributed between the vacuole wall and the cell membrane than in the thin hyperdistended rim of fat storage or of hypersecretion. This creates a thick rim of cytoplasm around the vacuole, between the wall of the vacuole and the cell membrane, rather than the virtually single membrane about a hyperdistended secretory vacuole. This thick wall of cytoplasm may represent more metabolic activity between the fat droplet and the cell cytoplasm, than the less metabolic activity in long-term fat storage. Thus, even though the storage vacuole is large and hyperdistended, it only appears paradoxical as a secretory cell that it does not thin out the cytoplasm overlying the embryonal or hibernational brown fat.

The true *secretory* type of *hyperdistended single vacuoles* (i.e.: signet-ring cell) occurs in irritation (e.g.: chronic irritants), inflammation (e.g.: chronic gastritis), infection (e.g.: chronic bronchitis, endometritis), hormonal imbalance (e.g.: endometrial atypia), hypersensitivity (e.g.: asthma), and neoplasia (i.e.: in situ or invasive adenocarcinoma). These are all conditions where the product of secretion is extremely rubbery and inspissated, and apparently difficult for the cell to release. Yet, this is the best evidence available on routinely-stained material that secretion is occurring.

4. Luminal Structures

Various types of columnar cells have *microvilli* on their luminal surface, outward finger-like projections of the cell membrane and cytoplasm (Figs. XIV-4 through XIV-7). While at times their function is unknown, they are usually present to increase interchange across the cell membrane (i.e.: absorption, secretion).

When they are numerous and of uniform length, they can be recognized by light microscopy as a *striated* border (e.g.: intestinal cells) or as a *brush* border (e.g.: renal tubule cells). The microvilli of these structures are short, uniformly spaced, of uniform length, and perpendicular to the cell surface. They have a rigid and distinct uniform periodicity.

Microvilli frequently branch, and may be very arborescent (Figs. V-3; XIV-4). They are not confined to columnar cells, but are formed on the surface of many cells having high levels of secretion or absorption as a function, such as mesothelial cells (Figs. VI-4 through VI-7).

Cilia, like microvilli, are also projections of the cell membrane. In addition, however, they possess internal longitudinal fibrils (Figs. V-3; XIV-4 through XIV-6) referred to variously as the axial filament complex, the axial microtubule complex, or the axoneme. This consists of 11 fibrils: two central unpaired fibrils and nine so-called paired fibrils, which are evenly spaced around the central two. The nine peripheral paired fibrils extend down into the cytoplasm to the basal bodies, which are apparently derived from the energized centrioles, and which then extend farther down into the cytoplasm as the striated rootlets and anchoring microtubules and filaments (Figs. XIV-4, XIV-6).

Cilia are larger and longer than the microvilli. Even though they bend, cilia retain a very uniform distance from each other, especially where they emerge from the cell surface, and thus impart a distinct *periodicity* in light microscopy (Figs. IV-12, XV-7). They usually stain eosinophilic or lavender – brilliantly, frequently – but they can stain virtually any color.

Sharply marking the luminal border of the cell in light microscopy is the *terminal plate.* This stains identical to the cilia and is actually the ciliary basal bodies and rootlets, just referred to above, embedded for a short distance in the body of the cell. When the luminal border of the cell is parallel to the line of sight of the microscope, the plate of rootlets is on end, superimposing the rootlets upon themselves, and appears as a brilliant line – the terminal plate. When the luminal border is angled to the line of sight, the rootlets do not superimpose on top of each other and the terminal plate becomes indistinct (Figs. IV-12, XV-7) or, even, disappears.

The presence of discriminable, identifiable, multiple cilia on exfoliated cells provides a most potent hallmark of *benignancy.* Brush borders, striated borders, and other forms of microvilli do not have the same diagnostic value. Thus *uniform periodicity, equal length within a given cell, countless numbers of cilia per cell, terminal plate,* and *staining qualities* take on extreme importance as features to unequivocally identify cilia; especially the *first three,* which are not mimicked by mucus strands and fibrin or other noncilia structures.

Malignant cells at times will have a few, abortive cilia, but not the countless massive numbers per cell with the above criteria. Only two isolated cases have been reported in the literature of cancer shedding cells with such identifiable cilia, both of these being ovarian carcinomas [67, 130]. By and large, however, it remains the best criterion of benignancy which is available.

The finding of true diagnostic cilia is especially of great diagnostic value in the respiratory tract, where cancer has *never* been described as exfoliating with such diagnostic cilia fulfilling the above three criteria. Conversely, and of great diagnostic importance, tissue fragments can be shed of benign, ciliated, extremely proplastic and atypical, hypersecretory, respiratory epithelium (so-called 'Creola bodies' [195]) aspirated at bronchoscopy in chronic bronchitis, or coughed up in sputum in asthma. These have been all too frequently miscalled cancer, where recognition of diagnostic cilia would obviate that grave error. To be usable, however, one must be certain that it is a ciliated cell (by the above criteria), and not just tenacious mucus stringing from the luminal border; furthermore, that the ciliated cell matches the rest of the cells in the tissue fragment in question, mainly by nuclear criteria, and that it is not a benign cell which has stuck to a cancer fragment (see Fig. XV-7).

Fig. XIV-6: Cilia and junctional complex, human bronchial columnar epithelium, transbronchoscopic biopsy. The *junctional complex* (jc), consisting of numerous zones, is the *terminal bar* of light microscopy (Fig. XIV-1). It runs across the lower right corner from the luminal surface as the tight junction (zonula occludens) along the intercellular border for about one-third of the electron micrograph, where the intercellular space appears. Where there is absence of intercellular space, there are areas of apparent membrane fusion with firm attachment of the two cell membranes. As the *cilia* reach the cytoplasm, the cell membrane is deflected off and the axial filament complex extends into the cytoplasm, becoming the basal body (b). The dark basal bodies with the striated rootlets (r) and the anchoring fibers (f), form the acidophilic or amphophilic *terminal plate* of light microscopy. Arborescent delicate glycocalyx (gc) extends in the lumen from the surface of the microvilli (mv).

Osmium/methacrylate.

XIV-6: × 63,000.

(Frost, Erozan, Donovan; unpublished electron micrograph.)

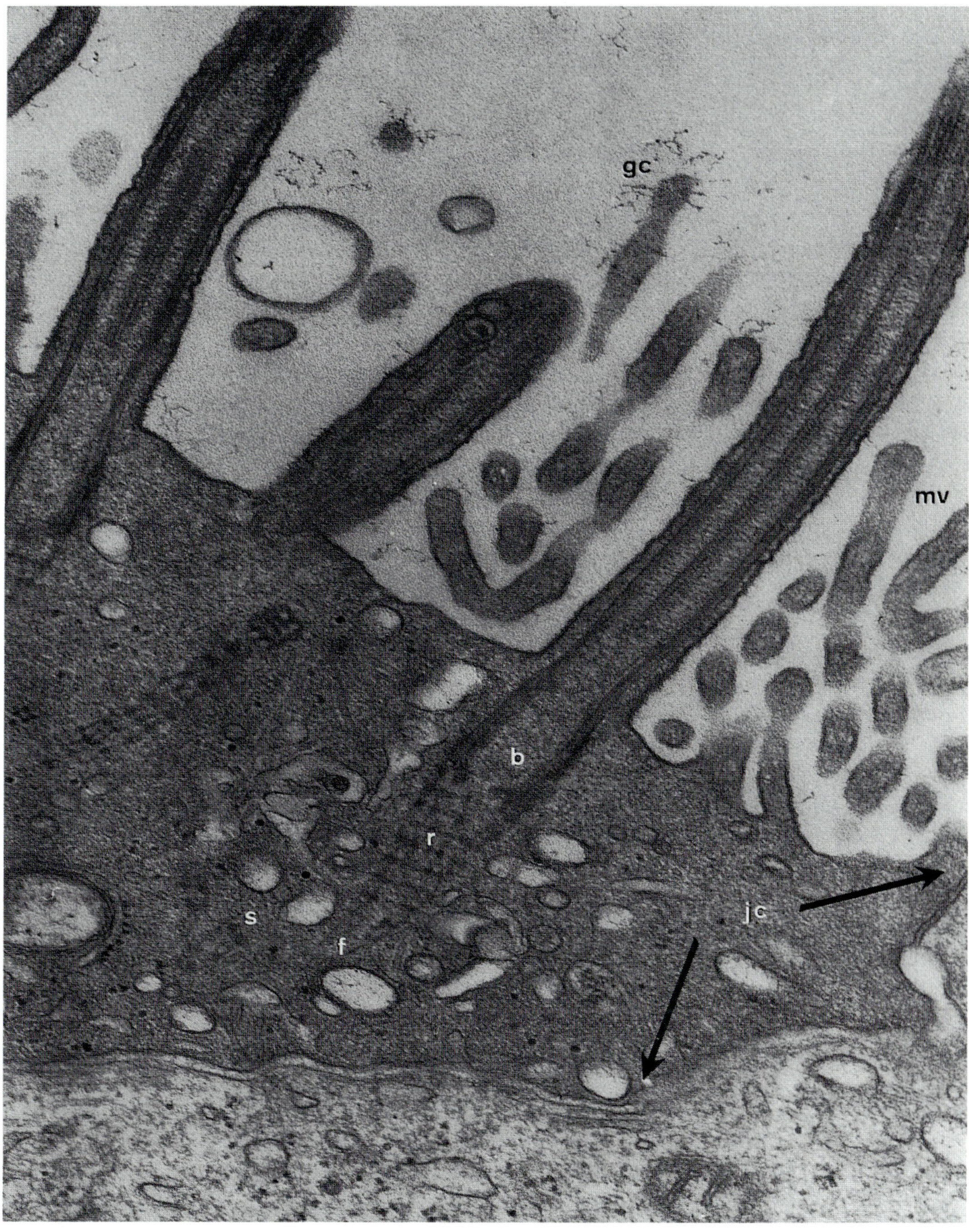

Cilia are usually the first cellular structures to disappear with degeneration, leaving nothing but a smooth columnar cell or, at times, leaving the terminal plate to be identified at the luminal end of the cell. At other times the terminal portion of cytoplasm of the cell pinches-off with the cilia (Fig. IV-12), such as the physiologically detached ciliated tufts (DCT) of endocervical cells cyclically changing from ciliated to nonciliated [144] and the pinched-off ciliated tufts of ciliocytophthoria in viral infection of the nose [50], the lungs [203] and the cervix [99].

D. The Exfoliated Cells of Columnar Epithelial Differentiation

1. Pseudostratified Columnar Epithelium

Each cell of columnar epithelium rests upon the basement membrane. In *pseudostratified* columnar epithelium, where two or more nuclei stratify, a thin tail of each cell theoretically reaches down to the basement membrane. Thus, the cell on the lumen has *cytoplasmic* processes extending from the basement membrane to the lumen. This is believed to be true even in the thick pseudostratified columnar epithelia (e.g.: trachea, major bronchi) where the tail can be extremely attenuated.

The rest of the cells do not reach the lumen. Nevertheless, these subluminal (e.g.: subcolumnar, subcylindrical, intermediate, parabasal, reserve, basal, germinal) cells reach to the basement membrane with a tail, and elongate with their major axis perpendicular to the basement membrane and the lumen. Their degree of elongation, or columnar maturation, is to a lesser degree than that of their luminal cells.

The basal or germinal cells lie on the basement membrane. They are the major source of replication and regeneration of the epithelium.

The *nuclei* of all columnar cell types, seek the basilar end of their cells. This is a characteristic orientation which is retained after exfoliation. The most mature cell lying on the lumen usually has a true columnar shape, with the nucleus near the tail or basement membrane end.

A cell of lesser maturation (e.g.: parabasal) has less cytoplasm and not such prominent nuclear eccentricity; still, it does demonstrate this tendency for the nucleus to be on the basilar side of the cell, which may be the only characteristic to suggest its columnar cell origin. With a foamy cytoplasm of multiple small vacuoles, and an eccentric nucleus indented toward the cytocentrum, it may very closely mimic a macrophage.

2. Simple Columnar Epithelium

In simple columnar epithelium (i.e.: cervix, stomach, bronchiole, renal tubules) all cells reach as a column from lumen to basement membrane, except for an occasional basal (germinal, reserve) cell. There is a definite orientation to each cell, with the nucleus basally located and with the more abundant cytoplasm extending to the lumen.

When their cellular activities and outside influences are the same from cell to cell, there can be the morphologic uniformity and monotony of a 'picket fence'. When the cells differ in function or reaction, this uniformity and predictability can be lost.

As a sheet of simple columnar cells land on a slide, they are oriented on their tails or on their faces (i.e.: *en face*). The intercellular borders are viewed along their length and appear well defined. The cells are polygonal, and the entire structure resembles a 'honeycomb'. When one moves to the edge of the sheet to observe the cells in profile, their columnar shape and orientation are identifiable (Figs. XIV-2, XIV-3). Other single cell sheets viewed *en face* can also appear as a honeycomb (e.g.: parabasal cells), so that this feature is not confined only to columnar cells.

3. Simple Cuboidal Epithelium

Virtually all cells reach from the basement membrane to the lumen. There is only an occasional basal (reserve, germinal) cell interdigitated upon the basement membrane, but not extending through to the lumen.

In cuboidal cells, there is much less cytoplasm per cell. In spite of this, however, the nucleus tends to orient below the center of the cell, especially after exfoliation.

Upon exfoliation, there may be little columnar or cuboidal orientation in evidence, with the nucleus occupying a central orientation. Some nuclear eccentricity can usually be discerned, however, with an occasional tail to bespeak cuboidal identity. Once again, with its scanty, foamy cytoplasm and its eccentric, indented nucleus, this exfoliated cell can closely mimic a macrophage (e.g.: from endometrium, lung, breast) from which it must be carefully discriminated at certain key times (e.g.: detecting 'early' adenocarcinoma of the endometrium and its precursors; separating adenocarcinoma of the lung from resolving pneumonia).

4. Simple Squamous Epithelium

In certain key areas of the body, columnar epithelium takes the form of simple squamous differentiation for specific functions. When it is physiologically necessary for columnar epithelium to thinly cover a structure (e.g.: respiratory exchange in pulmonary alveoli; blood filtration in renal glomeruli), a single cell attenuates (an active cell function) along a wide expanse of basement membrane and forms a thin polygonal simple squame.

This simple squamous epithelium is *not* keratinized and thus the cytoplasm is not rigid, but is rubbery and moldable. It can change back again, dramatically, in certain conditions and diseases. In states with fibrocellular proliferation involving the pulmonary alveoli (e.g.: asbestosis) or in collapse and inactivity of lung tissue (e.g.: atelectasis, scars, foreign bodies), the extremely thinned and actively attenuated alveolar epithelial cell changes strikingly. With *no* hard substance such as keratins to passively keep it squamous, such as in the intermediate cell or superficial cell of keratinizing stratified squamous epithelium, it ceases to attenuate into a simple squamous type of columnar cell, rounds up and becomes a cuboidal (so-called cuboidal metaplasia) or even simple columnar epithelium.

This simple squamous epithelium, therefore, is *not* a variant of stratified squamous epithelium (Chap. XI). In these situations, it is modified columnar epithelium. In others (e.g.: mesothelium, synovium, endothelium), it is modified mesoderm.

The cytoplasm is rubbery or elastic. Upon exfoliation, in contrast to the rigid keratinizing squame of stratified squamous epithelium, it rounds up into a sphere. Thus, in cellular spreads it may be difficult to definitely identify simple squamous cells, or to identify them as of columnar origin; however, they are easily distinguished from cells of keratinizing stratified squamous epithelium.

As with other columnar cells, the cytoplasm of this *exfoliated* cell is thick. The nucleus, however, is usually central rather than demonstrating the eccentricity hallmark of columnar cells. There is no tail for identification in these small, simple squamous cells.

5. Discrimination from a Macrophage

The ability to discriminate an exfoliated columnar cell from a macrophage, at times is very limited. When identification is of key diagnostic importance, it may be difficult, but certain features can be of assistance.

A columnar cell's nuclear chromatinic pattern is usually more uniformly granular than a histiocyte or, particularly, a more active macrophage. Cytoplasmic debris of phagocytosis is absent in the columnar cell, even though the cytoplasm may appear foamy from numerous small secretory vacuoles, or may lose its typical green and blue shading and become a somber gray, more typical of a macrophage.

Special stains for mucus do *not* help discriminate between mucus-secreting columnar cells and macrophages. The latter all too frequently phagocytose mucus and, thus, give a positive mucus reaction. Identification of monocyte/histiocyte antigens by immunocytodiagnostics, at times will help.

E. Tissue Fragments with Columnar Epithelial Differentiation

In many respects, fragments of columnar epithelium appear similar to the tissue from which they shed. Some important differences, however, do appear on exfoliation.

1. In Profile

Uniform columnar cells, when viewed in profile, appear as a 'picket fence' when they are rigidly held upright and parallel to one another by the forces from their adjacent neighbors. After exfoliation and removal from these lateral forces present in tissue, some columnar tissue fragments lose their typical columnar shape and ball-up, usually resulting in just a further unidentifiable ball of cells.

At other times, however, the lateral columnar cells of these balled-up tissue fragments curve and become semilunar, surrounding the more central cells in a somewhat 'pearly' fashion (pseudopearls). Thus, they can greatly resemble the pearls ('pearly bodies') of atypical squamous maturation (Fig. XII-5), with which they must not be confused. Careful up-and-down focusing at high power, reveals that these cells are *not* flattening and lying on top of each other in true stratified squamous fashion; but, rather, are columnar cells wrapping around each other.

Directed search for *luminal* borders on such tissue fragments will frequently identify them, at times with definite identification of *terminal bars* (Fig. XIV-1), and/or *terminal plates* and *discriminable cilia* (pp. 213–214) diagnostically emanating therefrom. Small tails from the opposite end of the cell help to identify the true columnar nature of these structures.

These are frequently encountered in the respiratory tract, especially in bronchoscopic specimens, in pulmonary sputum following bronchoscopy, or in the sputum of individuals with bronchial asthma [195]. It can also be encountered in gastric or in esophageal preparations where the specimen was obtained following intubation through the nose.

In such cases where the epithelium has extreme proplastic nuclear changes (i.e.: chronic irritation, repair), these balled-up columnar tissue fragments have been falsely mistaken for squamous cell carcinoma or large cell undifferentiated cancer by the unwary. At other times such proplastic, balled-up tissue fragments with large, single, hyperdistended, secretory cytoplasmic vacuoles from irritation and inflammation, have been mistaken for adenocarcinoma. The positive identification of unequivocal cilia, as previously discussed (pp. 313–314), allows one to recognize the benignancy of the tissue from which it arose.

2. En face 'Honeycomb'

Columnar cell sheets, when viewed from above or below (so-called '*en face*'), appear as a 'honeycomb'. With this viewing down the length of the cell, their cellular borders appear prominent. The shape of the cells is polygonal, mainly hexagonal, and the nuclei are centrally placed.

This is a finding of *simple* epithelium, a single layer of cells, no matter of what cell type. While it is most frequently encountered with cells of columnar epithelium, this does occur with other cell types stripped off as a single cell sheet (e.g.: parabasal cells in parabasal atrophy). The cell type can usually be determined by moving to the edge of the sheet, to carefully inspect cells lying in profile (Figs. XIV-2, XIV-3).

3. Identification of Cell-Type and Character

In a dense tissue fragment or cell ball, as in a sheet of simple epithelium, individual cell structures may not be discernible except along the periphery of the mass. In the depths of fragments, hyperchromasia is arti-

---

Fig. XIV-7: *Junctional complex* at the luminal border of two columnar cells with microvilli, human bronchial epithelium, transbronchoscopic biopsy. The junctional complex runs for about one-half of the electron micrograph, at which point the intercellular space appears. At numerous areas along the tight junction (tj), or zonula occludens, there are areas of apparent fusion of the two membranes, the rest appearing to be adherent along a most uniform clear region, which is apparently filled with a proteinaceous cement. The microvilli (mv) have a heavy coat of arborescent glycocalyx extending from their surface, a polysaccharide-rich layer.

Osmium/methacrylate.

XIV-7: × 61,500.

(Frost, Erozan, Donovan; unpublished electron micrograph.)

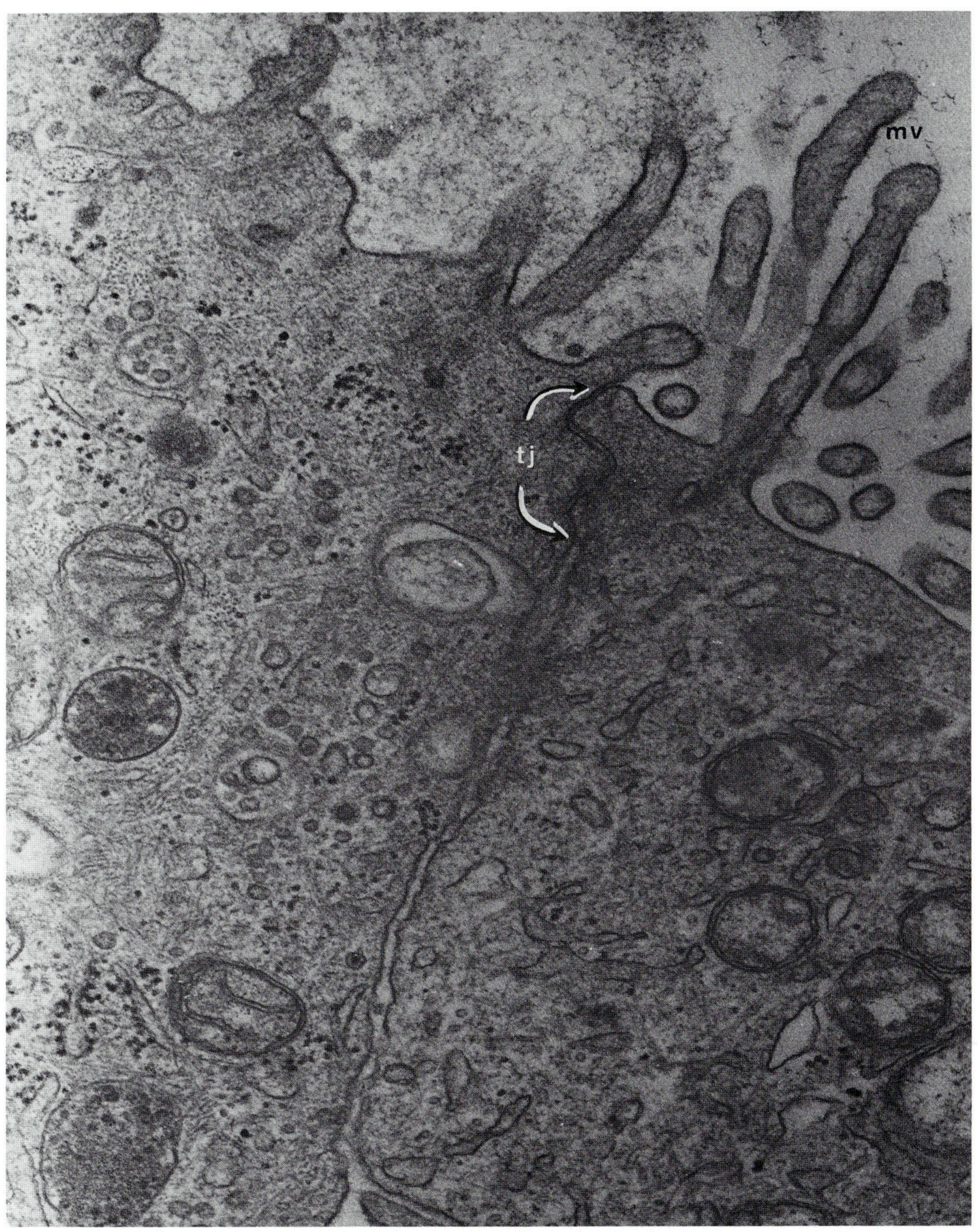
mv
tj

factual, nucleo-cytoplasmic ratio is distorted, and other apparent disturbances of morphology due to fragment density may be misleading unless meaningful peripheral cell examination and identity are made.

By moving to the edge of a sheet of cells or a tissue ball, one finds some lying in profile. Here their morphology can be carefully studied, their general activity determined, and their functional differentiation or cell type identified.

4. Polyps and Glands

Polypoid formations (e.g.: villi, polyps) and glandular formations (e.g.: glands, clefts) of columnar epithelium, in many ways allow for more effective epithelial surface utilization. This provides for greater absorption or secretion per unit of surface area (e.g.: stomach, intestine, endometrium, bronchus), and for physiological increase and distention of the lumen (e.g.: cervix uteri during delivery; mammary ducts in lactation).

Tissue fragments exfoliated from these, frequently retain the structure of a papillary frond or glandular fragment. When identified, this can identify the tissue-type of origin.

Glandular and polypoid formations are three-dimensional structures, whose morphology can best be appreciated by careful through-and-through focusing on the transparent, Papanicolaou-stained tissue fragment. By this maneuver, they can most accurately be discriminated from other tissue fragments or cell clusters which also have a three-dimensional structure but do not fulfill the criteria of an acinus, villus, etc.

In the latter conditions, while focusing through-and-through at high power, the microscopist notes that the lower and upper layers, of either a glandular or polypoid formation, present as *en face* sheets. The midportions of the structure come into focus as rings of cells *in profile,* facing either inward toward the lumen (e.g.: gland, acinus, duct), or outward from a connective tissue and vascular stalk (e.g.: villus, polyp, papillary frond).

Identification of these three layers is the first step. Then, by constructing a global mental image of the three-dimensional structure by careful observance during the through-and-through focusing, one perceives a true three-dimensional object for positive identification.

The surface of the lumen is recognized, on the inner or the outer surface of the structure (i.e.: gland or polyp, respectively), as a 'community' epithelial luminal border. Each cell undertakes, as a 'community effort', to make this epithelial luminal border a *uniformly straight* margin, frequently with identifiable terminal bars (Figs. V-2; XIV-1).

This is *not* the 'loop-the-loop' configuration of the borders of individual cells, having acute angles between them. On the contrary, it is a uniform, linear, fence-like structure of cells bound together at the lumen by a *tight junction* [74] (Figs. XIV-6, XIV-7), which is the *terminal bar* noted in light microscopy (Figs. V-2; XIV-1). In this way, their luminal surface is either linear or, if not under sufficient tension, there may be obtuse angles between the cells.

Exfoliation of glandular or polypoid forms, rather than of nondescript tissue fragments, is increased in various irritations (e.g.: cervix in inflammation), hypersensitivity (e.g.: bronchus in asthma), and endocrine states (e.g.: mammary duct in pregnancy). It may indicate a microacinar or micropolypoid inflammatory condition, or may only demonstrate the native glandular or villous appearance of the epithelium.

XV. Atypical Functional Differentiation: Columnar Epithelium and Adenocarcinoma

A. Atypical Columnar Differentiation

1. General Considerations

In general biologic activity states other than euplasia, columnar epithelial functional differentiation is frequently *a*typical. The probability for these atypical functional differentiating features to occur, increases in retroplasia (Chap. VII), is higher in proplasia (Chap. VIII), is even greater in the mixed (Chap. XVI), and occurs most frequently in malignant neoplasia (Chap. IX). As it is especially marked in the latter, atypical columnar differentiation will be discussed in this chapter, using adenocarcinoma as a model with reference to others where indicated.

It is to be clearly understood from the outset that the features of atypicalities in columnar differentiation do *not* differ essentially when found in retroplasia, proplasia, or malignant neoplasia, except in their frequency and severity. They, therefore, do *not* differentiate cancer from noncancer, and should not be used as malignant criteria. They are merely macabre indicators that all is not normal in the process of the cell differentiating to perform function.

First, cancer is to be *diagnosed* based upon the *malignant criteria* found mainly in the *nucleus* (Chap. IX); then, and *only then,* are typical and atypical functional differentiating features entered in to help determine cell type. Functional differentiating features, no matter how atypical, do not discriminate from nonmalignant atypicalities and are *not* to be used as malignant criteria.

The cells of adenocarcinoma, thus, are *primarily* cancer by malignant criteria (Chap. IX) and, only *secondarily,* functionally differentiating toward columnar. Again, as with other cancers, this order of approach is important in diagnostic work in order to obviate atypical differentiations in nonmalignant states being incorrectly diagnosed as cancer.

2. A Cardinal Rule:

First, determine that cancer is present (mainly by *nuclear* features, Section C);

Second, determine the cell type of cancer present by its functional differentiation (mainly by *cytoplasmic* features, Section D).

Reversing this order, by using functional differentiating features as malignant criteria no matter how atypical, is not only usually meaningless, but it may be misleading and invite diagnostic disaster.

In some adenocarcinomas (e.g.: endometrial) more individual cell exfoliation occurs in the early stages or low grades, then the number of cells shed in true tissue fragments increases in the later stages. In the *very* late stages or high grades, the proportion of single tumor cells exfoliated may again increase, somewhat; however, adenocarcinoma generally sheds more cells in tissue fragments than individually, in contrast to the single cell shedding of squamous cell carcinoma.

Usually the cells of adenocarcinoma are basically large cell undifferentiated carcinoma (pleomorphic cell carcinoma) in type, with some of them displaying sufficient columnar differentiating characteristics to recognize them as being from adenocarcinoma. The rest of the tumor shows only the features of large cell undifferentiated carcinoma (Chap. XVIII), the 'wastebasket' tumor of those non-small cell carcinomas without recognizable functional differentiation, but a few aggressive tumors are small cell.

A lower percentage of cells from adenocarcinoma, than from squamous cell carcinoma, sheds with identifying characteristics of functional differentiation. Squamous cell carcinoma usually sheds a profusion of cells with identifying squamous differentiation, while the greatest number of cells from adenocarcinoma has equivocal or no discriminable differentiating features.

B. Nuclear Features in Atypical Columnar Differentiation

1. Shape and Orientation

Adenocarcinoma nuclei have a great tendency to be *rounded* and to appear *vesicular,* as though they are holding copious amounts of fluid under pressure or as if the nuclear membrane is pulled tightly about the nuclear contents. In certain adenocarcinomas (e.g.: endocervical) (Fig. XV-1) this tendency is present even when a nucleus is molded by pressure from secretory vacuoles, and even with the irregularities and extremes imposed upon each nucleus by its malignant nature.

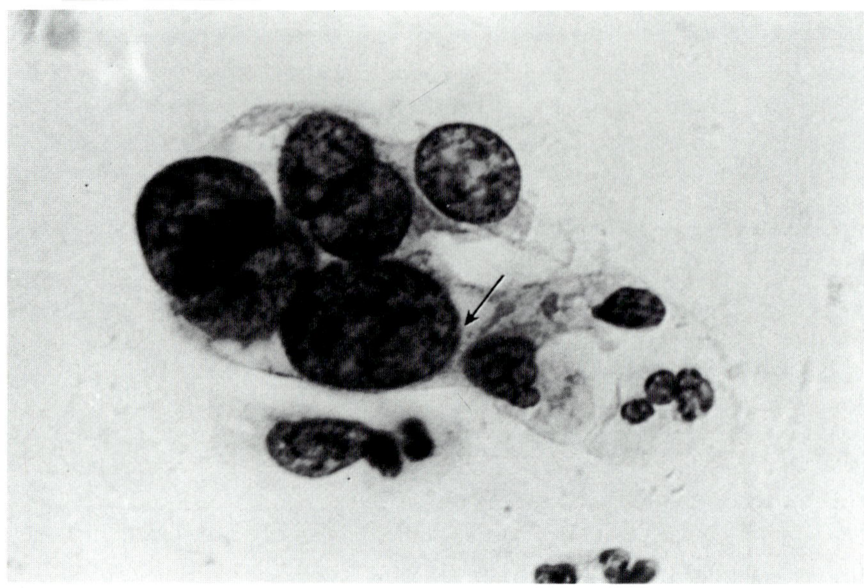

Fig. XV-1: Vesicular nuclei in adenocarcinoma, cervix uteri. Fast (vagino-pancervical) smear. The nuclei are extremely vesicular, appearing to be distended as though the fluid contents are exerting great pressure upon the nuclear membrane, or the latter is contracting tightly about its contents. While this usually irons out nuclear membrane irregularities of cancer, when they do occur even small ones are of greater significance than in a shrinking, wrinkled nucleus. Note the small 'bite' irregularity in the nuclear membrane of the largest nucleus (arrow). The hyperchromasia, the extremely but unevenly cleared parachromatin, the irregular distribution of the chromatinic net, and the unpredictable extremely thick/extremely thin chromatinic rim are all good malignant criteria. The intact neutrophil in the cytoplasm (4-o'clock) and the vacuolated cytoplasm are both suggestive, but not diagnostic of secretion. The extreme nuclear vesiculation is highly suggestive of an endocervical origin.
Papanicolaou stain.
XV-1: × 1,500.

Nuclei usually lie eccentrically within the cytoplasm near the basement membrane (tail) end of the cell. The greater volume of cytoplasm is at the opposite (luminal) pole. Cytoplasmic contents (e.g.: vacuoles) are nearly always in this latter area, even in extreme atypical formation.

Indentation of the nuclear membrane frequently occurs on the side which is oriented toward the center of the cell. This is from the great metabolic activity in that region of a columnar cell (e.g.: centriole, secretory products packaging in the Golgi complex) (Fig. XIV-1). Off-round also occurs from extreme growth (i.e.: molding from adjacent nuclei, Chap. VIII) and from degenerative changes (e.g.: dehydration, shrinkage, wrinkling, Chap. VII). Nuclear roundness and vesiculation, even with these irregularities, however, are potent hallmarks of columnar differentiation (Figs. XV-1, XV-4).

2. Chromatin Pattern

The nuclei of atypical columnar cells react to the causes of the atypia in a more extreme degree than cells of the squamous series. Thus, they show injury and degenerative changes of retroplasia (Chap. VII) to a great degree, with massive rounded chromatin clumping, extreme parachromatin clearing, and fearsome hyperchromasia – but with varying degrees of blurring of the chromatin/parachromatin interfaces – due only to retrogressive processes (e.g.: viral pneumonitis). Likewise, even a modest amount of cell stimulation results in much more extreme nuclear changes of proplasia (Chap. VIII) than occurs in squamous cells under the same stimulation. This includes marked hyperchromasia, large chromatin granules, parachromatin clearing, enlarged nuclei, mitoses, and multinucleation (e.g.: chronic cervicitis, bronchitis).

These are potent sources of false identification of cancer. If great care and caution in interpretation are not taken, a false diagnosis of cancer can be encouraged in the unwary, unfortunately, by atypical functional differentiating features.

The nuclei of adenocarcinoma cells run the gamut in extremes in quality of their malignant morphology (Chap. IX). Many have some of the *best* malignant criteria (Figs. I-1, V-1), such as sharply pointed and irregular huge chromatin clumps and nucleoli, prominent abnormally cleared areas of parachromatin, extremely thick next to extremely thin chromatinic rims, and macabre shaped nuclear membranes.

In sharp contrast, however, many other adenocarcinomas have some of the *poorest* malignant criteria of all cancer cells. Classic in the latter category is a nondescript 'ground glass' appearance of extremely finely divided, powdery chromatin in a perfectly uniform dispersion within a rounded and vesicular nucleus. 'Ground glass' can be present in simple proplasia, and is thus *not* a criterion of malignancy; *nor* is it a feature for recognizing columnar differentiation. While adenocarcinoma can have prominent 'ground glass' nuclei, so can squamous cell carcinoma (Chap. XII) as well as many other cell types of cancer.

3. Nucleolus

Nucleoli are associated with *protein* production, either for intracellular use (e.g.: repair, rapid growth) or extracellular production (e.g.: secretion). It is *not* a criterion for recognizing adenocarcinoma.

Any cancer cell can have prominent and multiple nucleoli due to *rapid growth.* This is particularly true in the poorly differentiated forms of squamous cell carcinoma, transitional cell carcinoma, and virtually all sarcomas.

Additionally, columnar cells can have prominent nucleoli related to their *secretory* activity. While adenocarcinoma thus has an added *possibility* of having more prominent and multiple nucleoli than do other cell types of cancers, this is a very *weak* criterion. It has been used frequently in the past, and has led to many a *false* identification of poorly-differentiated 'anything' as an adenocarcinoma.

It is best *not* to consider prominent nucleolar material as a functional differentiating criterion of secretion in columnar differentiation. It should be accepted for what it truly is – strong evidence of protein production.

Characteristics of nucleoli, at times, provide valuable evidence suggesting the type of cancer or the site of origin of an adenocarcinoma. A very *prominent,* single *nucleolus* in cancer cells exfoliating predominantly as *single cells,* should make one think of melanoma. Various adenocarcinomas can shed with equally *prominent* one or two *nucleoli* in cells of diagnostic true *tissue fragments,* including hepatic cell adenocaricinoma (i.e.: hepatoma), renal cell adenocarcinoma (i.e.: hypernephroma, Grawitz tumor), ovarian adenocarcinoma, choriocarcinoma (esp. gonadotropic hormone secretors).

At times, ovarian adenocarcinoma will shed cells having a prominent, peculiarly punched-out clearing of the parachromatin, symmetrically placed about an extremely cherry-red nucleolus. This perinuclear halo, first described and associated with ovarian adenocarcinoma by W. Howdon [145], increases the nucleolar prominence and, while not exclusively ovarian, mandates consideration to rule this in or out as the site of origin.

C. Cytoplasmic Features in Atypical Columnar Differentiation

1. Shape and Orientation

There is definite columnar shape and orientation in most cells with atypical columnar differentiation, and discriminable cells shed from adenocarcinoma. On profile viewing, the polarity is that of columnar maturation, with scanty cytoplasm at one end forming, perhaps, a tail; then lies the nucleus, with the major amount of cytoplasm at the other end, presenting to the luminal surface (Fig. XV-2).

The cell membranes and intercellular borders are frequently hazy and indistinct (Figs. XV-1, XV-2). This is in sharp contrast to the crisp and well-defined borders of keratinizing squamous cell carcinoma (Figs. XII-12, XII-15). If the cell borders are viewed through a long expanse, however, as in *en face* presentation (Fig. XIV-2), they can appear prominent.

2. Secretion

Most of the atypical morphologic changes of functional differentiation which appear in the cytoplasm are related to cellular secretion. These are related to any of the three major phases of the secretory process: the *production* of secretory material, intracellular *storage* of secretion and extracellular *release*. Any of the phases or all can be present.

a. Cytoplasmic Basophilia. The cytoplasm of cells which are actively *producing* secretory products becomes highly basophilic (Fig. XV-2). This is associated with increased numbers of ribosomes in the cytoplasm, especially those of the granular endoplasmic reticulum [76]. This basophilia of the RNA-associated protein of the mature cytoplasmic ribosomes is in striking contrast to the acidophilia of the RNA-associated protein of the maturing ribosomes of the nucleolus, both representing increased protein production. This increased RNA with its basophilia, is easily demonstrated as the 'flaming' red in the cytoplasm of cells stained with acridine orange and activated by deep blue or ultraviolet light in cellular spreads [180, 283] and in cell flow-through automated systems [64, 110, 282, 284].

The cytoplasmic basophilia at times is biphasic, being deeper in the perinuclear and central areas and paler in the periphery of the cells (Figs. XV-2 through XV-4). This is frequently noted in single cells shed from adenocarcinoma; most particularly, the single cells of 'early' endometrial adenocarcinoma, which are so easily confused with 'active' histiocytes. When this biphasic cytoplasmic basophilia is present, it helps somewhat to differentiate such single adenocarcinoma cells from macrophages; but the latter, at times, can show the biphasic tendency also.

This biphasic cytoplasm of a secretory cell is to be distinguished from the 'ecto-endoplasm' of a cell with atypical cytoplasmic keratinization of stratified squamous differentiation (Figs. XII-16, XII-17). The line of demarcation in the former secretory biphasic cytoplasm, between the central basophilic cytoplasm and the peripheral lighter cytoplasm (Figs. XV-2 through XV-4), is not sharp like it is at the ecto-endoplasmic border of atypical keratinization (Figs. XII-16, XII-17). Furthermore, the texture of the cytoplasm of the former appears foamy or vacuolar in *both* of the cytoplasmic phases with, in fact, this texture carried over from one to the other across an indistinct, irregular boundary; conversely, the foamy textured endoplasm in an atypically keratinizing cell, contrasts sharply with its hyaline ectoplasm (Figs. XII-16, XII-17) and the two are separated by a distinct, sharply defined ecto-endoplasmic border.

Extreme cytoplasmic basophilia or secretory production, at times is restricted to the center of the cell, and adjacent to the area of nuclear indentation. Electron microscopy reveals that rough endoplasmic reticulum aggregates in this region around smooth endoplasmic reticulum with the Golgi complex, 'packaging' secretory material for secretion (Fig. V-3). In routine light microscopy, the center of this basophilic area may become 'cleared' where the Golgi complex and smooth endoplasmic reticulum concentrate (Fig. V-2).

b. Cytoplasmic Granules. Frequently, granules appear in the cytoplasm representing those products of secretion which are visible to routine processing and light microscopy. They usually make their first appearance in the region of the center of the cell, frequently about the periphery of the cleared area. They then move toward the luminal surface, at times being diffusely spread throughout the cytoplasm.

The granules can stain either acidophilic, basophilic, or amphophilic (i.e.: enzymes of pancreatic adenocarcinoma). At times they have their own pigment color, such as the golden-brown to green color of bile (Figs. XV-3 through XV-5). The latter usually can be differentiated from hemosiderin, by its greater loss of gleam in lowering the microscopic condenser out of good Koehler illumination (Table IV-1, and text). It may be difficult to differentiate bile (Figs. XV-3, XV-4) from melanin (Figs. XVII-1, XVII-3; XVIII-4), however, without cytochemical reactions (e.g.: Stein's reaction); however, melanoma tends to shed as single cells (Figs. XVII-1, XVII-2), while bile producing hepatocellular adenocarcinoma (Figs. XV-3, XV-4) sheds mainly in diagnostic true tissue fragments (DTTF, Table VI-1), as do most adenocarcinomas.

c. Inspissated Secretory Material. At times, large secretory products dehydrate and shrink, appearing to be an amorphous aggregation of inspissated proteinaceous material. These occur anywhere from the area of the cytocentrum to the lumen, and are an atypical presentation of secretory material. It frequently appears hard and deeply chromatophilic, unmold-

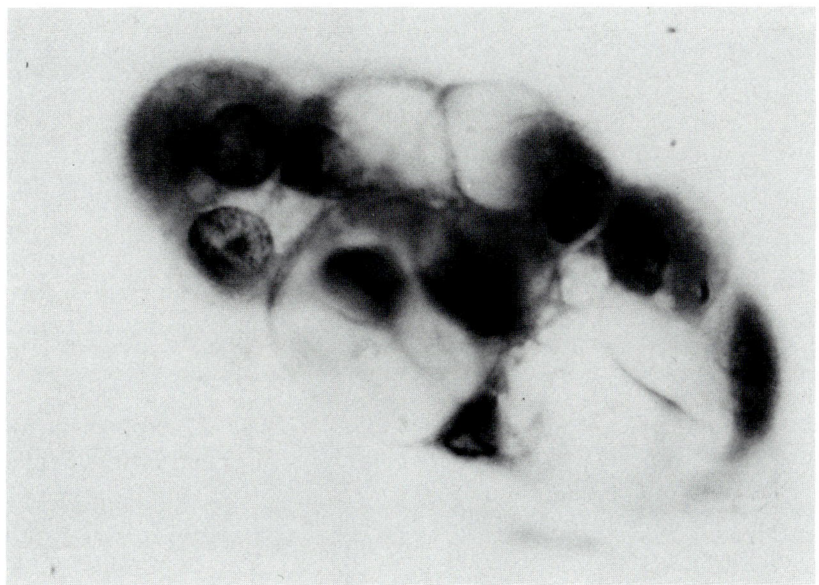

Fig. XV-2: Secretion in adenocarcinoma, ovary. Fast (vagino-pancervical) smear.
First, this is *cancer* by cellular malignant criteria, mainly nuclear. *Second,* regarding functional differentiation, the nucleoli are prominent, indicating active manufacture of ribosomes preparatory for *protein* production, and the cytoplasm is markedly chromatophilic (blue-green), indicating a high ribosome content of the cytoplasm, also for *protein* production. While both of these indicate increased protein production, *neither* of them discriminates protein production for *secretion* from that of proteins for other uses, such as for increased cell growth which can be found in any type of cancer, and thus do not help to positively identify functional differentiation. Nuclei are eccentric in the cells, even when there are no large vacuoles pushing them there (upper left cell), indicating columnar differentiation and suggesting columnar cancer (i.e.: adenocarcinoma). There is a peculiar clearing of the outer portion of the cytoplasm in the cells at 2-o'clock and 10-o'clock, suggesting adenocarcinoma. The best evidence of secretion, however, is the large, thin-walled, hyperdistended, bubble gum, secretory type of single vacuoles at 5-o'clock and 6-o'clock. Their huge size and their *pouching out* of this papillary-like tissue fragment, further suggest ovary as a primary. This suggestion of an ovarian site is strengthened by a single, cherry-red nucleolus in each of the nuclei, and the characteristically 'punched-out' clearing about the one on the left.
Papanicolaou stain.
XV-2: × 1,200.

able, at times irregular, and may well be more difficult for the cell to secrete than when the product is in a more normal watery phase. It may completely fill the lumen of an acinus (Fig. XV-8).

Apparent inspissation is most frequently present in the mucoid-producing adenocarcinomas. While it is usually found in the company of cells containing large, obviously secretory vacuoles, it can be the only real evidence of a secretory cell.

d. Foamy Cytoplasm. A markedly foamy cytoplasm is frequently a hallmark of a secretory cell. This may be difficult at times to distinguish, however, from a phagocyte or from a cell undergoing degeneration (Fig. XV-6).

The foamy cytoplasm of a secretory cell tends to be mainly on the luminal side of an eccentric nucleus (Fig. XV-2), as also occurs with phagocytosed vacuoles of a macrophage (Fig. IV-1). In contradistinction, small vacuoles of cell degeneration usually occur anywhere within the cytoplasm, leaving the nucleus to lie where it was before degeneration, or to be shoved without meaningful direction (Figs. VII-1, VII-7). This feature of nuclear eccentricity with secretion can be of little value, however, with foamy cytoplasm of secretion at times appearing to disperse all about the nucleus (Fig. XV-6) (e.g.: epithelial 'foam' cell of nipple secretions) and may even extend into the tail (Figs. XIV-2, XIV-3).

The cytoplasm of a secretory cell does not contain phagocytosed debris and degenerating nuclei, as does that of a macrophage. However, secretory

---

Figures XV-3 through XV-5: Hepatoma (hepatocellular adenocarcinoma) with bile production, metastatic to lung.

Figs. XV-3, XV-4: Sputum. Golden-brown granules of bile (arrows) are present in foamy cytoplasm which contains many small vacuoles. The nucleoli are prominent in rounded, vesicular nuclei. Heavily basophilic cytoplasm is slightly biphasic with a thin pale periphery. The cells have exfoliated in a diagnostic true tissue fragment. In Figure XV-3, the cell on the left demonstrates a peculiar close relationship and uniform distance between the cell membrane and nuclear membrane, maintained uniformly between 6- and 11-o'clock, in spite of fairly abundant cytoplasm. As this does not appear to be due to secretion and is more extreme and uniform than simply columnar orientation, it is a criterion favoring malignancy.

Fig. XV-5: Tissue section of needle biopsy of liver. Hepatoma.

XV-3, XV-4: Papanicolaou stain; XV-5: Hematoxylin and eosin stain.

XV-3, XV-4: × 2,000; XV-5: × 500.

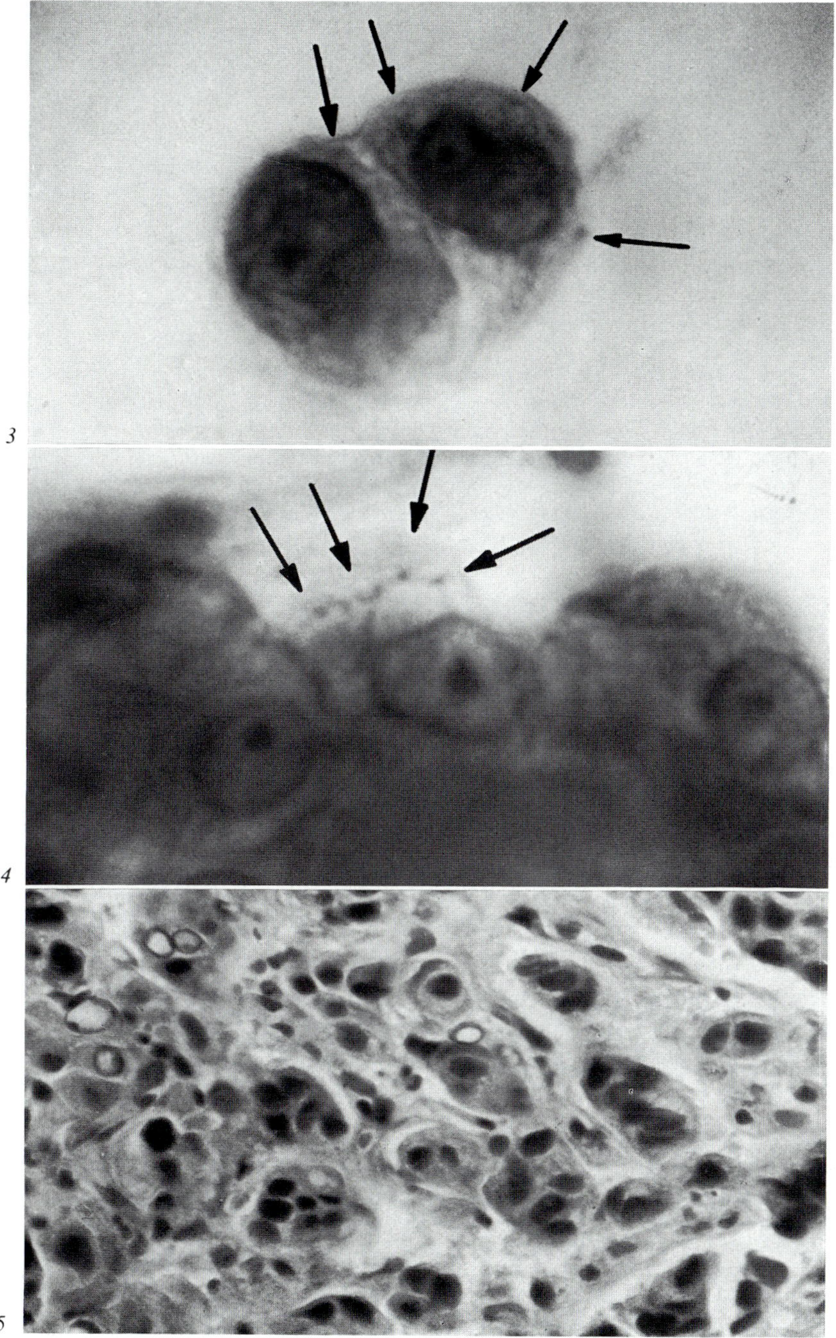

3

4

5

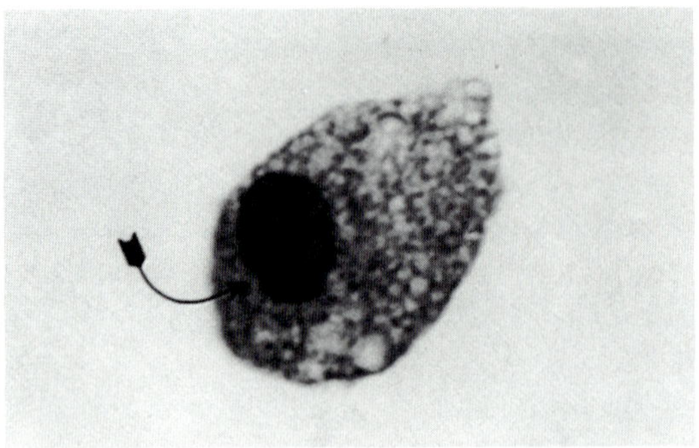

Fig. XV-6: Foamy, multiple vacuoles in histiocyte vs. adenocarcinoma, endometrium.
Fast (vagino-pancervical) smear. The cytoplasm is extremely foamy, with small, multiple
vacuoles extending throughout the entire cytoplasm. This is not a good criterion for secretion, as it could be found in either secretion, degeneration, or phagocytosis. There is no
evidence of phagocytosed tingible particles in the foamy cytoplasm, which speaks against a
macrophage. The nucleus is 'kidney bean' shaped and eccentric, as can be present in either
a secretory cell or a macrophage. If the cell is a macrophage, then nuclear indentation is
expected toward the center of the cell, at the hof (Fig. IV-1). In this cell, however, the
irregularity of the nuclear membrane occurs *opposite* from the cell center, in the 'anti-hof'
region (arrow), a valuable malignant criterion (see Fig. IX-12). 'Early' adenocarcinoma of
the endometrium.
Papanicolaou stain.
XV-6: × 1,150.

cell cytoplasm is attractive to neutrophils, which frequently move inside of
the cell membrane and phagocytose the secretory cell from within – endo-
phagocytosis. As a result, secretory cells frequently contain many *well-preserved* neutrophils (Fig. XV-1, XV-8), in contradistinction to the partially
digested neutrophils in a macrophage (Fig. IV-3). Degenerating cells can
also have well-preserved neutrophils in their cytoplasm.

A cell with foamy cytoplasm, therefore, is suggestive of, but is not a
dependable sign of secretion.

e. Cytoplasmic Vacuoles. Many times these small foamy secretory
structures within the cytoplasm coalesce (Fig. XIV-5), or otherwise become
large enough to be recognized as secretory type vacuoles (Fig. XV-7). Their
association with secretion then becomes more obvious. They are more
numerous on the luminal side of the cell, but can occur throughout. When
only a few secretory vacuoles are present, they usually are restricted to the
immediate paranuclear cytocentral area. As the vacuoles coalesce (Fig. XV-

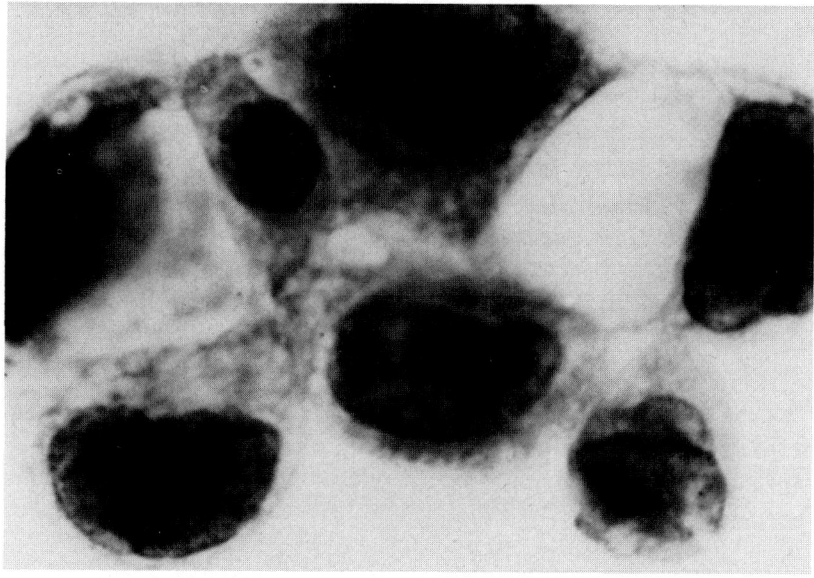

Fig. XV-7: Foamy and singly vacuolated adenocarcinoma cells. Sputum. This diagnostic true tissue fragment (DTTF) of cancer cells shed from a primary adenocarcinoma of the lung, has one ciliated cell (11-o'clock). Does this signify a benign tissue fragment? Note that the nuclear structure of the ciliated cell is benign and is in striking contrast to that of the other six nuclei, which reveals it as a benign cell stuck in among malignant cells. Many of the multiple vacuoles of the malignant cells are coalescing, with the production of single hyperdistended, secretory-type vacuoles, which are much better criteria of secretion than the foamy cytoplasm.

Papanicolaou stain.

XV-7: × 1,920.

7) they form larger but fewer vacuoles, that are more easily recognized as being secretory in nature (Fig. XV-2).

Multiple cytoplasmic vacuoles as with the smaller 'foamy' structures, are also not clearly discriminable from phagocytosis or degeneration. While *phagocytic* cells occasionally have large vacuoles, they usually are uniform in size and contain degenerating products of phagocytosis. The nucleus is not pushed to the side, as with a secretory cell with large vacuoles, but tends to occupy the moderately eccentric position of a macrophage.

Degenerating cells can also contain multiple vacuoles which tend to be fairly evenly dispersed throughout the cytoplasm. They spread around the nucleus without appearing to displace it.

Single and *hyperdistended,* secretory-type vacuoles (Figs. XV-2, XV-13) are much more significant indicators of secretion. As the vacuoles become larger and more filled with abnormal, apparently hard-to-get-rid-of secretory material, the cell wall is markedly distended and thinned. The huge vacuole tends to pouch out past the usual dimensions of the cell, carrying the cellular border out as a thin rim about the vacuole.

The degree of *hyperdistension* and of the intravacuolar pressure, is accurately displayed by the decreasing amount of cytoplasm present in the nucleo-vacuolar cytoplasmic triangle (Figs. XV-2, XV-13). In this triangle between the nuclear membrane, the cell membrane, and the wall of the highly distended vacuole, the cytoplasm becomes almost nonexistent. The nucleus can be drawn down to a fine point in this area. The vacuolar wall, over the rest of the distended vacuole, is pressed against the cell membrane so that they appear to be one under light microscopy, as do the nuclear membrane and the cell membrane along the basilar aspect of the nucleus.

Such hyperdistended, single vacuoles are extremely good evidence of cellular secretion when found in the usual mucoid specimens. In materials which are watery, however (e.g.: body cavity fluids, cerebrospinal fluids, cyst contents, tissue culture fluids), vacuoles of cell degeneration can be single and markedly hyperdistended (Fig. VII-7). Other than in these very watery specimens, however, single hyperdistended secretory-type vacuoles are extremely good evidence of cell secretion.

f. Cellular Stickiness. The presence of extremely sticky and viscous secretory material is evidenced by great adherence of cells to each other (Figs. XV-2, XV-7). In adenomatosis (bronchiolo-alveolar type of adeno-carcinoma of the lung) and in some very mucoid adenocarcinomas (e.g.: some colonic, gastric) the cells do not fall away from each other easily, but maintain themselves as tight, cohesive groups, or as long streams of adhering cells.

g. Intracellular Phagocytic Leukocytes. The cytoplasm of certain secretory cells contains well-preserved polymorphonuclear leukocytes. These have invaded the host cell as phagocytes (microphages) and are digesting the cell from within (Figs. XV-1, XV-8).

Frequently, but not invariably a vacuole forms about the leukocyte, either individually or in groups. It appears that the leukocytes are attracted to the interior of secretory cells and of cells undergoing degeneration. At times, they go into secretory vacuoles; at other times, they feast off of the general cytoplasm.

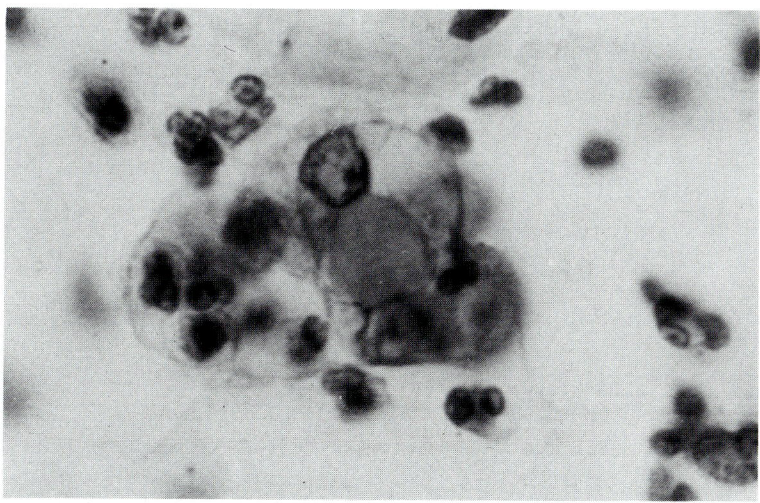

Fig. XV-8: Well-preserved neutrophils in the cytoplasm of acinus-forming adenocarcinoma, endometrium. Fast (vagino-pancervical) smear. Note that the neutrophils within the cancer cell are as well preserved as those outside, evidence that they are the phagocytes (microphages) and the cancer cell is being eaten from within. This is found in secretory cells principally, but also in degenerating cells of any type. These tumor cell nuclei are well preserved, favoring secretion. There is a large mass of inspissated secretion filling the lumen of an acinus which is formed by this tissue fragment of cancer cells.
Papanicolaou stain.
XV-8: × 960.

In this process, the polymorphonuclear cells appear *healthy* (Figs. XV-1, XV-8). This is in striking *contrast* to the fragments of *degenerating* leukocytes and nuclear debris found in macrophages (Fig. XV-3), and is valuable evidence in favor of the host cell being secretory.

Especially is this true if there is no overt evidence of cell degeneration. Neutrophils do go into degenerating cells of all types, so that their presence usually signifies a certain amount of cellular degeneration. If, however, the host cell shows no overt evidence of degeneration and the intracellular leukocytes are healthy, this speaks strongly in favor of a secretory-type cell.

3. Luminal Structures

Cancer cells virtually do not produce abundant, normal, well-formed cilia (Figs. V-3; XIV-4 through XIV-6). Occasionally, under electron microscopy, a few cilia occur in obviously cancerous cells. However, the large numbers of uniform cilia per cell found in normal ciliated columnar cells virtually are not observed in cancer cells.

Most of the so-called cilia, recognized on the luminal surfaces of adenocarcinoma by light microscopy in either tissue or cell spreads, are actually mucous strands or, at times, fibrin mistaken for cilia. They do not fulfil the strict criteria of discriminable cilia, however (Chap. XIV, C, 4). Two adenocarcinomas of the ovary are the only cancers to have been described in the literature as shedding malignant cells with diagnostic cilia [67, 130]. None have been described in the lung, and the careful identification of cilia is a most valuable criterion for *that* cell; however, one must be sure that all of the cells of the tissue fragment are of *that* type, and are not different from it and, thus, possibly malignant (Fig. XV-7).

Thus, the finding of discriminable cilia on the luminal end of the cell (Figs. IV-12, XIV-5, XV-7) which are of *uniform length, numerous,* of *uniform periodicity,* and with a prominent *terminal plate* (rootlets) virtually assures that the cell is *not* from cancer.

D. Tissue Fragments in Atypical Columnar Differentiation

Cells from nonmalignant epithelium with atypical columnar differentiation, as well as cells from adenocarcinoma, generally shed in clumps or tissue fragments. This is in contradistinction to the usual individual cell shedding from atypical stratified squamous epithelium and from squamous cell carcinoma.

1. Glandular Acini and Tubules

Acini shed from adenocarcinoma and from atypical columnar epithelium yield the three-dimensional 'tennis-ball' arrangement of glands in columnar differentiating epithelium (Chap. XIV).

Three distinct planes of focus with all layers between, can be delineated by carefully focusing up and down at magnification sufficiently high for adequate resolution. An *en face* sheet of cells lies on the glass slide. As the plane of focus moves up, the tissue unfolds into an equatorial rim or ring of cells formed about a lumen (Fig. XV-8). With further upward shifting of the focus, this ring moves into a second *en face* sheet of cells covering the top of the acinus.

Some acini shedding from cancer have enough cytoplasm in their cells at the equatorial ring to orient the nuclei outward toward the basement membrane, and the larger amount of cytoplasm inward toward the lumen. Here the central luminal end of the cells, with their continuous 'community' cell border and terminal bars (Fig. XIV-1), identify a true epithelial luminal border (Fig. XV-8).

The duct of a gland is identified as an elongated, open-ended tubule. The lower sheet of cells lying *en face* on the slide, gives way by focusing upward to a tube of cells in profile with their luminal ends facing inward, and then to a top sheet of cells *en face.*

2. Polyps, Papillary Fronds, and Stalks

Two *en face* sheets of cells are again present, the lower on the glass slide and the upper on the top of the structure. In the middle equatorial ring, however, orientation of the basement membrane and lumen is reversed if there is adequate cytoplasm. The luminal ends of the cells, with their greater amounts of cytoplasm and their 'community' luminal border are located facing outward about the periphery of the structure (Fig. XV-10). The basal ends of the cells with their nuclei are identified toward the center of the structure. At times stroma can be discerned lying in the center, inside of the basement membranes.

Papillary fronds frequently occur as solid fingers of tumor cells in a branching, fern-like configuration (Figs. XV-9, XV-10). The outer surface of these fronds may present a 'community' luminal border, signifying it as an epithelial surface; but the basement membrane and connective tissue stroma may not be discernible.

Thin interglandular villi at times appear as papillary fronds, both histologically and cytologically. Conversely, the interpolypoid clefts of a papillary structure may be morphologically indistinguishable from acini and tubules.

Stalks of polyps or papillary fronds of carcinoma are shed as elongated tubules of cancer. The lower and upper *en face* sheets appear similar to ducts, but at the midplane the orientation is reversed. The lumen is on the outside of the structure, while the basement membrane and connective tissue stroma form a central core (Fig. XV-9).

3. Psammoma Bodies

Round psammoma (*psammos:* Gr. = sand) bodies occur in some papillary adenocarcinomas. The psammomas appear *no* differently in benign conditions than they do in cancer. The diagnosis of malignancy lies exclusively in the morphology of the cells (Chap. IX).

Chiefly present in ovarian papillary adenocarcinoma, they at times occur in adenocarcinomas from other sites (e.g.: endometrium, tubal, thyroid, pancreas, stomach). Similar-appearing structures are found in other sites (e.g.: meningioma, mesothelioma, prostate concretions, pulmonary microlithiasis alveolaris, ovarian clefts and inclusion cysts, brain 'sand' from degeneration) and in extremely small forms occurring intercellularly (e.g.: Michaelis-Gutman bodies of malakoplakia). These all have in common a three-dimensional structure of concentric shells of proteinaceous mineral-bearing material (Figs. XV-11, XV-12).

Psammoma bodies of *papillary adenocarcinoma* are found as a part of the cancer, embedded well within the tissue fragments (Figs. XV-11, XV-12). If there is only a near association with the tumor, it is not sufficient to label it a psammomatous-bearing tumor. If it is a *part of the cancer,* however, that is good evidence of the cancer being a papillary adenocarcinoma.

The presence of psammoma bodies is *not* a criterion of malignancy. They are found in cul-de-sac aspirations and in vaginal pool material in benign conditions.

Psammoma bodies are golden-brown, lavender, or amphophilic with darker concentric shells. These features discriminate them from somewhat similar-appearing artifacts (i.e.: talcum powder, spores, pollens, diatoms). Discriminable psammoma bodies of adenocarcinoma, however, have a distinctive 'tree-ring' lamellation in two dimensions (Fig. XV-12); but they are *three-dimensional spheres,* with concentric shells resembling those of a 'jaw-breaker' (British: 'gob-stopper'), which are easily defined for identification by carefully focusing up and down at high magnification.

To obviate misidentification, two major identifying features should be present before identifying psammoma-producing papillary adenocarcinoma: their morphologic appearance (e.g.: concentrically lamellated shells) *and* their definite incorporation *within* diagnostic true tissue fragments (DTTF) of papillary *cancer.*

E. Morphologic Hallmarks of Biologic Behavior

The correlation of morphology with grade and biologic behavior of adenocarcinoma, resembles that of squamous cell carcinoma. In general, the more advanced is the grade of adenocarcinoma, the more striking are

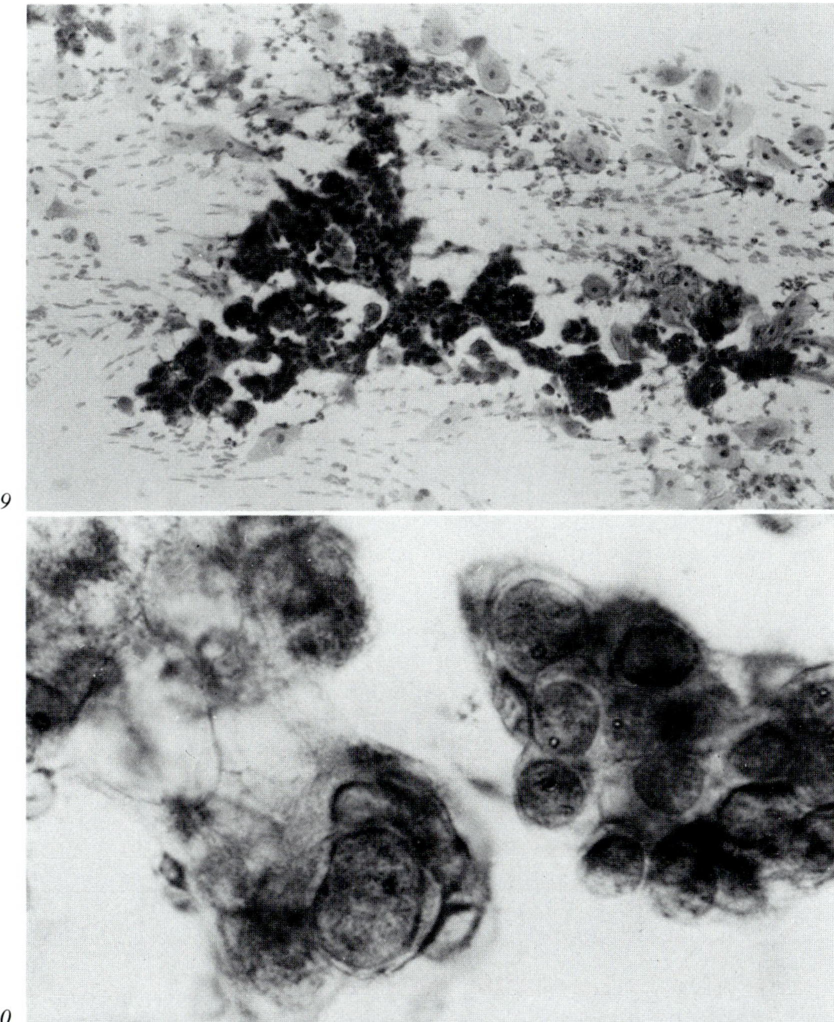

Figures XV-9, XV-10: Papillary adenocarcinoma. Fast (vagino-pancervical) smear.

Fig. XV-9: A papillary frond with many high order branches composed of cancer cells, indicates a papillary adenocarcinoma.

Fig. XV-10: The midplane reveals a luminal surface on the outside of the diagnostic true tissue fragments (DTTF, Table VI-1), substantiating the papillary nature of the structure. Many malignant criteria lead to a diagnosis of cancer. These malignant cells wrap around each other in a pseudopearl configuration not to be confused with a true pearl arrangement of squamous differentiation.

Papanicolaou stain.

XV-9: × 108; XV-10: × 1,620.

the malignant criteria (Chap. IX) and the less are the functional differentiating features.

Early adenocarcinoma of the endometrium (*in situ* and Grade I) usually sheds many single cells with a moderate to coarsely granular chromatin, moderate hyperchromasia, and occasional small nucleoli. There is usually abundant cytoplasm which is often biphasic [38, 94, 99, 228]. The nucleus is eccentric, and the cell closely mimics an 'active' histiocyte.

As the grade increases these features generally change, so that the advanced endometrial adenocarcinomas (Grade III or IV) have more of the classical criteria of invasive malignant neoplasia (Chap. IX). There is chromatin clumping and angularity, larger areas of abnormally cleared parachromatin, increased hyperchromasia, more irregular and pointed macronucleoli, and larger nuclei.

Bronchial carcinoids, even when metastatic to a lymph node, tend to have a uniformly dispersed coarsely granular chromatin, and abundant columnar cytoplasm. Most adenocarcinomas of the lung, which are encountered in present clinical practice, are advanced in grade analogous to endometrial Grades III and IV. The few 'early' types encountered, however, have nuclear morphology similar to the 'early' endometrial cancers.

Cells shed from bronchiolo-alveolar adenocarcinoma (adenomatosis) which have vesicular nuclei with granular chromatin and abundant cytoplasm, are associated with tumors whose cells line the alveoli, do not

---

Figures XV-11 through XV-13: Psammoma bodies in a papillary serous cystadenocarcinoma. Ovary.

Fig. XV-11: Peritoneal fluid. Papillary fragments of adenocarcinoma with psammoma bodies. Note the smaller psammoma bodies on the right, deeply embedded in the fragment of cancer tissue.

Fig. XV-12: Greater detail of the delicate arrangement within the concentric shells of the psammoma body, with cancer cells closely applied all about it.

Fig. XV-13: Hyperdistended secretory cells, each with a single pouched-out vacuole. In the right upper cell, the scanty cytoplasm in the nucleo-vacuolar triangle and about the vacuole, along with the nuclear molding, are all evidence of great pressure within this hyperdistended secretory-type vacuole.

Papanicolaou stain.

XV-11, XV-12: Fast (vagino-pancervical) smear; XV-13: Peritoneal fluid.

XV-11: × 320; XV-12: × 800; XV-13: × 1,320.

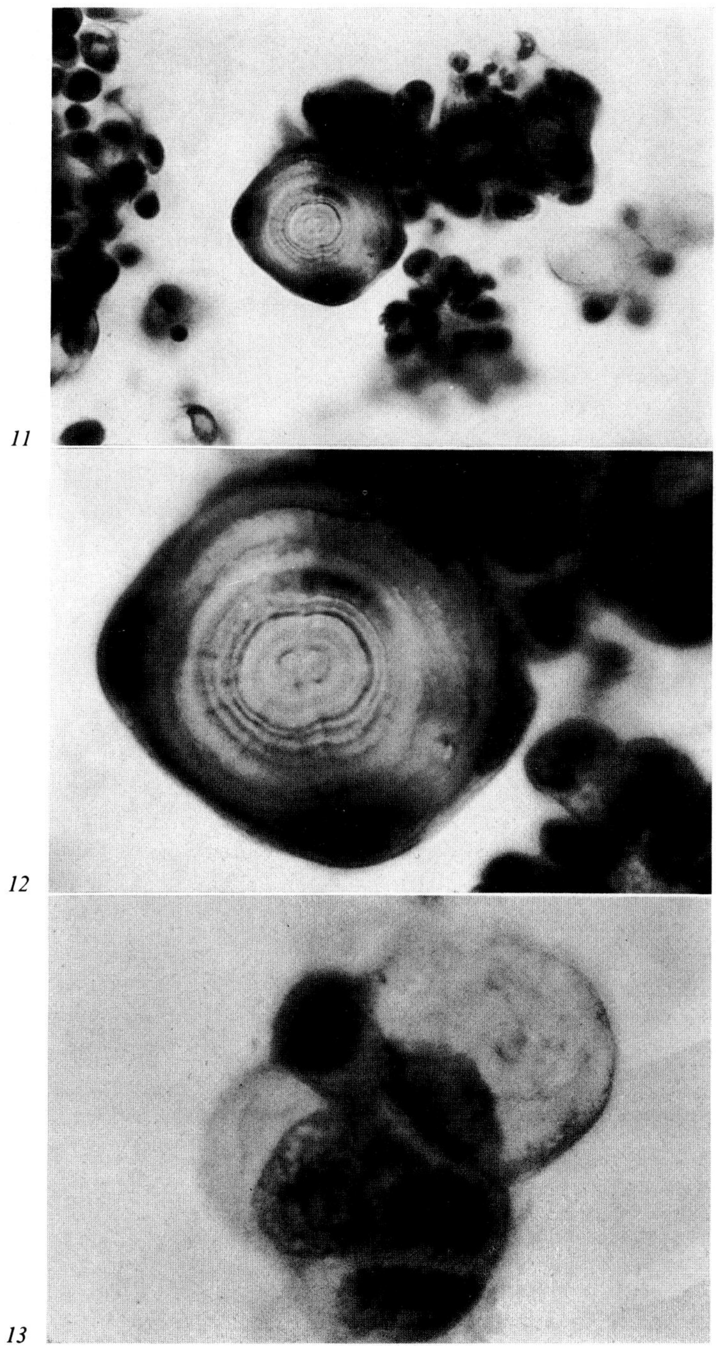

11

12

13

destroy pulmonary structure, and permit long years (10–30) of patient survival. On the other hand, when bronchiolo-alveolar adenocarcinomas shed cells with nuclei having sharp, irregular chromatin aggregates, angular macronucleoli, cleared parachromatin, irregular nuclear membranes and scanty cytoplasm, the tumor tends to have foci of tissue destruction and invasion, areas of solid tumor formation, and a shorter patient survival, approximating that of the more classical bronchogenic adenocarcinoma.

XVI. Functional Differentiation during Squamous Metaplasia: Transformation from Columnar Epithelial Functional Differentiation to Keratinizing Stratified Squamous Epithelial Functional Differentiation

A. General Considerations

Metaplasia (*meta-*: Gr. = change; *-plasis:* Gr. = molding, development, formation) denotes a change in the development of a tissue *from* producing one type of functional differentiation *to* producing another type of functional differentiation.

In health and disease, numerous types of metaplasia are encountered (e.g.: connective tissue *to* cartilage *to* bone, urothelium *to* keratinizing stratified squamous epithelium, gastric epithelium *to* intestinal epithelium) with many factors involved (e.g.: endocrine, nutritional, chronic irritation, hereditary).

The term *metaplasia,* however, is most frequently employed to designate that change in functional differentiation *from columnar epithelium to keratinizing stratified squamous epithelium,* usually as a result of longstanding chronic irritation from untoward stimuli. This *process* of *squamous metaplasia* will be used herein as a model for detailed consideration. This will allow development of basic cellular concepts involved in the overall *processes* of *metaplasia.*

Assessment of *biologic behavior* in this spectrum of metaplastic development, is greatly aided by evaluating *morphologic patterns*, particularly of the nucleus. Aspects of these patterns are derived from evaluating the cytoplasm to determine the *type* of differentiation, the *degree* of maturation attained, its *atypicality,* and the *severity* and *quality* of general activity.

1. Type of Functional Differentiation

Morphologically, the type of differentiation which a cell or its tissue undergoes to perform its function is evidenced mainly by *cytoplasmic* changes, with minor information provided from its nuclear structure. In the model of metaplasia under consideration, two major types of differentiation are encountered in the most mature surface cells: those which mature as *columnar* cells and orient *perpendicular* to the luminal surface and basement membrane (Chap. XIV); and those which mature as *squamous* cells and orient *parallel* to the latter structures (Chap. XI).

2. Degree of Maturation

The degree to which a cell or its tissue has matured, toward being able to fully perform its functions, is evaluated mainly from cytoplasmic morphology. It is generally a rule concerning a cell and its parent lesion that, as its *proplasia* and *neoplastic potential increase*, its *immaturity* also *increases* (i.e.: its ability to mature, decreases).

3. Atypicality of Functional Differentiation

As maturation changes from columnar to squamous, varying degrees of atypicality appear, first in the columnar (Chap. XV) and then in the squamous (Chap. XII). Another general rule is that the *severity* of the untoward *stimulus* is directly proportionate to the *severity* of the *atypical* forms of maturation.

4. Severity and Quality of General Activity

The more the epithelium reacts to the situation which is bringing forth the metaplastic process, the more proplastic the cells become. The severity and quality of increased general activity in developing neoplasia, are reflected mainly in the degree of *proplasia,* by nuclear morphology, but also in cytoplasmic *immaturity.*

Marked proplastic changes (Chap. VIII) can thus bespeak abnormal and alarming general activity. Injury and degeneration are reflected in retroplastic changes (Chap. VII). Retroplasia and proplasia frequently occur *together* in squamous metaplasia (Figs. II-9, II-10, II-12).

When the stimulation decreases, the nuclei show decrease in their proplastic changes and, eventually, again become euplastic. Thus, nuclei of euplasia (Chap. VI) occurring within the metaplastic spectrum, attest to the waning of the stimuli and 'settling-down' of the epithelium, while retaining cytoplasmic changes for a longer period of time.

Conversely, when *proplasia* is marked with very *little* general inflammatory reaction or degeneration within the epithelium, it is disturbing. This paradoxical finding increases the possibility that initiation has occurred, and that neoplasia is present.

B. Acute Injury and Repair

When euplastic epithelium (Figs. II-1 through II-3) is subjected to a harmful agent sufficient to produce injury and degeneration (Fig. II-4), it enters a retrogressive phase which can lead to death (Figs. II-5, II-6). If the cells do not die, but recover, they pass from retroplasia through proplasia during repair (e.g.: squamous inflammatory atypia, columnar inflammatory atypia), and then return to euplasia upon healing (Fig. II-8).

In *columnar* epithelium, the nuclear changes yield most valuable morphologic evidence of this activity, while the cytoplasm retains its columnar differentiation, throughout, with varying degrees of maturity (Fig. VIII-8). During injury, degeneration is evidenced mainly in nuclear changes of retroplasia (e.g.: parachromatin clearing, chromatin clumping, chromatin/parachromatin interface blurring); but also in features of its cytoplasm (Chap. VII).

During repair, the exfoliated proplastic columnar cell has prominent eosinophilic nucleoli (Chap. VIII), while its cytoplasm becomes basophilic, both from the increased ribosomal RNA and associated proteins. Eventually, with recovery, the cell returns to health and settles down into a euplastic level of biologic activity (Chap. VI).

In *keratinizing stratified squamous* epithelium, the changes of acute inflammatory atypia are much *less* severe than in columnar epithelium for the *same* type and degree of injury. When the injury does become more *extreme,* however, *columnar* epithelium dies and ulcerates, or undergoes squamous metaplasia (below); while stratified *squamous* epithelium undergoes marked retroplastic and proplastic changes to squamous atypia. If the injury is sufficiently severe, however, even squamous epithelium will eventually die and ulcerate.

From the same stimulus, multinucleation occurs less often in squamous atypia than in columnar atypia, and to a lesser degree. Furthermore, nucleoli are absent or not prominent in squamous atypia; conversely, they are very prominent in even slight proplasia of columnar epithelium, and of nearly every cell type other than squamous (e.g.: myocytes, fibrocytes, macrophages, urothelial cells).

In rare exceptions to this rule, *extreme* reparative reactions may produce nucleoli in squamous atypia. However, even though the nucleoli can be large and hideously irregular [114, 207], their corners and projections are *rounded* and there are varying degrees of degenerative *blurriness* of the nucleolar/parachromatin interfaces, as well as the general chromatin/parachromatin interfaces (Chap. VII). If one carefully examines these nucleoli, they are thus different from the thinly spiculated and sharply angled diagnostic nucleoli at times found in well-preserved cancer (Chap. IX).

C. Chronic Injury and Repair: The Process of Squamous Metaplasia

1. Presquamous Squamous Metaplasia: Columnar Differentiation Phase

If a second injury occurs during the active healing phase of regenerating epithelium, more injury and then delayed regeneration result (Fig. II-9), which is added to an already actively regenerating epithelium. If such injuries continue to occur to a columnar epithelium over a period of time, as in chronic irritation, varying degrees of degeneration can be superimposed upon ever-increasing cellular activity of regeneration (Figs. II-10, II-12).

The surface cells at the lumen attempt to continue to mature as columnar during this period of irritation and resulting increased activity (Fig. XVI-1). As the chronic irritation continues, there is proliferation of the less

---

Figures XVI-1 through XVI-3: Squamous metaplasia in human bronchial epithelium.

Fig. XVI-1: Presquamous squamous metaplasia: columnar differentiation with mild inflammatory atypia. Orientation of luminal cells remains perpendicular to both the lumen and the basement membrane. Thin delicate goblet cells (Fig. XIV-5) contain watery mucus, with only an occasional atypical signet-ring cell in the epithelium. Ciliated cells on the luminal surface contain *many long* cilia per cell. There is some proliferation of the subluminal or parabasal (reserve) cells with moderate nuclear features of proplasia indicating moderately increased general activity.

Fig. XVI-2: Transformational squamous metaplasia. Transformational phase of differentiation. There is great loss of orientation of the cells, with most facing in all directions. Pouched-out, hyperdistended secretory cells (signet-ring cells) with rubbery mucus are numerous and disoriented, some are even facing toward the basement membrane. Only a *few short* cilia per cell remain on the luminal surface. The nuclear membranes have the marked undulation of proplasia found in extremely increased activity (note right midfield), with granular chromatin and prominent nucleoli.

Fig. XVI-3: Stratified squamous metaplasia. Surface cells are maturely thinned squames, but they are *smaller* than normal, nonmetaplastic squames. They are completely oriented parallel to the lumen and to the basement membrane. Secretion and cilia are absent. The nuclei are less proplastic (dyskaryotic) than those of Fig. XVI-2, as this epithelium is settling down toward euplastic squamous metaplasia.

Hematoxylin and eosin stain.

XVI-1 through XVI-3: × 950.

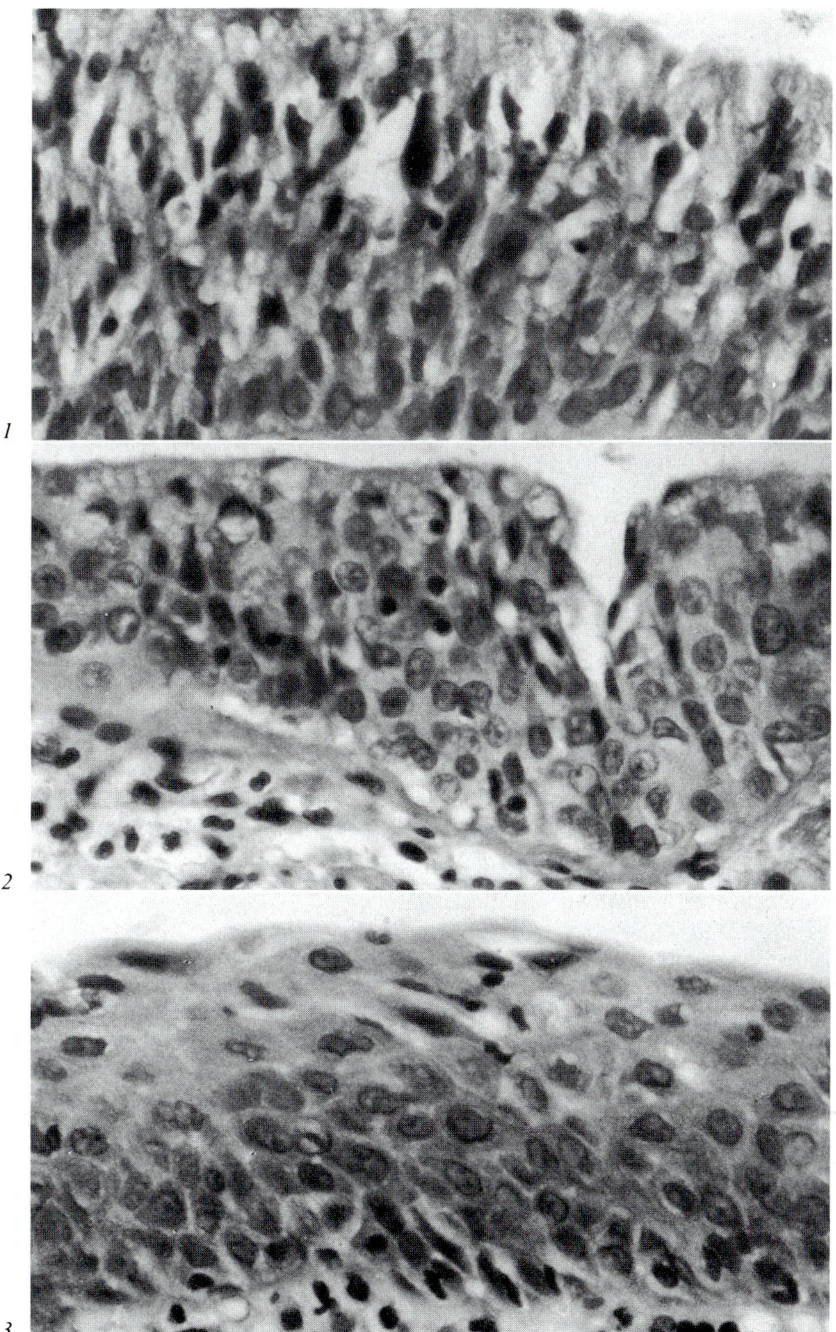

mature subluminal cells (also called: subcolumnar, subcylindrical, parabasal, basal, reserve, germinal cells).

The columnar atypias or dysplasias thus produced are designated by a babel of numerous morphologic terms (e.g.: subcolumnar cell proliferation, subcylindrical cell atypia, subluminal cell hypertrophy, basal cell hyperactivity, parabasal cell atypical hyperplasia, reserve cell hyperplasia). Nuclear features of proplasia or dyskaryosis (Chap. VIII) (Figs. VIII-3, VIII-8) are a hallmark of this regenerating epithelium, which is reacting to irritation by increasing its general activity. The *severity* of this nuclear morphologic pattern of *proplasia* and the *level* of cytoplasmic *immaturity,* directly reflect the *degree* of *increased* general activity (Fig. XVI-1).

Surface cells from these lesions shed as mature columnar dyskaryotic cells (Fig. VIII-8). Their dyskaryotic nuclei usually contain prominent nucleoli from increased protein production, in contradistinction to the virtual absence of nucleoli from nonmalignant squamous cells under increased general activity.

Cells shed from the *subluminal* layers (e.g.: subcolumnar, parabasal, reserve cells) also contain prominent nucleoli. Their cytoplasm exhibits only partial maturation, however, with a high nucleo-cytoplasmic (N/C) ratio in the cells shedding as moderately mature dyskaryotic cells or immature dyskaryotic cells (Fig. VIII-8), the latter especially in severely reacting epithelium.

With removal of the irritation, the lesion becomes quiescent and eventually returns to a nuclear morphology of euplasia. If the injurious and stimulatory situation continues, however, a high degree of increased general activity is maintained with marked nuclear proplasia (i.e.: dyskaryosis) resulting and, eventually, the process of metaplastic alteration actively progresses toward squamous metaplasia.

If the stimulation and the resulting degree of increased activity are of a *modest* grade, the tissue passes through a relatively brief transformational phase (Transformational Squamous Metaplasia, C.2., below) into a keratinizing stratified squamous metaplasia (C.3., below). If the stimulation and the degree of increased activity are marked, however, it passes into an extended phase of transformational differentiation (C.4., below).

2. Transformational Squamous Metaplasia Phase

As metaplasia progresses from columnar epithelial functional differentiation (Fig. XVI-1), the cells progressing upward from the germinal epithelium toward the lumen change or lose their characteristics of typical columnar epithelium. Thus the typical *watery* mucus of the *goblet cell* (Figs.

XIV-5, XVI-1) thickens and is changed to the *rubbery*, globoid mucus of the hyperdistended *signet-ring cell* (Fig. XVI-2). The *orientation* of the cells becomes *disturbed,* so that some are secreting sideways, while others have their nuclei on the luminal end of the cell with their secretory vacuole facing downward toward the basement membrane (Fig. XVI-2).

The *ciliated* cells reaching the lumen, which *typically* produce hundreds of long, active cilia per cell (Fig. XVI-1), now *atypically* produce shorter and fewer cilia per cell which are less active or are inactive [25], or else produce no cilia at all [147] (Fig. XV-2). Thus, there are not sufficient efficient cilia to move the mucus and its debris out of the bronchi and the trachea.

Regarding the *mucus*-producing cell, its rubbery glob of mucus is finally released from an atypical signet-ring cell and sits upon the surface like a wad of clay or of chewing gum. In this configuration and consistency it does not spread out to form the protective watery mucus blanket which is produced by the healthy goblet cell. The result is that the very effective 'mucociliary blanket', which in health covers and protects the epithelium, is lost. Also gone or greatly impaired from the unhealthy cilia and the suboptimal mucus blanket, is the 'mucociliary escalator' which, in health, effectively moves microorganisms, toxins and carcinogens up and out of the respiratory tract. These harmful substances thus gain greater access to the epithelium.

Prekeratins increase in the lower cells as metaplasia progresses, which become more mature as their cells mature and move toward the lumen. Rather than columnate, these more mature cells become globular or cuboidal (Fig. XVI-2). Eventually, as the process of metaplasia progresses, the cells begin to *thin* but, rather than being columnar with their long axis perpendicular to the lumen (as with the typical columnar cell), they squamify and shift their orientation 90 degrees, so that they are parallel to the luminal surface and to the basement membrane.

This *transformational phase* is not to be confused with the truly functionally differentiated *transitional epithelium* of the urinary tract – the *urothelium.* The latter is a very plastic, adaptable, highly specialized tissue. It can rapidly change from a squamous-type epithelium, which is stratified but nonkeratinized and covers a markedly distended structure (e.g.: urinary bladder, ureter), *to* a columnar epithelium lining a contracted, virtually emptied lumen.

3. Keratinizing Stratified Squamous Metaplasia Phase

As metaplasia progresses from its columnar presquamous phase and through the transformational phase, it moves into this keratinizing stratified squamous phase. Here the cells, which mature and lie at the surface, have no cilia and cease production of mucus. They flatten (squamify) and become thin squames which are, on the average, appreciably smaller than normal intermediate or superficial cells (Fig. XVI-3).

They orient laterally, parallel to the lumen and to the basement membrane. They stratify, produce keratin, form intercellular bridges, and develop karyopyknosis. Features of columnar epithelial functional differentiation have been replaced by those characteristic of stratified squamous epithelium.

Nuclear changes of proplasia reflect the level of continuing activity, which can range from extreme increased activity to normal baseline activity. Thus the former are varying degrees of proplastic, dysplastic, or atypical squamous metaplasias, while the latter are euplastic squamous metaplasias. The nuclear morphology, therefore, bespeaks the degree of continuing reaction.

A variety of histologic terms (e.g.: atypia, atypical metaplasia, basal cell hyperplasia, dysplasia, intraepithelial neoplasia, spinal cell atypia) have been applied to these epithelial lesions with surface maturation of stratified squamous differentiation. Some denote different phases in the process, while others connote biologic features which may be felt to be present and significant to characterize.

The greater is the degree to which the epithelium is able to mature on its surface, the greater is the percentage of mature squamous dyskaryotic cells which are shed therefrom. Conversely, the less is the ability of the epithelium to mature, the more immature are the shed dyskaryotic cells (Fig. VIII-8). The immature dyskaryotic cells and the moderately mature dyskaryotic cells which are shed from this epithelium lack prominent nucleoli, as do the mature squamous dyskaryotic cells (Chap. VIII). Even with modest diagnostic scraping and brushing of the stratified squamous epithelium, the deeper metabolic cells which have nucleoli (showing protein production) do not exfoliate, as the cells of this epithelium are held so firmly together. The prominent exception to this is the occasional extremely edematous repair epithelium which, with heavy scraping, will shed deeper cells. This is in contrast to the *frequently* present nucleoli in cells shed from squamous cell carcinoma, which can otherwise appear identical.

Atypical keratinizing stratified squamous functional differentiation (Chap. XII) is more atypical in squamous metaplasia than in euplastic keratinizing stratified squamous epithelium. However, it is less atypical than in squamous cell carcinoma functional differentiation, but of the *same* type.

a. Size. The overall cell area of squames shed from metaplasia tends to be smaller than normal intermediate cells or superficial cells. In a metaplasia which has become quiescent and euplastic – euplastic metaplasia – the nucleus may appear perfectly normal; yet, this small squame size of the cytoplasm may be the sole feature to identify that it comes from squamous metaplasia.

b. Thinning. Cytoplasmic thinning usually occurs uniformly about the nucleus, as in typical differentiation. At times, expecially in the more severe metaplasias, it may occur asymmetrically, producing cytoplasmic tails. This atypical thinning may occur unilaterally (i.e.: 'tadpole' cells) (Figs. XII-1, XII-3), or bilaterally (i.e.: 'spindle' cells) (Figs. XII-2, XII-4), as in cells from squamous cell carcinoma (Figs. XII-10 through XII-12, XII-15).

c. Keratinization. Keratinization is usually typical in metaplasia but, at times, can be extremely atypical. As previously discussed in detail (Chap. XII), atypical keratinization includes hyalinization, orangophilia, keratinization particles (Fig. XII-6), 'hard' cellular borders (Fig. XII-15), development of ectoplasm and endoplasm (Figs. XII-7, XII-12), and refractile ringing (Figs. XII-16, XII-17). In some advanced metaplasias, a distinct ecto-endoplasmic border appears fibrillary as it twists down the cytoplasmic tail (Fig. XII-12), eventually ending in a bulb. While acidophilia is associated with keratinization, it is not specific and can often be misleading. It frequently is present as an artifact of degeneration.

d. Stratification. As with typical differentiation, sheets of overlying metaplastic squames typically shed as single cells or as indiscriminate clumps of squames. At times, however, atypical small plugs of the superficial layers of stratified squamous epithelium shed together as 'pearls' or 'pearly bodies' (Figs. XII-5, XII-14, XII-15). A pair of such cells shed in tight stratification may appear as 'cannibalism', one cell having engulfed another (Fig. XII-14); in actuality, however, one lies embedded on top of the other, as a cup in a saucer, or as a baseball in a catcher's mitt. These are features of squamous *functional differentiation* and are *not* malignant criteria, no matter how atypical. When these cells are shed from squamous metaplasia, they may have severe dyskaryotic nuclear changes of proplasia, and the presence of the atypical pearl formation must never sway one towards a malignant diagnosis.

Native stratified squamous epithelium, which did not arise through metaplasia, can reach this same state of atypia under chronic injury and repair (e.g.: oral, esophageal, vaginal, ectocervical). This is referred to as *atypical* or *dysplastic* squamous epithelium. If one cannot be certain whether atypical cells have come from atypical native squamous epithelium or from atypical squamous metaplasia (e.g.: sputum with oral and lower respiratory admixture, cervical with endocervical and ectocervical admixture), a less specific term is in order, such as *squamous atypia.*

If the irritation and the resulting proplastic reaction continue and become more intense, squamous metaplasia can pass back into a transformational phase of differentiation (Phase 2., p. 250). This can be morphologically indistinguishable from transformational squamous metaplasia originating directly from columnar epithelium.

4. Atypical Exaggerated Transformational Metaplasia:
Continuing Severe Transformational Differentiation Phase

If the injury or stimulus is so great that the increased general activity (i.e.: proplasia) continues at an extremely high level, the metaplastic process remains at the transformational phase (2., p. 250). Furthermore, if it already has reached the squamous phase of metaplasia (3., p. 251) and then the general activity increases to a sufficient degree, it can revert back to the transformational phase at this exaggerated level.

The morphology can become extremely atypical. It is in this phase of exaggerated transformational metaplasia that most of the clinically borderline lesions are found (e.g.: dysplasia, atypical metaplasia, severe atypia, CIN) and the possibility of intraepithelial neoplasia is greatest. Many, frequently subtle, morphologic changes occur in this transformational phase of metaplasia, which offer significant assistance in the assessment of its true biologic nature.

Nuclear changes are of greatest value in assessing the biologic meaning of morphologic changes, and in understanding their clinical importance. Chronic irritation has both stimulative (proplastic) and degenerative (retroplastic) features. Furthermore, features of retroplasia frequently mimic or mask those of proplasia. Key to assessing the true biologic process(es) present is a careful evaluation of features of proplasia (Chap. VIII) and of retroplasia (Chap. VII), to determine roles they are playing and detect neoplasia.

For a given degree of severe proplasia, an increase in retroplastic features decreases the neoplastic significance of apparent proplastic features and, thus, the likelihood of neoplasia to be present. Conversely, with the same degree of severe proplasia, a virtual absence of manifestations of retroplasia evidences greatly in favor of a neoplastic process.

Cytoplasmic changes and atypical functional differentiation, on the other hand, are of less value in assessing biologic and clinical significance than they are in recognizing enviromental stresses playing upon the tissue. Morphologic features of *atypical columnar* functional differentiation appear similar to those frequently found in adenocarcinoma (Chap. XV). The number of *cilia* per cell is greatly decreased, as is the percentage of cells which bear cilia; or, cilia can be completely lacking from the surface of the epithelium.

The morphologic character of *secretion* alters markedly, or completely disappears. Cytoplasm appears less foamy, becoming more homogeneous and basophilic. At times, an occasional highly vacuolated cell will remain, with a single, hyperdistended vacuole containing abnormal rubbery, inspissated mucus (i.e.: 'signet-ring' cells). Special stains and reactions will at times identify secretory products (e.g.: mucus) or their precursors, especially when their forms are atypical in routine diagnostic stains.

The *orientation* of cells is variable (Fig. XVI-2) and, in the extremely exaggerated transformational differentiation, can become completely disrupted. While some cells retain normal polarity, this state of orientation becomes less frequent with increasing severity of the reaction.

Cells can become completely inverted with nuclei at their luminal end and secretory vacuoles facing toward the basement membrane. Others lie in varying degrees of disorientation between these two extremes, with squames appearing which orient parallel to the luminal surface and are interspersed between the atypical cuboidal cells. These squamous cells are smaller than normal intermediate or superficial squamous cells, and are recognizable hallmarks of the squamous metaplastic process, even of this transformational phase.

Features of *atypical stratified squamous* differentiation appear, which are similar to those occurring in squamous cell carcinoma (Chap. XII). Atypical *keratinization* (dyskeratosis) in its many forms, is very frequent. Special stains (e.g.: -SH, -SS-) [211] and immunodiagnostic procedures [132] help identify this process, as do the more classical morphologic features (e.g.: ecto-endoplasm, ringing, hyalinization). Cytoplasmic *thinning* (e.g.: tail formation) and *stratification* (e.g.: pearl formation) bespeak the extreme atypicality in functional differentiation of these markedly stressed cells.

Karyopyknosis is frequent. When this form of nuclear degeneration develops in a large proplastic or dyskaryotic nucleus, it can present extreme diagnostic difficulties for the unwary, and may be incorrectly called cancer. The resulting pyknotic nucleus, even though shrunken from its original size, is still abnormally enlarged and hyperchromatic, and cannot be accurately morphologically discriminated from one of squamous cell carcinoma which has become pyknotic and lost its discriminating nuclear features of malignancy. Such 'India ink nuclei' have been a serious source of false identification of cancer [99].

If injury and recovery are properly balanced, epithelium can be maintained in this transitional state indefinitely (Figs. II-10, II-12). If this state of incessant stimulation is sufficiently severe, little or no recognizable functional differentiation may be present.

In a few cases it is extremely difficult, or impossible, to draw the fine morphologic line to differentiate the biologic behavior of extremely atypical transformational metaplasia from *in situ* carcinoma, either histologically or cytologically. For most lesions, this is possible from a practical sense [69, 94, 99, 103, 155, 206, 229, 237, 273]. However, expert opinions on a few, given, borderline lesions will continue to vary widely [224, 258].

Some of these lesions already are neoplastic. Thus, if one is certain, intraepithelilal neoplasia is a logical term to use. However, the term cervical intraepithelial neoplasia (CIN) is employed much too lightly in clinical usage today to be interpreted as an unquestionable *denotative* identification of biologic behavior, and must be accepted as a *connotative* term designating an *implied* possibility or probability (i.e.: increasing from CIN I to CIN II to CIN III) that there is neoplasia present.

Other lesions are becoming cancer [72], are cocarcinogenic sets (Fig. II-13) or, in fact, may be innocuous reactions. Various exciting studies (e.g.: detecting developing tumor antigens in these atypical cells) and assistance from automated systems (e.g.: DNA ploidy) are promising to add additional parameters to aid in the assessment of the biological behavior of such borderline lesions [32, 41, 64, 70, 84, 102, 107, 110, 111, 128, 132, 156, 180, 194, 198, 217–219, 221, 231, 238, 243, 246, 284, 298, 305].

At the present time, the most accurate assessment of a given lesion and the prognosis of its biologic behavior, is achieved by careful evaluation of all available data – especially *both histological* and *cytological*. This is carefully weighed with any additional evidence (e.g.: ploidy).

5. Outcome and Pattern Variations of Squamous Metaplasia

To gain key insight into what the morphology of those borderline lesions has to offer in assessing their biologic behavior, experimental and animal studies have been invaluable [4, 28, 55, 103, 134, 185, 219, 230, 231, 292]. Experimentally, some agents (e.g.: podophyllin) produce this severe reaction [148, 251] (Figs. XVI-4 through XVI-7) as a promoter or cocarcinogen without, by itself, apparent progression to malignancy. In others (e.g.: methylcholanthrene, benzpyrene), this severe epithelial reaction (Figs. XVI-8, XVI-9) is a step in the development of cancer [4, 55, 134, 230–232].

The more severe dysplasias shed the more immature dyskaryotic cells (Fig. VIII-8). The less mature is the cytoplasm (i.e.: less differentiation, less amount) and the more severe is the dyskaryosis (i.e.: more hyperchromasia, larger chromatin granules, clearer parachromatin, more undulated nuclear membranes, increased nuclear size), the more potentially sinister is the lesion (Figs. VIII-1 through VIII-3; XIII-4 through XIII-15; XVI-4 through XVI-9).

As the injury and stimulation *diminish,* the epithelium tends to mature and differentiate more toward stratified squamous with, also, a lessening of the degree of dyskaryosis and an approach toward euplastic metaplasia. If the injury and stimulation are withdrawn completely in those sites where columnar epithelium previously existed (e.g.: bronchi), there is good evidence that complete reversal *from* squamous *to* columnar differentiation can occur with time.

There are many variations on this theme of columnar epithelium undergoing transformation to squamous metaplasia [99, 185, 194]. In some epithelia during the process of squamous metaplasia, a single sheet of the older overlying columnar cells remain attached to the underlying epithelium, while definite transformation and maturation toward squamous is taking place in these lower portions of actively growing epithelium [99]. Frequently a slit formation occurs between the sheet of older columnar cells and the newer squamifying epithelium. Eventually, the overlying columnar cell sheet exfoliates, resulting in exposure of the underlying proliferating squamous epithelium which then lines the lumen [194].

While the covering columnar cells remain on the surface, the underlying proliferating epithelium frequently shows poor squamification. With natural exfoliation and loss of this columnar sheet cover, squamification of the new underlying epithelium progresses. An artificial removal of the cover, however (i.e.: abrasion), leaves a poorly-differentiated, or undifferentiated, transformational epithelium which may be most difficult histologically to discriminate from *in situ* carcinoma [99]. Cytologically, however, the immature dyskaryotic cells which are shed therefrom have significantly more cytoplasm than the discriminable *in situ* cells (Figs. VIII-3; XIII-1, XIII-3).

An ingrowth of adjacent squamous epithelium may occur, with replacement of previously existing columnar epithelium. This process of epi-

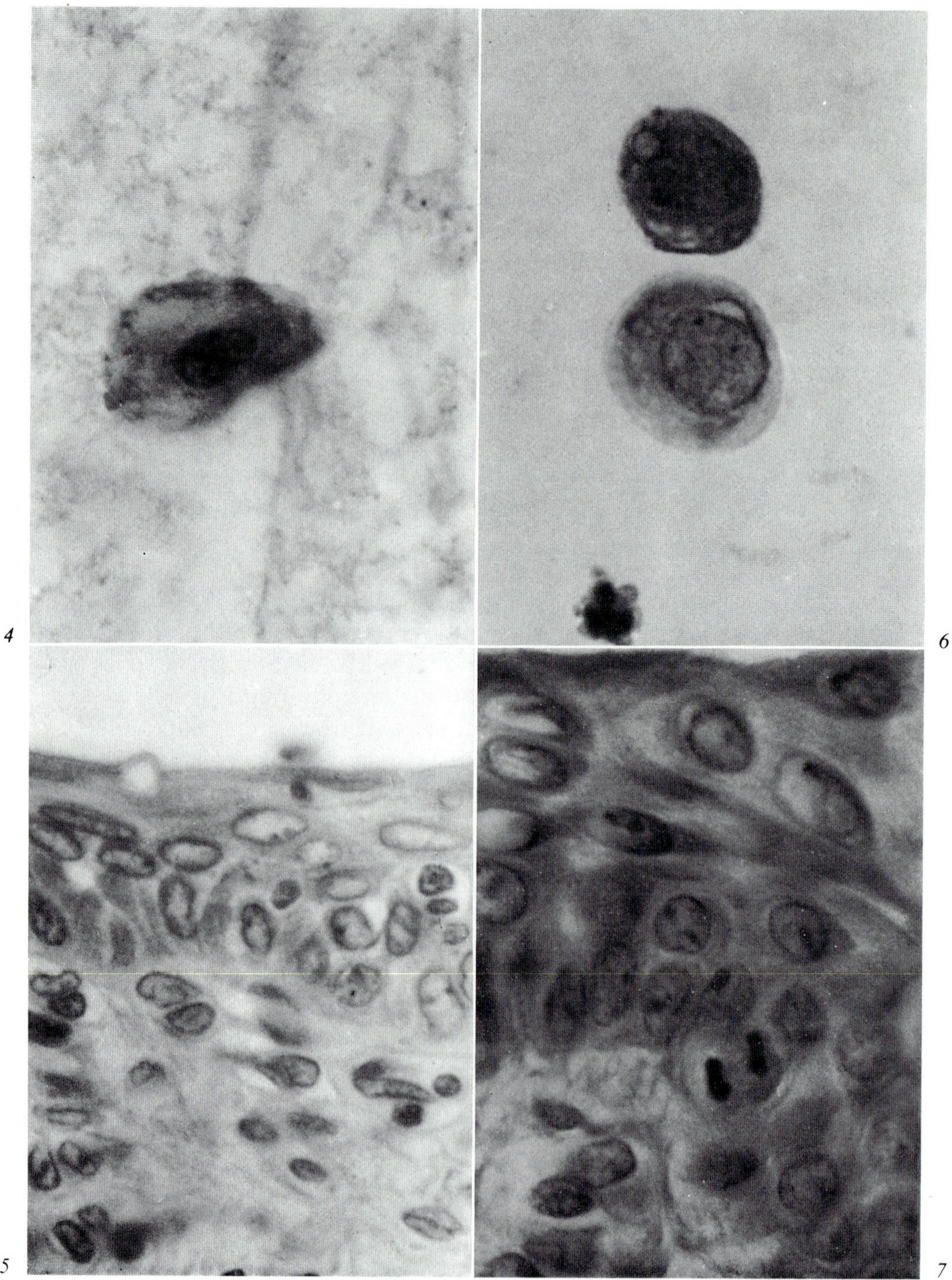

4
6
5
7

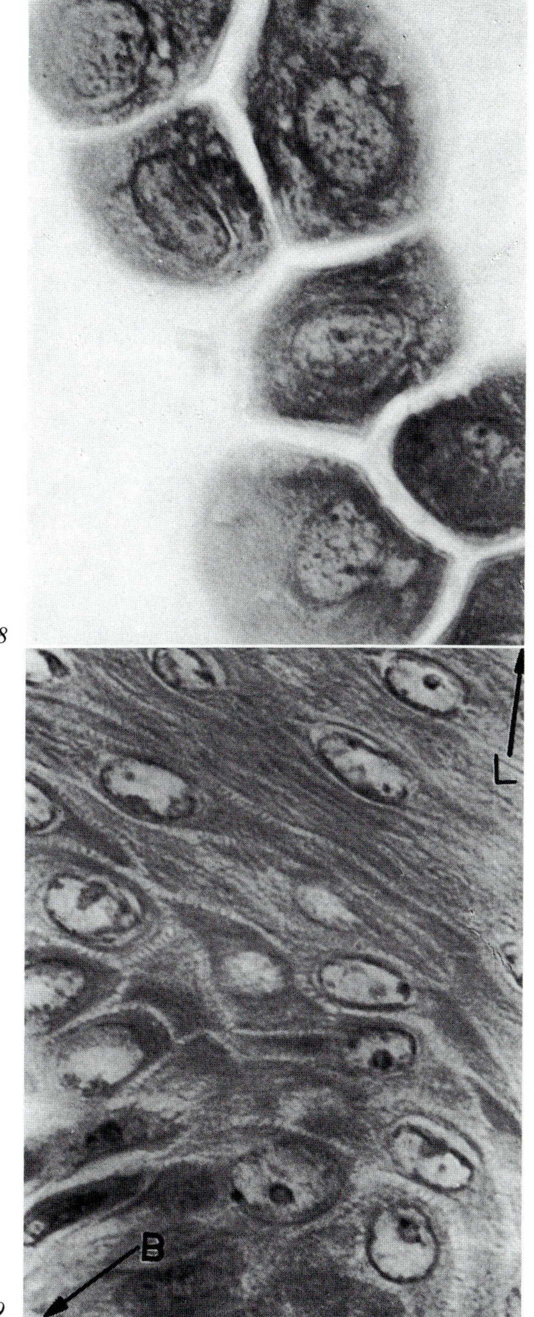

Figures XVI-4 through XVI-9: Chemically induced proplasia, *Endocervix, Wistar rats* [28, 292].

Figs. XVI-4, XVI-5: Normal control, treated with mineral oil, four weeks. The exfoliated cells (Fig. XVI-4) are normal-appearing squames. The endocervical tissue section (Fig. XVI-5) reveals the stratified squamous epithelium which is characteristic of the normal endocervical canal of the rat.

Figs. XVI-6, XVI-7: Treated with *podophyllin* in mineral oil, four weeks. Exfoliated cells (Fig. XVI-6) and endocervical tissue (Fig. XVI-7). Note the neutrophil for size. There is marked proplasia with increase in cell size, nuclear size, nucleo-cytoplasmic ratio, epithelial thickness (2 ×), nuclear membrane undulation, granular chromatin, parachromatin clearing, hyperchromasia, mitosis (i.e.: telophase), cytoplasmic basophilia, and cytoplasmic immaturity. Nucleoli are prominent in the tissue section of this stratified squamous epithelium (Fig. XVI-7), but they decrease as the surface is neared and are absent in the exfoliated cells (Fig. XVI-6).

Figs. XVI-8, XVI-9: Treated with *methylcholanthrene* in mineral oil, four weeks. There is extreme proplasia with marked dyskaryosis and basophilia of the cytoplasm. The hyperplastic endocervical epithelium is markedly (4 ×) thickened (Fig. XVI-9). Nucleoli are prominent but decrease toward the surface. B = Basement membrane; L = Luminal surface.

XVI-4, XVI-6, XVI-8: Papanicolaou stain; XVI-5, XVI-7, XVI-9: Hematoxylin and eosin stain.

XVI-4 through XVI-9: × 1,320.

dermidization may be difficult morphologically to distinguish from that of metaplasia, especially when it undergrows the columnar epithelium before the latter is exfoliated. It is not rare for these two processes to be found together, and for many other related lesions to coexist at a given time.

D. Multiplicity of Coexisting Related Lesions

The entire spectrum of squamous metaplasia is enormous, *from* the differentiation of a highly specialized columnar epithelium (e.g.: mucus-producing, ciliated) *to* the differentiation of a highly specialized stratified squamous epithelium (e.g.: keratinized, anucleate). There are many intermediate stages or shades of this spectrum through which a given process may pass, whence it may start, or where it may end.

The direction and the rapidity with which epithelia go through the spectrum vary with many factors. Stimuli differ in their capability for bringing epithelium to certain stages of the spectrum of metaplasia, and vary in the speed with which given stages are reached. Cells and tissues themselves, due to many factors (e.g.: type, site, age, nutrition), differ in their capabilities for reaction to stimuli.

The complex process is not only that of gaining increased passive protection by going to mature stratified squamous epithelium. The great factor of stimulation and injury to these tissues, with regeneration and repair wherever the situation allows, produces many intermediate steps in the transformational phases of the process which are more highly susceptible to mutation from carcinogens, and which afford many opportunities for genetic misinformation in the frequent mitoses of chronic irritation and repair.

Different areas of a given columnar epithelium do not necessarily proceed through metaplasia at the same speed (quantitative) nor in the same fashion (qualitative). In addition, some areas arise from initiated cells which have established clones of their progeny. Other areas represent clones from cells which have undergone further mutations along the line of neoplastic progression.

As a result, therefore, when a specimen is examined at a given time in the process, multiple different stages of the metaplastic process and of the neoplastic process usually have been attained by different areas of the epithelium. In addition, there are numerous variations in cell types within each stage [70, 250] as well as varying degrees of atypicality.

Thus a comprehensively obtained cytologic sample will contain different cell types from different areas, as well as different cell types from the same lesion (Fig. XIII-3). This morphologic heterogeneity bespeaks the biologic heterogeneity which is present in neoplasia.

It has become a 'diagnostic way-of-life' to recognize that the *most severe lesion* (the one with the gravest clinical significance) is the one whose diagnosis will be *the diagnosis* of the disease processes present. In reality, however, hosts of related lesions in this spectrum exist at the same time.

The causative forces of these related lesions frequently are labile and, in many respects, are difficult to identify or we are unable to determine them with accuracy. Morphologic alterations which result from them, fluctuate accordingly. Structural patterns present at any given time (e.g.: at biopsy or cellular specimen), therefore, may be *transformational* in either direction (i.e.: on the way *to* complete squamous metaplasia, or reversing *back* to the preexisting columnar type) or may be *static* at that time.

A more severe stimulus does not necessarily cause the epithelium to proceed farther toward complete metaplasia. In fact, some of the most significant stimuli appear to hold the epithelium *at the atypical exaggerated transformational phase* (C.4, p. 254) with minimal degrees of cytoplasmic maturation (i.e.: squamous or columnar differentiation), increased numbers of the less mature subsurface cells (e.g.: subcolumnar, parabasal), and intensified biologic general activity (i.e.: proplasia) which is reflected in the nuclear morphology (Figs. XV-6 through XV-9).

Marked increase of general activity and excessive decrease of maturation, frequently occur together. So extreme do they become in certain reactions to injury, that morphological (both histological and cytological) distinction from *in situ* carcinoma is most difficult, and further features (e.g.: DNA ploidy) are of additional assistance. There exists a great discrepancy of opinion regarding the biological significance of many of these morphologically borderline – shades of gray – lesions. Repeated cytological and histological study and correlation with DNA ploidy are invaluable to better understand the *true biologic behavior* of these biologically and clinically extremely important lesions [41, 64, 110, 111, 128, 150, 156, 194, 219, 243, 305].

XVII. Cancers with Other Cell Types of Functional Differentiation

The *same approach* for recognition and identification which was used for squamous cell carcinoma (Chap. XII) and adenocarcinoma (Chap. XV), is employed in cancers of other cell types, namely:

First, that it is cancer (by malignant criteria, mainly nuclear: Chap. IX), and

Second, into what cell type is the cancer trying to differentiate (by functional differentiation criteria, mainly cytoplasmic: Chap. X).

A few models hereinafter serve to exemplify the second part of this approach, the recognition of functional differentiation.

A. Pigmented Cancers

1. Malignant Melanoma

Single cancer cells are usually shed from malignant melanoma (Fig. XVII-1). At times loosely held together tissue fragments do appear (Fig. XVII-2), particularly in needle aspirations or brush samples where tissue fragments are more likely to be cut out or torn off. When this occurs, however, the cells are very loosely held together, appearing to separate easily from each other.

The cells tend to be *large* with *modest* to *marked* variation in size, and varying from poor (Fig. XVII-1) to extremely good (Fig. XVII-2) nuclear malignant criteria. Nucleoli are prominent (Fig. XVII-2); indeed, at times, they are gigantic. Giant cancer cells in malignant melanoma can be multinucleated and most pleomorphic (Fig. XVII-3).

Some cells contain no melanin (amelanotic) while others in the same tumor are loaded down with it. When melanin is present in the cytoplasm, it is usually characteristic and recognizable (Fig. XVII-3). It ranges from a fine dusting of golden-brown pigment, which may be extremely difficult to detect (Fig. XVII-1), to massive, irregular, golden, yellow-brown clumps.

In the same cell, the size of pigment granules may vary markedly (Fig. XVII-3). While this variance may be greater than that which one usually

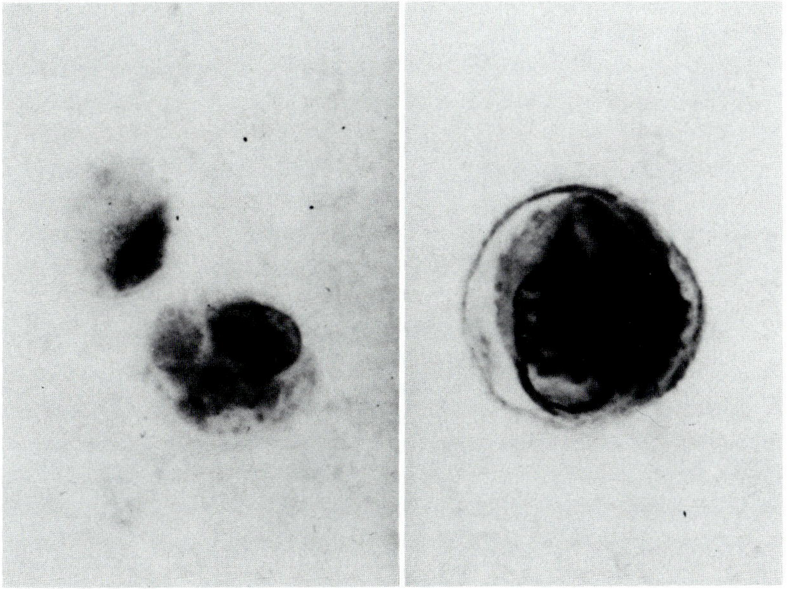

12

Figures XVII-1, XVII-2: Cancer cells containing pigment in their cytoplasm, pleural fluid. First, these cells are shed from cancer, as evidenced by malignant criteria (Chap. IX), especially of the nuclear features of the two cell DTTF in Fig. XVII-2. *Second,* their cell type is determined by features of functional differentiation, mainly cytoplasmic.

Fig. XVII-1: The cell on the left has a fine dusting of golden-brown pigment in its cytoplasm which is compatible with hemosiderin, bile or melanin. There are three large accumulations in the bottom cell which are rounded and could also be either one of the three pigments. They darken as the condenser is racked down from Koehler illumination about one-third of its excursion (Table IV-1), which favors bile or melanin. The cells are shed singly, predominantly, which favors melanoma over a bile-producing hepatoma (hepatocellular adenocarcinoma), which sheds cells mainly in diagnostic true tissue fragments (DTTF; Table VI-1) (cf. Figs. XV-3 through XV-5).

Fig. XVII-2: There are multiple, prominent nucleoli in the inner nucleus. This two-cell tissue fragment, one of the few diagnostic true tissue fragments (DTTF) in the specimen, is *not* a pearl of stratifying cells (Fig. XI-6), but is one cell appearing to be completely enveloped by a second cell. This cell-within-a-cell configuration is characteristically encountered at times with malignant melanoma involving body cavity fluids.

Papanicolaou stain.

XVII-1, XVII-2: × 1,500.

finds with hemosiderin, bile, carbon, lipofuscin, and other cytoplasmic pigments, it is *not* a good diagnostic feature, discriminating poorly between these pigments.

Bile may also appear as fine dust or granules in hepatic cell adenocarcinoma (i.e.: hepatoma, hepatocellular carcinoma) (Figs. XV-3, XV-4), but larger amounts of it usually appear as *amorphous* masses (intracellular, or as intercellular plugs) rather than the irregular, more *crystalline* granules of melanin. In addition, hepatoma cell cytoplasm frequently is vacuolated, appearing secretory (Figs. XV-3, XV-4) and with the cells tending to exfoliate in diagnostic true tissue fragments (DTTF; Table VI-1) as do other adenocarcinomas, rather than singly as do melanomas.

Special stains and reactions are available. Those for bile and melanin, however, are extremely harsh on cellular specimens, and the diagnostic cells in question frequently wash off during the staining procedure. On the other hand, the Prussian blue reaction to identify hemosiderin is mild and ideally suited for cellular spreads. Use of this technique can rule out phagocytosis of the hemosiderin by cells from a poorly-differentiated cancer of any type or an undifferentiated cancer, which frequently mimics melanin in the cytoplasm.

A rapid way to tentatively classify this group of golden pigments, is by moving the condenser down from Koehler illumination (Table IV-1). Melanin and bile darken and do *not* continue to 'gleam' as much as do hemosiderin, cigarette pigments and lipochromes, when one lowers the condenser away from Koehler illumination. This rapid determination helps discriminate melanin from hemosiderin, but not from the bile of hepatoma (i.e.: hepatocellular adenocarcinoma) (Figs. XV-3 through XV-5).

Figures XVII-3, XVII-4: Cancer cells producing melanin pigment in their cytoplasm.

Fig. XVII-3: Malignant melanoma. Numerous golden brown melanin granules are present in the cytoplasm of this giant malignant cell, ranging markedly in size and shape from fine dusting to large granules. Nucleoli are modestly prominent.

Fig. XVII-4: Pigmented basal cell carcinoma. This is a diagnostic true tissue fragment of malignant cells, whose nuclei have mainly granular chromatin and prominent nucleoli. Nucleo-cytoplasmic ratios are high. A few, fine, golden-brown pigment granules form a row adjacent to, and above the top nucleus, and scatter in the cytoplasm to the right.

Papanicolaou stain.

XVII-3: × 2,000; XVII-4: × 3,200.

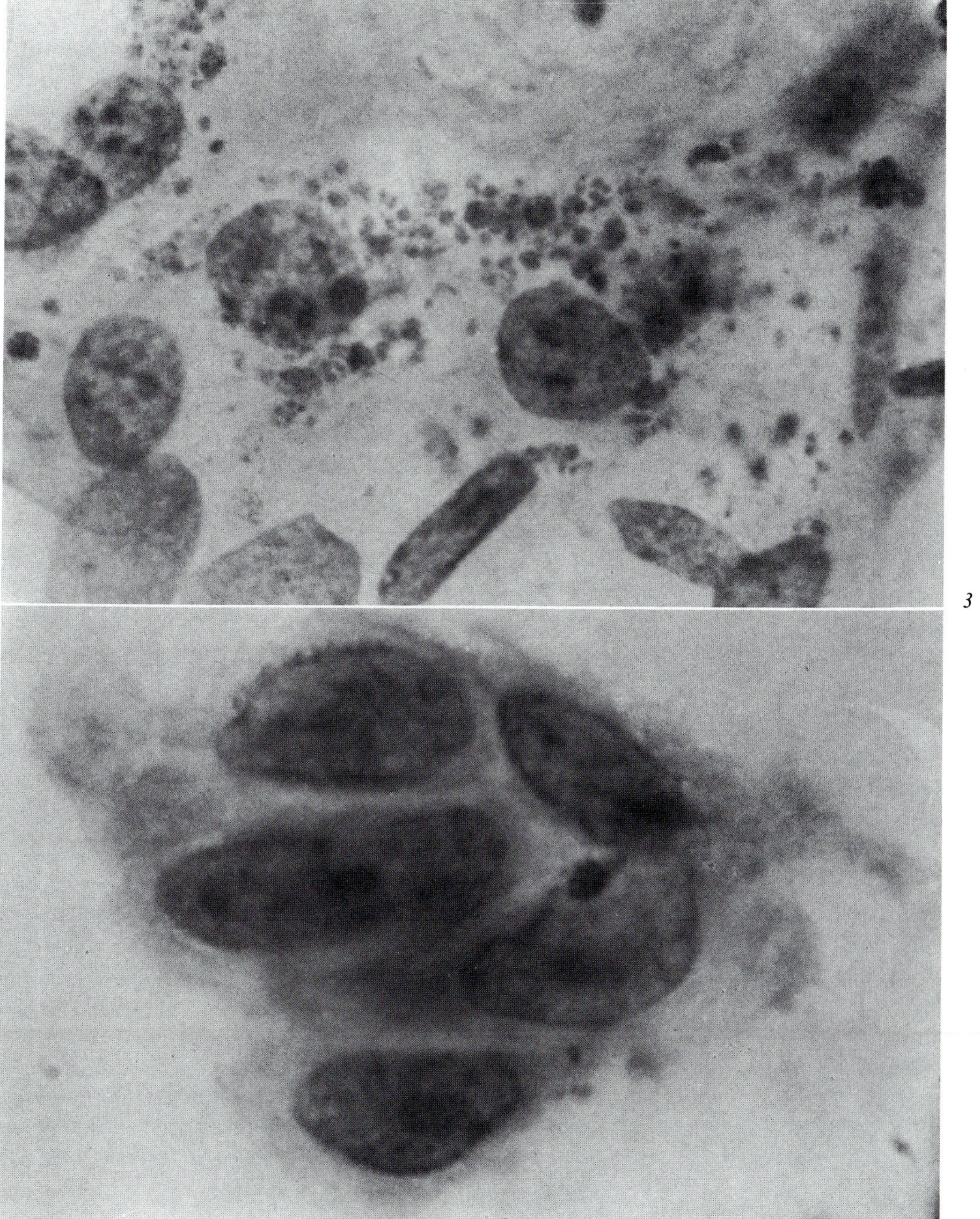

3
4

In melanomas of the uveal tract, cytologic recognition of the Callander type is possible in about half of the cases of aspiration of the anterior chamber [125]. The cellular shape in spindle A type and spindle B type is fusiform, while the nucleus of the latter has a much more irregular chromatin pattern, and the nucleoli are prominent. These are discriminated poorly from the fascicular type, but the epitheloid type sheds more rounded cells with very irregular, active nuclei.

2. Pigmented Basal Cell Carcinomas

Basal cell carcinoma tends to shed cells with fairly poor criteria of malignancy, as compared to the malignant melanoma. They mainly have large oval nuclei with a granular chromatinic net, slightly irregularly distributed throughout cleared parachromatin (Fig. XVII-4). The chromatinic rim thickness is generally uniform. The nuclear membrane has occasional malignant-type irregularities (Chap. IX) which may be key to the diagnosis, as may a rare sharply irregular nucleolus. Hyperchromasia is modest in well-preserved nuclei. They frequently exfoliate in diagnostic true tissue fragments (DTTF) in contradistinction to the single-cell tendency of melanoma. There is an increased nucleocytoplasmic ratio which is much more uniform, from cell to cell, in a DTTF than in a melanoma. At times intercellular bridges may be present.

Occasionally, basal cell carcinoma forms pigment in its cells. The melanin granules appear identical to those in melanoma, but frequently have a strikingly close association with the nuclear membrane (Fig. XVII-4). This paranuclear rowing of melanin granules, a greater tendency to shed in diagnostic true tissue fragments rather than single cells, the more subtle malignant criteria of the vesicular nucleus, and intracellular bridges favor an identification of a pigmented basal cell carcinoma rather than that of melanoma.

B. Sarcomas

1. Fibrosarcoma

These cancers often shed elongated cells, both nucleus and cytoplasm. If they have been in watery fluids (e.g.: body cavity fluids, cysts), they will round up. In mucoid material they can resemble the spindle cells (i.e.: 'fiber cells') of squamous cell carcinoma, but their cell borders usually are lacy and *indistinct,* rather than being sharply defined as with a keratinizing squamous cell (Chap. XII).

Most frequently their delicate cytoplasm has been stripped off, leaving behind bare, elongated nuclei lying parallel with their neighbors as logs in a log raft. This is frequently a prominent feature of meningioma, or meningeal fibrosarcoma. In neurofibrosarcoma (malignant Schwannoma) the elongated nuclei are characteristically wavy, like a jammed and crumpled cigar.

At times one can recognize smooth, thin fibrils or cytoplasmic furrows running lengthwise along the *outside* of the cell borders. These possible collagen fibers are not to be confused with longitudinal muscle fibers lying *within* the cytoplasm.

2. Leiomyosarcoma

The cells are elongated with soft, textured cytoplasm and extremely delicate cell membranes. The nuclei are also elongated, tending to have a somewhat more pointed nucleus than those of fibrocytes or spindle cells.

The fibrils in leiomyosarcoma are not distinguishable from those in fibrosarcoma by routine stain, except that they lie within the cytoplasm while those of the latter lie outside of the cell membrane and extend throughout the extracellular fluid. Therefore, in watery fluids (e.g.: pleural, peritoneal) the fibrosarcoma cells tend to round up very early, while the leiomyosarcoma cells remain elongated for an indefinite period of time, if their intercellular nonstriated myofibrils are sufficiently thick and rigid enough to serve as a backbone structure holding the cytoplasm elongated. With careful inspection (e.g.: 100 $\times$, oil with Koehler illumination, Table IV-1), the nonstriated fibers frequently can be resolved running longitudinally within the cytoplasm of the leiomyosarcoma cell.

3. Rhabdomyosarcoma

The discriminable cells are also elongated, with indistinct and delicate cell borders. The nuclei tend to be elongated, but are frequently rounded. At times the nuclei are multiple and line up tandemly within the elongated cell, as a so-called 'strap' cell.

Within the cytoplasm are longitudinal myofibrils, as in leiomyosarcoma cells. In contradistinction, however, these intracytoplasmic myofibrils have prominent nodes along them, which are resolvable with the light microscope (100 $\times$ oil, with Koehler illumination) at uniformly periodic intervals. The nodes occurring in adjacent fibrils, line up with each other to form the effect of intracytoplasmic cross-striations. As these cross-striations are valuable identifying features, they are to be carefully discriminated from mucus strands or other background material drawn across the cell, and from wrinkling of the cell membrane.

C. Leukemia and Lymphoma

Malignant lymphocytes *also* have the morphologic features of malignant neoplasia. It is *not* true, as has been alleged and often repeated over the decades, that the malignant lymphocytes of leukemia and lymphoma look no different than benign lymphocytes, just that there are more of them and they are more immature. This misconception has been carried over scores of years of hematologic work. Even in the air-dried Romanowsky stain, they have malignant criteria which are usually improved upon by immediate wet fixation and proper Papanicolaou staining.

In nearly any given cancer there are *deadly* cancer cells, in the tissue and exfoliated, which morphologically appear *normal* (Chap. VI) and do *not* bespeak their malignant behavior. Next to them may lie morphologically diagnostic malignant cells (Chap. IX). As with other sarcomas and carcinomas, diligent *search* through *adequate* and properly prepared material for the important discriminating *criteria,* will yield a diagnosis of cancer in most, or nearly all of the leukemias and lymphomas.

The nuclear chromatin/parachromatin pattern in an air-dried smear stained with a Romanowsky stain (e.g.: Wright's, Giemsa) is extremely bland, masking much of the nuclear criteria of malignancy (Chap. IX) which can be brought out more clearly in the properly wet-fixed smear stained by Papanicolaou's method. Here the irregular and sharply angled large chromocenters and macronucleoli are readily demonstrated in many of the malignant lymphocytes, as is their irregular distribution throughout the abnormally cleared parachromatin.

The air-dried Romanowsky stained nucleus, however, does *excel* in demonstrating the *shape* of the *nuclear membrane.* Where it may appear perfectly rounded and regular in the less flattened, wet-fixed, Papanicolaou-stained smear, the air-dried Romanowsky preparation frequently accentuates the unpredictably irregular and angled indentations and spicules of the nuclear membrane. Where this could be an important diagnostic feature, recourse to an air-dried Wright's-Giemsa smear is of great help.

One or more prominent *large nucleoli* are a most striking feature of malignant lymphocytes (Fig. XVII-5). This is in contrast to their absence

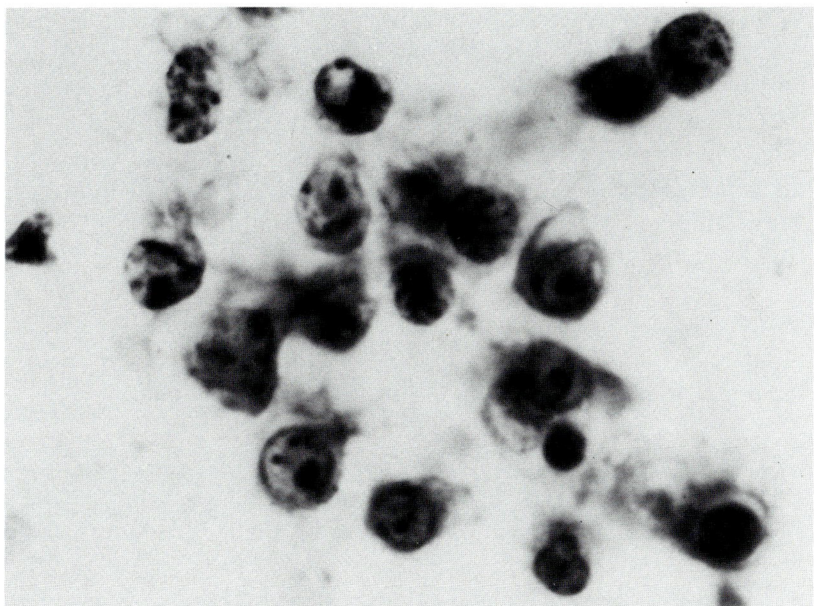

Fig. XVII-5: Malignant lymphoma, intermediate and mixed cell, small cleaved. *Pleural fluid.* These moderately large lymphocytic cells have an extremely high nucleocytoplasmic ratio, with scanty and frequently whispy cytoplasm. There are a number of malignant criteria, including moderate nuclear membrane irregularities (note the square 'bite' with sharp angles at 1-o'clock in the nucleus, below, at 7-o'clock of the picture). Markedly cleared parachromatin is irregularly distributed throughout the chromatin granules. Nucleoli are moderately prominent and vary markedly in size and shape.
Papanicolaou stain.
XVII-5: × 1,500.

or inconspicuous presence in normalcy. In certain marked reactions (e.g.: viral infection) extremely reactive (proplastic) nonmalignant lymphocytes (i.e.: immunoblasts, transformed lymphocytes) may have nucleoli, but they are small or moderate, and occur in large cells.

Prominent nucleoli, particularly irregular and sharply angulated ones, are frequently the only feature present to indicate a malignant process in mature, small, and otherwise undramatic lymphocytes. This extremely discriminating diagnostic aid is lost entirely when the preparation has been *overstained* with *hematoxylin;* or *understained* with weak, overused *counterstains –* or *both.* The resulting hematoxylin stained nucleoli are unable to be discriminated from chromocenters.

This is best remedied by preparing new fresh counterstains, and properly staining a *new* cell spread. However, the original preparation's diagnostic value can be improved, by: 1) removing the coverslip, 2) hydrating to water, 3) destaining the hematoxylin in weak hydrochloric acid (to where it is light enough so that the dark neutrophil nucleus' chromatin/parachromatin pattern is clearly discernible), 4) blueing, 5) washing thoroughly, 6) dehydrating to 95 percent ethanol, and 7) restaining with *freshly prepared* counterstains (i.e.: Orange G & EA) [124].

Once cancer has been established, *then* some assistance from cellular morphology can frequently be called upon to help determine a) whether one is dealing with a *leukemia* or a *lymphoma,* and b) the cell type into which the tumor cells are attempting to functionally differentiate.

1. Leukemia

Leukemias passively shed only *single* cells. Simple cell clusters may be encountered, where degenerating and sticky single cells simply adhere to one another, but there are *no* well-preserved *diagnostic true tissue fragments* (DTTF; Table VI-1) exfoliated *passively* in leukemia.

The more blastic and acute the leukemia is, the more it consists of larger and more pleomorphic cells with better malignant criteria. The more chronic forms shed mainly cells which more typically differentiate into the mature-appearing leukocytic cell types (i.e.: lymphocytic, granulocytic), best recognized morphologically under a Romanowsky stain. Prominent nucleoli in a few of these mature cells can be of value in establishing a diagnosis of malignancy. Cytoimmunodiagnosis using monoclonal antibodies and other techniques to recognize cellular antigenic markers [83, 132, 176] assist in positive cell type identification.

2. Malignant Lymphoma

In contradistinction to leukemias, *lymphomas occasionally* shed cells in diagnostic true tissue fragments (DTTF; Table VI-1). The percentage of cells shed in DTTF is so *very* low, however, that both lymphomas and leukemias are discriminatingly different in this feature from adenocarcinomas and from squamous cell carcinomas. On the other hand, the difference between lymphomas and leukemias is, at times, sufficient to be of help in discriminating between the two of them.

However, it is exceedingly difficult to discriminate lymphoma from leukemia on this, the *only one* feature available, unless *adequate* amounts of *fresh* and *properly* prepared material is carefully *screened* for *well-preserved, diagnostic true tissue fragments* (DTTF). This can be of value in body cavity fluids, where hundreds of thousands of tumor cells are available for careful screening and evaluation, but rarely in other sites. After thorough screening for DTTF, identifying a few well-preserved exfoliated DTTF, favors lymphoma over leukemia.

The quantity of DTTF involved, is very small. To provide a better appreciation, for example, body cavity fluids which contain lymphoma cells will have, roughly, only *five* to *thirty* DTTFs present in a well-populated cellular preparation under a 1 × 2 inch coverslip. In contrast, in *leukemia* there are *no* well-preserved DTTFs to be found within the same large number of cells (lymphocytes merely sticking or clumping together in groups are not DTTF, see Table VI-1). In further striking contrast, tens of thousands of DTTF will be present in the usual adenocarcinoma. In this way, the definitive recognition of *well-preserved* DTTF (Chap. IV; Table VI-1) can be of great diagnostic importance.

As with leukemia, a more exact cell type can be determined by cytoimmunodiagnostic identification of antigenic marker production, using suitable monoclonal antibodies [83, 132, 176, 178]. Furthermore, the formation by the tumor of nodules or follicles *vs.* a diffuse pattern, is able to be evaluated only by tissue pattern in histologic preparations. Even with these appreciable limitations, however, it is frequently extremely useful to arrive at a tentative identification of certain lymphoma groupings which frequently is possible with careful cytologic examination.

a. Non-Hodgkin's Lymphomas, General. Groupings of non-Hodgkin's lymphomas identifiable in routine cytopathology, fit into *tentative* categories related to those of the International Working Formulation of Non-Hodgkin's Lymphomas [197] and other prognostic histologic classifications (e.g.: Rapaport, Lukes-Collins, Kiel, Dorfman, British, WHO). *Cytopathologic groupings:*

1) *malignant lymphoma, small lymphocytic* (e.g.: small uniform, small noncleaved, small cleaved);

2) *malignant lymphoma, intermediate or mixed cell* (e.g.: small and large cleaved, mixed, histiocytic) (Fig. XVII-5);

3) *malignant lymphoma, large cell* (e.g.: large cleaved, large noncleaved, lymphoblastic, immunoblastic, plasmacytic, convoluted, transformed, polymorphous, histiocytic, lymphoepitheloid) (Figs. XVII-6 through XVII-8).

Then, following immunologic investigations for expression of functional differentiation, and histologic study of tissue pattern (e.g.: follicular (i.e.: nodular) *vs.* diffuse; fibrosis), *definitive* classification can be made based upon all available information.

The latter two can have convoluted nuclei, particularly the last, whose nuclei can be severely lobulated and form 'nose-like' protuberances (Fig. XVII-7). *Nucleoli* also are most prominent in the last, usually being large, multiple and irregular [24]. While most lymphomas functionally differentiate as B-lymphocytes, one cannot accurately differentiate neoplastic B-lymphocytes, neoplastic T-lymphocytes, and neoplastic monocytes/histiocytes by pure morphology without immunodiagnostic marker studies.

b. Malignant Lymphoma, Plasmacytoid. Plasma cells are B-lymphocytes which have further differentiated to produce more immune globulins, the antibodies of the blood and tissues in the body's defense against infection. Many B-cell lymphomas exhibit plasmacytoid features.

The typical plasma cell *nucleus* has multiple prominent rounded chromocenters distributed throughout, principally about the periphery as a part of the chromatinic rim, producing its characteristic 'clock-face' pattern. These characteristics are retained in the malignant plasma cell, with prominent nucleoli and irregular, pointed chromatin clumps.

The cytoplasm is highly basophilic, due to great numbers of ribosomes along the endoplasmic reticulum where protein (immune globulin) is actively being synthesized. An area of cytoplasmic clearing, corresponding to a prominently active Golgi complex, is in the vicinity of the cytocentrum, forcing the nucleus to be *eccentric.* At times the clearing is filled with a chromatophilic proteinaceous material, which stains with a different texture than the surrounding cytoplasm.

c. Malignant Lymphoma, Burkitt's Type. This also is *first* morphologically identifiable as cancer then, *second,* as a lymphoma and then, *third,* as

---

Figures XVII-6 through XVII-8: Malignant lymphoma, large cell, large non-cleaved (i.e.: histiocytic lymphoma; reticulum cell sarcoma).

Figs. XVII-6, XVII-7: Pleural fluid. The chromatin is coarsely clumped, the parachromatin is cleared and irregularly distributed, and there are prominent acidophilic nucleoli. The nuclei are minimally convoluted, with a small nose-like projection in the one in the upper right (Fig. XVII-7). Occasional cells have well-preserved, intact cytoplasm. There is a two-cell tissue fragment with nuclear molding (11-o'clock, Fig. XVII-7) evidencing a lymphoma rather than a leukemia.

Fig. XVII-8: Tissue section with the same pleomorphic nuclear pattern and prominent nucleoli.

XVII-6, XVII-7: Papanicolaou stain; XVII-8: Hematoxylin and eosin stain.

XVII-6 through XVII-8: × 800.

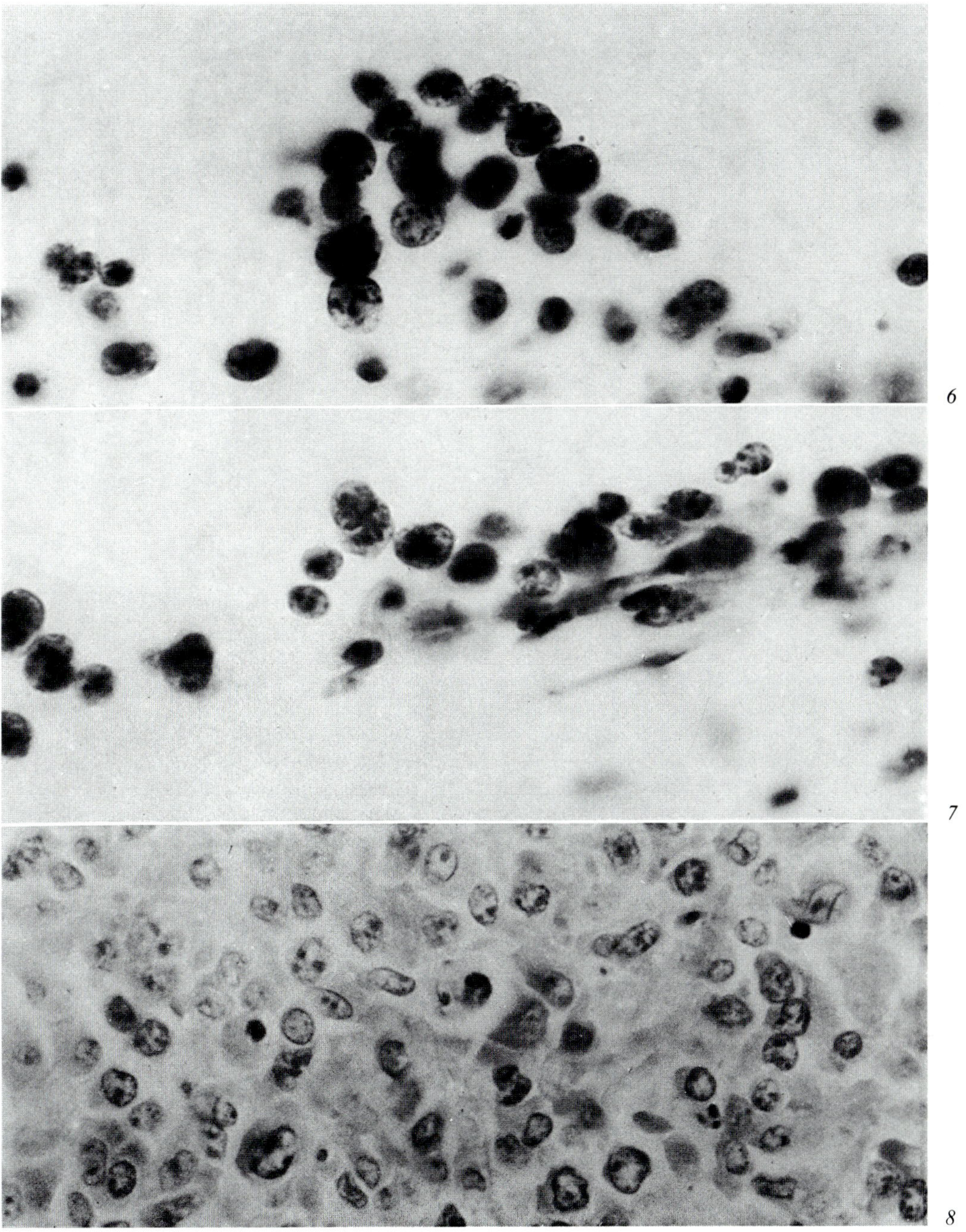

6

7

8

Burkitt's type. The neoplastic cells of this lymphoma are medium in size, of the non-cleaved variety, with or without prominent nucleoli and minimal convolution (Figs. XVII-9, XVII-10).

In addition, a second cell type occurs, the apparently benign, 'starry-sky', phagocyte and reactive histiocyte. These are fewer in number than the neoplastic cells, and scattered throughout them. Their nuclei are bland and they have abundant, pale cytoplasm which usually contains phagocytosed nuclear debris (Fig. XVII-9).

These pale cells are difficult to discriminate from macrophages or (in body cavity fluids) from mesothelial cells. They contain partially digested nuclear debris. Occasionally one will encounter these pale cells in a two- or three-cell diagnostic true tissue fragment (Table VI-1; Fig. XVII-10). This evidence identifies them as a phagocytic reticulum cell of Burkitt's lymphoma, the pale cells which create the 'starry sky' on tissue section (Fig. XVII-11), rather than mesothelial cells.

d. Mycosis Fungoides. The Sezary cells of mycosis fungoides and of Sezary's syndrome, have an extremely convoluted (cerebriform) nucleus. While most functionally differentiate as T-lymphocytes by cytoim-

---

Figures XVII-9 through XVII-11: Malignant lymphoma, Burkitt's type.

Figs. XVII-9, XVII-10: Most of the smear was made up of the smaller round cells with hyperchromatic nuclei (small, non-cleaved), lying about the periphery in these two fields. They are malignant by usual criteria (Chap. IX) with prominent nucleoli and a high nucleo-cytoplasmic ratio. They are moderate in size and have no, or minimal convolutions (note the small nose-like lobulation at 6-o'clock, Fig. XVII-10). Their cytoplasm contains clear vacuoles which stained with oil red-0. They are the malignant lymphoma cells. The three larger, pale cells in Figure XVII-9 have phagocytosed nuclear debris in their cytoplasm, and well could be single mesothelial cells or macrophages which have come together secondarily and adhered to each other. The two pale cells in Figure XVII-10, however, have nuclear molding from each other, are well preserved, and one can clearly identify a well-defined line extending the entire distance between the two nuclei representing the cell membranes of the two cells in this diagnostic true tissue fragment (DTTF). These are the benign, 'starry-sky' tingible macrophages, the reactive (proplastic), phagocytic reticulum cells of Burkitt's lymphoma.

Fig. XVII-11: The 'starry sky' pattern of Burkitt's lymphoma. The large, pale phagocytic reticulum cells sit within the massive array of moderate-sized, neoplastic, small, non-cleaved follicle center cells of this malignant lymphoma.

XVII-9, XVII-10: Pleural fluid; XVII-11: Biopsy.

XVII-9, XVII-10: Papanicolaou stain; XVII-11: Hematoxylin and eosin stain.

XVII-9, XVII-10: × 1,200; XVII-11: × 300.

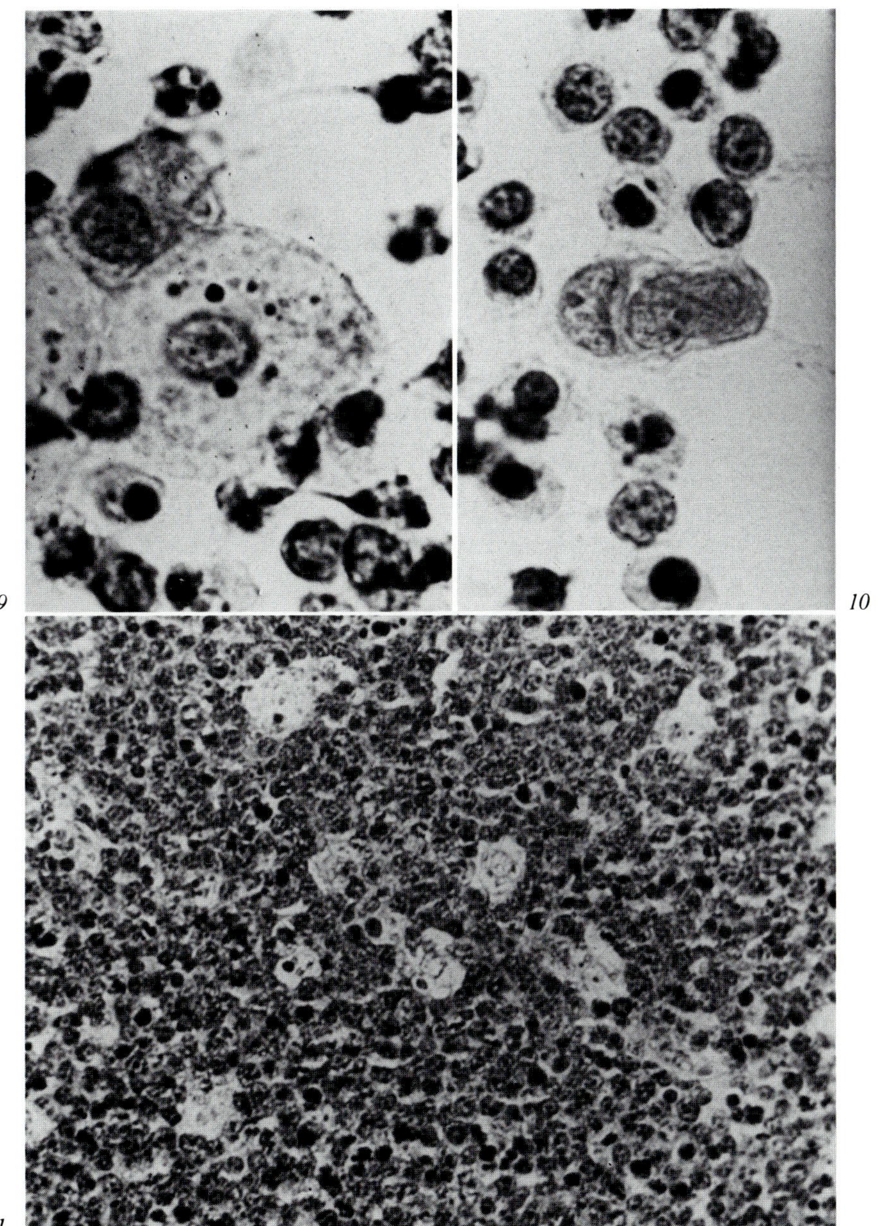

munodiagnostic studies, they cannot morphologically be discriminated accurately from the malignant B-lymphocyte of large convoluted lymphoma (histiocytic), or of malignant histiocytosis.

e. Malignant Histiocytosis. At times malignant monocytes/histiocytes can be recognized morphologically, *first* as single cancer cells resembling histiocytes, and then *second* by the occasional phagocytosis of erythrocytes by these malignant macrophages. The monocyte/histiocyte nature of these malignant cells is most accurately identified, of course, by cytoimmunodiagnosis [83, 132].

f. Hodgkin's Disease. While the large, binucleated Reed-Sternberg cell is frequently considered to be the most characteristic cell of Hodgkin's disease, it is *not* diagnostic of that lymphoma by itself, and must be present in the proper cellular pattern for the diagnosis to be established. Cells morphologically identical to it are found in many poorly-differentiated cancers or in pleomorphic large cell undifferentiated cancers.

Cytologic evidence for Hodgkin's lymphoma includes:

first, a *diagnostic lymphoma,* consisting of the Hodgkin cell (i.e.: a malignant immunoblast) which is usually of moderate cell size.

second, identification and quantification of the *Reed-Sternberg cell;*

third, recognition and quantification of small *lymphocytes,* as well as of eosinophils, neutrophils, plasma cells, and histiocytes.

The *diagnostic lymphoma cell* (i.e.: the Hodgkin's cell) consists of a moderate, mononuclear, immunoblastic-type sized cell with minimal nuclear convolutions (Fig. XVII-12). Nucleoli are prominent, and numerous. Malignant criteria are usually good involving parachromatin clearing, chromatin clumping, and nuclear membrane irregularity (Chap. IX).

Characteristically, the *Reed-Sternberg cell* has an equally bilobed nucleus which appears as two nuclei (Fig. XVII-13). Each lobe, or nucleus, is of equal size and shape, with virtually the same chromatin pattern with prominent, irregular chromatin clumps and marked parachromatin clearing. A prominent acidophilic nucleolus is present in each lobe, or nucleus, creating a 'mirror-image' or twin nucleus appearance.

Frequently *eosinophils,* plasma cells, histiocytes, *lymphocytes,* and, to a less extent, neutrophils accompany the diagnostic lymphoma cells (i.e.: Hodgkin's cells) and the large Reed-Sternberg cells (Figs. XVII-12, XVII-13). The lower grades of Hodgkin's lymphoma are associated with a higher number of lymphocytes (i.e.: lymphocyte predominance). The moderate grades have greater numbers of associated eosinophils, neutrophils and plasma cells (i.e.: mixed cellularity, nodular sclerosis). The higher grades

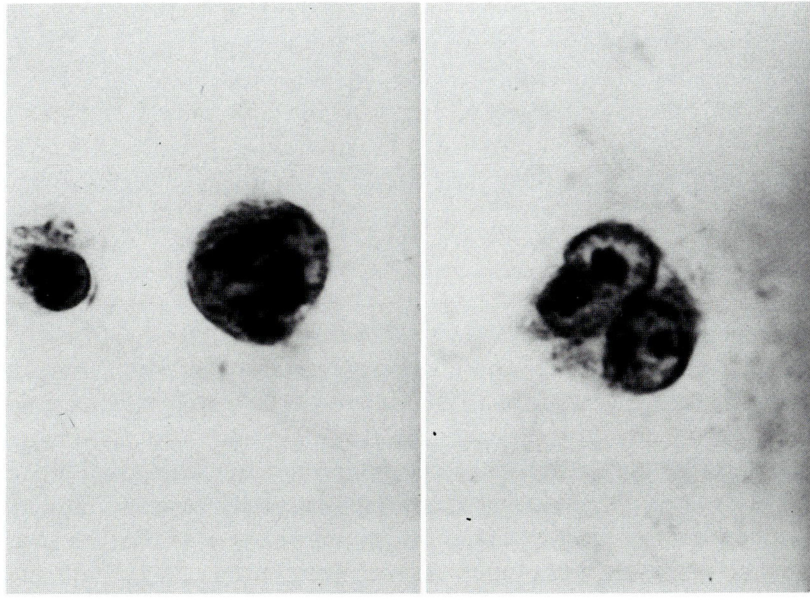

12   *13*

Figures XVII-12, XVII-13: Malignant lymphoma, Hodgkin's type.

Fig. XVII-12: The leukocyte on the left (note size) is an eosinophil, while the larger cell to its right is a Hodgkin's cell (i.e.: malignant immunoblast). It is characteristic of the malignant lymphoma cells present in this specimen, with an extremely high nucleo-cytoplasmic ratio, irregular nuclear membrane, marked hyperchromasia, massive chromatin clumping, irregular parachromatin clearing, and a large, prominent nucleolus.

Fig. XVII-13: A Reed-Sternberg cell present in the same specimen. This large cell is bilobed or binucleated, each lobe or nucleus is a 'mirror-image' of the other with identical features, e.g.: nuclear notching, same size, same degree of chromasia, one prominent eosinophilic nucleolus with an extreme clearing of the parachromatin about it which gives it an 'owl-eyed' appearance.

Papanicolaou stain.

XVII-12, XVII-13: × 2,000.

shed fewer of the normal appearing inflammatory cells, with the anaplastic malignant lymphoma cells (i.e.: Hodgkin's cells) and Reed-Sternberg cells predominating (i.e.: diffuse fibrosis, reticular, Hodgkin's sarcoma, lymphocyte depletion). A Romanowsky stain clearly demonstrates the eosinophils, while they are poorly shown in routine Papanicolaou stain.

It is generally believed that the Reed-Sternberg cell derives from the mononuclear Hodgkin's cell. The nature of the Hodgkin's cell, however, is not at all clear. Conflicting evidence points to either B-cell, T-cell, or monocyte/histiocyte origin or, even, mixtures thereof. In a recent study [83] the malignant lymphocytes of most of the cases studied expressed T-cell phenotypic markers, while some reacted with anti-B-cell antibodies. The Reed-Sternberg cells failed to react positively to any of the anti-B- and anti-T-cell antibodies, but some cases reacted to anti-monocyte antibodies. The disease thus appears to be heterogeneous in its expression of functional differentiation.

D. Choriocarcinoma

These cancer cells are rarely present in diagnostic clinical specimens from the female genital tract. When they are, however, they morphologically functionally differentiate either as cytotrophoblasts or as syncytiotrophoblasts. At times, both cell types are in the same specimen.

The cytotrophoblast has one nucleus per cell, centrally located in a modest amount of cytoplasm and with prominent nucleoli. The syncytiotrophoblast has clusters of multiple malignant nuclei gathered into one area of an abundant cytoplasm.

E. Cancers of Mixed Differentiation

As with choriocarcinoma, many carcinomas have multiphasic patterns where one area functionally differentiates one way (e.g.: adenocarcinoma), while another area matures differently (e.g.: squamous cell carcinoma, sarcoma). The resulting mixed tumor (e.g.: adeno-squamous carcinoma, mixed mesodermal tumor, mucoepidermoid, adenoacanthoma) can shed cells with different characteristics [99, 145], some reflect columnar differentiation (Chap. XV), while others have squamous differentiation (Chap. XII). This is frequent in bronchogenic and in uterine tumors.

Mixed functional differentiation of different cell types occurs in numerous sites throughout the body. Mixed mesodermal tumors (e.g.: fat, striated muscle, smooth muscle, fiber, epithelial cells) are most common in the genital and urinary tissues, while Wilms' tumor (e.g.: epithelial tubules, glomeruli, muscle, connective tissue) arises in the kidneys.

With the formation of metastases, one or all types can be found in the secondary sites. It is most frequent for adenocarcinoma, sarcoma, or the most poorly-differentiated areas of a tumor to metastasize, rather than for the squamous cell carcinoma; but this is not a predictable occurrence.

XVIII. Cancers without Identifiable Functional Differentiation

Cancers which do not bear morphological evidence of functional differentiation, are referred to as being *undifferentiated.* In many instances this is a *misnomer,* as some 'undifferentiated' tumors *do* functionally differentiate, but in a way which cannot be morphologically identified by routine staining procedures.

Some differentiation can be demonstrated by special stains and reactions (e.g.: mucus, melanin) [211] and immunocytodiagnostic techniques (e.g.: anti-keratin, anti-calcitonin) [58, 132, 214, 267]. Others produce significant biologic effects through their endocrine secretion (e.g.: carcinoid syndrome or Cushing syndrome in some poorly and undifferentiated pulmonary tumors). When they are not marked enough to produce signs and symptoms, they are most difficult or impossible to identify without biological, biochemical, or cytoimmunodiagnostic techniques [132].

Poorly differentiated tumors shed *mainly* undifferentiated cells. The cells which do have evidence of differentiation are so *few* or *poorly sampled* that the functional differentiation may not be identifiable, and it may be very poor or at times misleading (Tables X-1 through X-5).

Cancers showing no *morphological* evidence of differentiation, whatsoever, tend to fall into two main groups morphologically as well as biologically: the nonuniform, heterogeneous, pleomorphic, *large cell undifferentiated cancer* or *non-small cell undifferentiated cancer,* and the uniform, homogeneous, monotonous, *small cell undifferentiated cancer.*

A. Large Cell Undifferentiated Cancer (Non-Small Cell)

1. The Cells Generally
Completely undifferentiated cancer consisting of large pleomorphic cells or, most frequently, large, small, and other heterogeneous sizes, are conveniently referred to as large cell undifferentiated cancer. This is in contradistinction to the monotonous uniformity present in the small cell undifferentiated cancers. Usually, the cells of the large cell type of undifferen-

tiated cancer, are extremely pleomorphic (Fig. XVIII-1) with some of the largest, as well as the smallest, cancer cells encountered in the same tumor.

When an adenocarcinoma is very poorly differentiated and does not mature sufficiently for it to be recognized, it most frequently sheds as large cell undifferentiated carcinoma. Sarcomas, likewise, are usually essentially large cell undifferentiated cancers if they are not differentiated sufficiently for their cell type to be recognizable. In the cervix uteri, if there is the least bit of cellular functional differentiation toward squamous (Chap. XII), a large cell keratinizing squamous cell carcinoma [297] is suspected and to be ruled out by careful search for further features of squamous differentiation. Transitional cell carcinoma usually sheds as large undifferentiated, at times with papillary trends or atypical squamous cell differentiation.

2. Nucleus

The nuclei vary markedly in size, shape and configuration. While they may tend to be vesicular, they are usually markedly irregular. They usually contain some of the best malignant criteria with an unpredictably irregular nuclear membrane and large, jagged, angular chromatin masses in prominent and abnormally-cleared parachromatin (Chap. IX). The nucleoli are frequently multiple, massive, and sharply angular or spiculated (Fig. XVIII-1). Multinucleation with marked variation and bizarre mitoses are not uncommon.

3. Cytoplasm

The *amount* of cytoplasm per cell characteristically varies markedly from cell to cell, from extremely scanty to very abundant. The same variation is true for the nuclear size so that, in the same tumor, a given cell may have a huge nucleus in wispy cytoplasm with an extremely high nucleo-cytoplasmic ratio, one adjacent cell may have a normally adequate nucleus and cytoplasm, while another adjacent cell may have a very small nucleus in an excessive amount of cytoplasm resulting in an extremely *low* nucleo-cytoplasmic ratio (Chap. XVIII-1).

The cytoplasm is usually fluffy, without indicating secretion. Small intercellular vacuoles of degeneration, may resemble intercellular bridges, unless one examines them carefully (Figs. XII-8, XII-9, XII-18, XII-19) and cautiously interprets them (Chaps. XI, XII).

The cellular borders are usually indistinct in large cell undifferentiated cancer, so that it may be impossible to unequivocally discern them by light

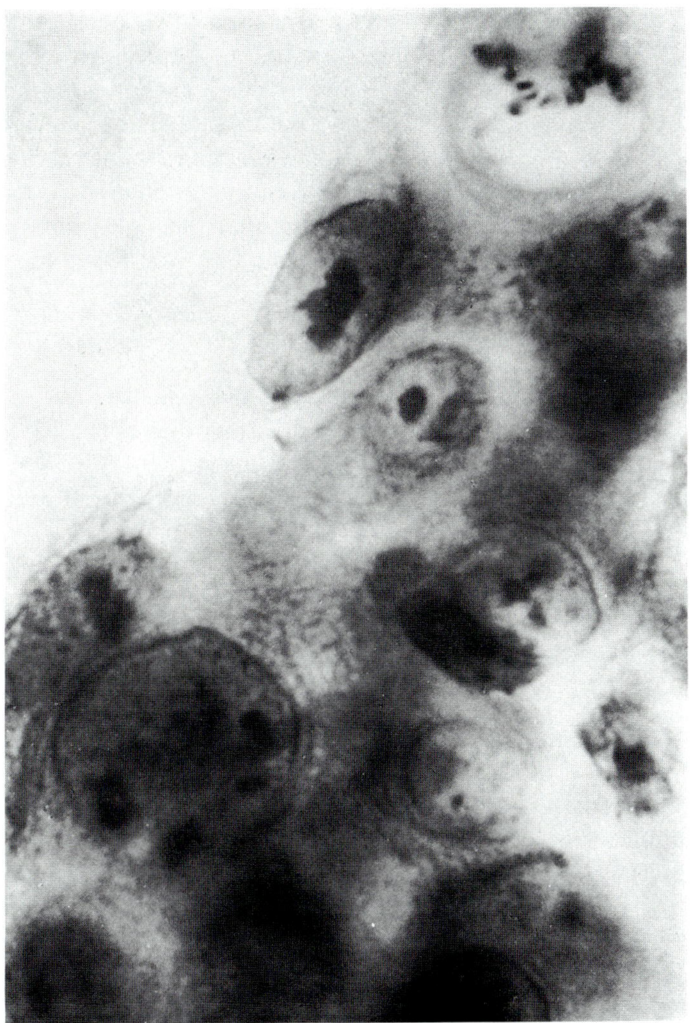

Fig. XVIII-1: Large cell undifferentiated cancer. Very pleomorphic cells with varying amounts of cytoplasm and nuclear size. Malignant criteria are usually quite good, note the sharply angulated nucleolus, the variation in the numbers of nucleoli per nucleus within this diagnostic true tissue fragment (DTTF), and the abnormal mitosis in the upper field. The foamy cytoplasm is common and may suggest secretion, but does *not* indicate it.

Papanicolaou stain.

XVIII-1: × 2,000.

microscopy, arousing false suspicions of a syncytium. On careful examination along the stretched edge of a diagnostic true tissue fragment of such a large cell undifferentiated cancer, a strong intercellular junction may be accentuated as a pronounced dot. In a poorly-differentiated adenocarcinoma, this may signal a terminal bar of a tight junction (zonula occludens) (Figs. XIV-1, XIV-6, XIV-7). In other large cell undifferentiated cancers, however, it merely indicates a robust intercellular junction which is not giving way to the lateral pulling forces of cells in a true tissue fragment under tension.

4. Tissue Fragments

Large cell undifferentiated cancer usually sheds in true tissue fragments. They typically hold together fairly well, rather than appearing to fall apart easily as with small cell undifferentiated cancer.

The cells exhibit great tendencies for nuclear molding from nuclei of the adjacently growing cells, with modest crowding from a moderately high average nucleo-cytoplasmic ratio. The marked variation in nucleo-cytoplasmic ratio from cell to cell, however, results in the high degree of cellular pleomorphism which is demonstrated well in exfoliated tissue fragments (Fig. XVIII-1).

B. Small Cell Undifferentiated Cancer

Cells from small cell undifferentiated cancer tend to be uniformly small and monotonous. This is in striking contrast to the mixture of large and small cells present in large cell undifferentiated cancer.

Small cell undifferentiated cancer in the *adult* is frequently associated with squamous cell carcinoma (e.g.: small cell nonkeratinizing carcinoma of

---

Figures XVIII-2 through XVIII-4: Small cell undifferentiated cancer: Small cell nonkeratinizing carcinoma, cervix uteri. Fast (vagino-pancervical) smear.

Fig. XVIII-2: Stringing out along cervical mucus, some cells break off from the loosely held together tissue fragments.

Figs. XVIII-3, XVIII-4: The cytoplasm is scanty, yet intact. There are frequent sharp indentations of the nuclear membrane with moderate nuclear molding, monotony of nuclear size and pattern, parachromatin clearing, and inconspicuous nucleoli. Note, for size, two lobes of a neutrophil nucleus along the lower border of Figure XVIII-4.

Papanicolaou stain.

XVIII-2: × 500; XVIII-3, XVIII-4: × 2,000.

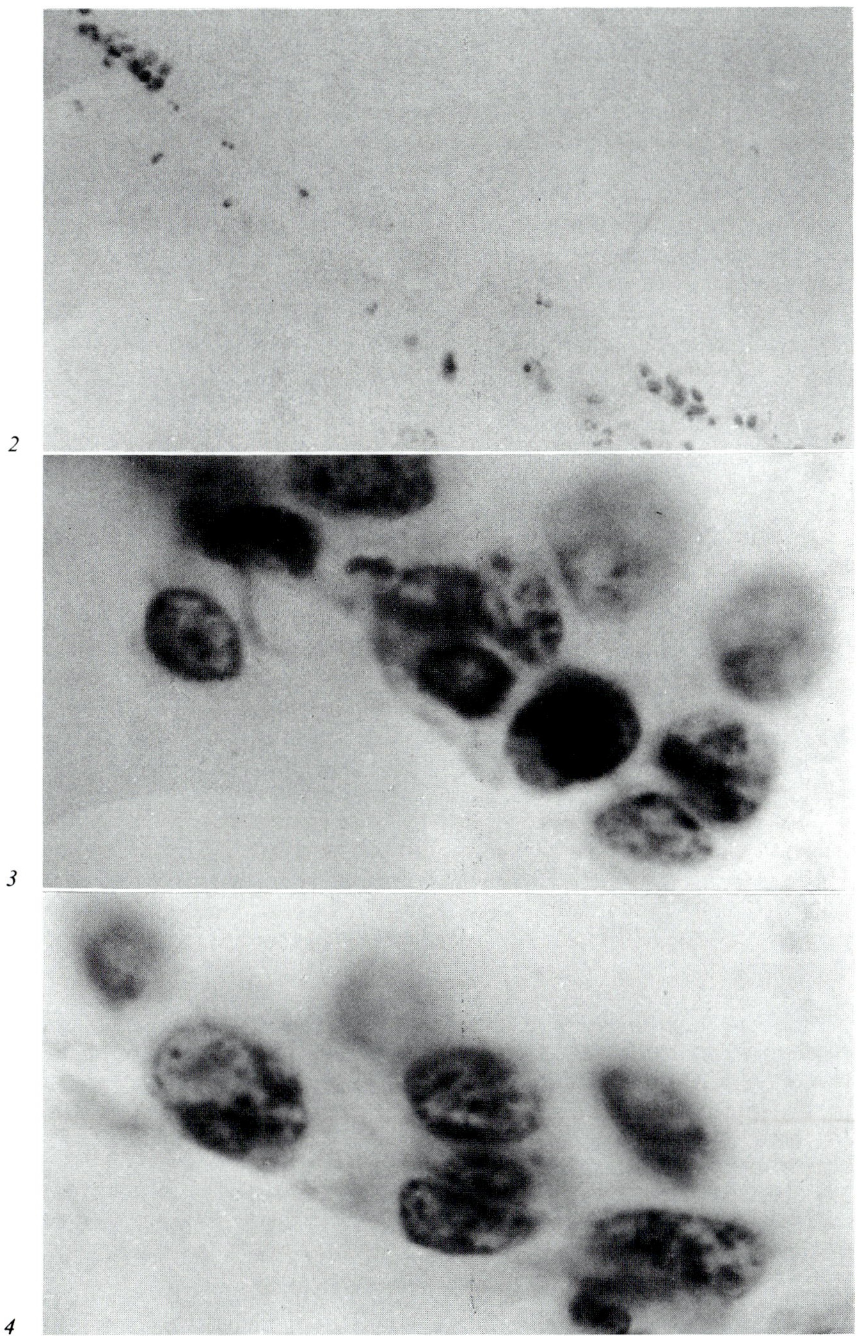

the cervix (Figs. XVIII-2 through XVIII-4); or 'oat' cell or small cell carcinoma of the lung (Figs. XVIII-5, XVIII-6); or esophagus). It is so rapidly growing that it quickly overgrows its sluggish precursors and is usually the sole presenting tumor. Here the nuclear membrane is equidistant from the cell membrane, as the nucleus attempts to occupy a central position.

At times, poorly-differentiated adenocarcinomas (e.g.: endometrium, breast, stomach, pancreas, prostate) can also shed only as small cell undifferentiated cancer. When this occurs, adenocarcinoma frequently can be suspected by the nucleus lying slightly eccentrically in the extremely scanty cytoplasm. The cell membrane may be hard to discriminate from the nuclear membrane along one side of the nucleus, while there is a bit more cytoplasm at the opposite side of the nucleus where it frequently flattens or even indents.

Small cell undifferentiated cancer can also be found in other cell types. When it occurs in a *child,* one considers mainly the blastic tumors (e.g.: nephroblastoma (Wilms' tumor), neuroblastoma, retinoblastoma, medulloblastoma, malignant lymphoma, Ewing's sarcoma, rhabdomyosarcoma, ependymoblastoma, esthesioneuroblastoma) [2, 95, 155].

1. The Cells Generally

Small cell undifferentiated cancer usually sheds in loosely adhering tissue fragments, from which cells easily separate and appear to 'stream out' from the fragment along mucus or fibrin strands (Fig. XVIII-2). This

---

Figures XVIII-5, XVIII-6: Small cell undifferentiated cancer: Small cell carcinoma of the bronchus.

Fig. XVIII-5: Sputum. The extremely small cells are nearly uniform in size and shape (for size, note the neutrophil stuck to the upper surface of the large true tissue fragment). Nuclear molding is marked. Note the sharp, deep, thin, V-shaped notched irregularity of the nuclear membrane at 11-o'clock in the left-most and the right-most cells in the large true tissue fragment. Nucleoli are present, but are small and are not prominent. There remains a well-preserved, thin rim of cytoplasm equidistant about each nucleus, best seen in the right lower and left upper.

Fig.XVIII-6: Tissue section of the same patient. The small, uniform cells have the same pattern identical to those exfoliated (Fig. XVIII-5). The cytoplasm is scanty, and frequently in wisps.

XVIII-5: Papanicolaou stain; XVIII-6: Hematoxylin and eosin stain.

XVIII-5, XVIII-6: × 2,000.

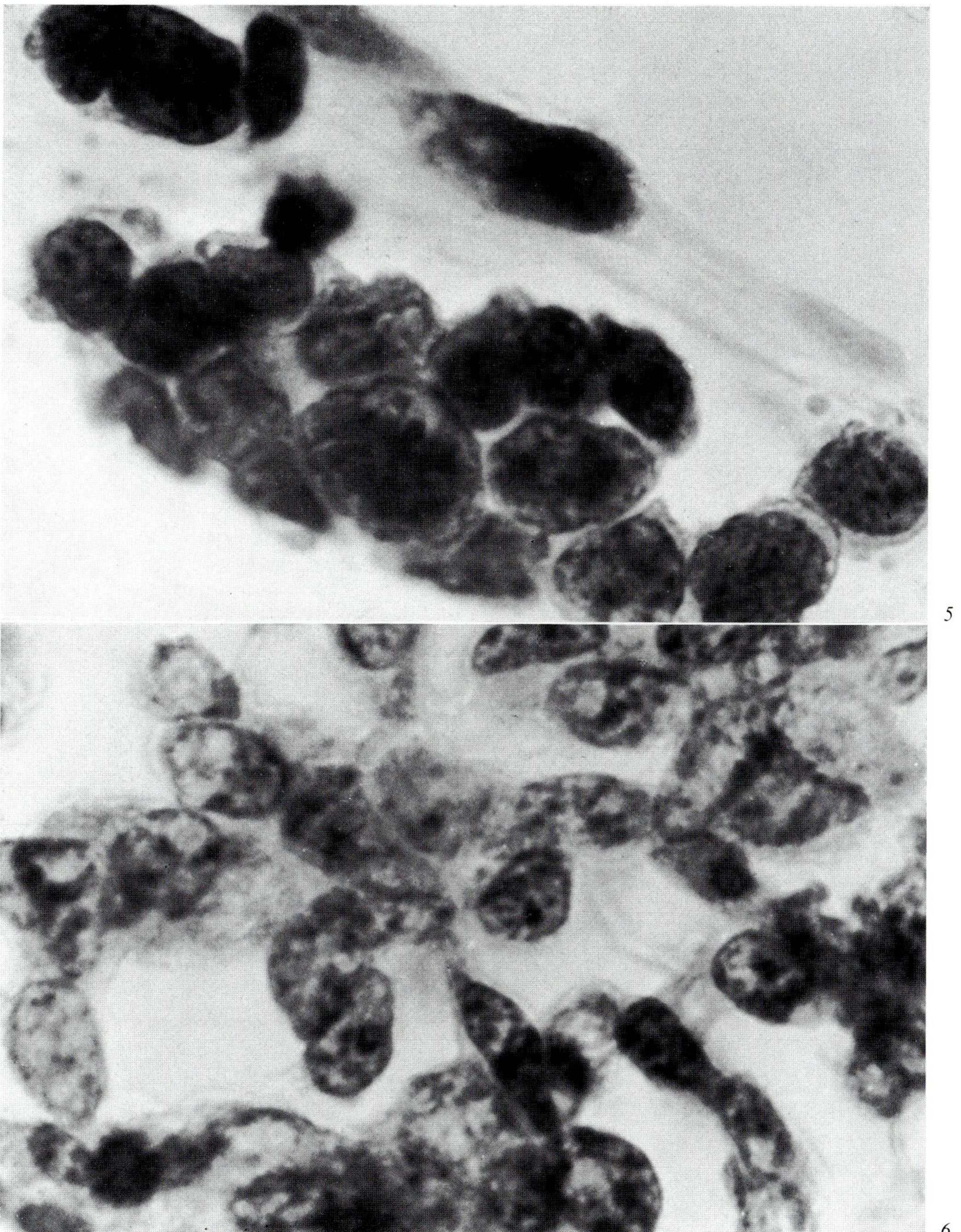

5

6

is most characteristic in small cell carcinoma of the lung (Fig. XVIII-5), but occurs also with small cell undifferentiated cancers in other sites. While the cells are small, they are larger than lymphocytes and usually range about the size of small histiocytes (cf.: Figs. IV-1; XVIII-3 through XVIII-5).

2. Nucleus

The nuclei of cells in small cell undifferentiated cancer are small and monotonously uniform in size. However, they are large for the overall size of the small cell, with a resulting extremely high nucleo-cytoplasmic (N/C) ratio. The nucleus tends to be somewhat vesicular (Figs. XVIII-3 through XVIII-5), but there is definite folding and irregularity. In addition, there is severe nuclear molding from the adjacent cells, indicating crowding in tissue, which is retained even after the cell exfoliates (Fig. XVIII-5). This is in striking contrast to the single, rounded nuclei of *in situ* carcinoma (Figs. XIII-9, XIII-13).

Truly bare nuclei should not be identified as small cell undifferentiated cancer. While tumors can degenerate and their cells lose cytoplasm, so do other very active tissues (e.g.: columnar cells) (Fig. VII-1). Such degenerated proplastic cells can appear morphologically very similar to small cell undifferentiated cancer, especially when their degenerated, bare nuclei stick together and mold. To be *diagnostic,* a cell must be *intact.*

The intact cell of small cell undifferentiated cancer has an extremely high nucleo-cytoplasmic ratio. So scanty is its cytoplasm that, at times, only by using highest resolution (e.g.: Koehler illumination, Table IV-1) and the highest meaningful magnification (e.g.: 63X dry; 100X oil) can one identify a thin rim of intact cytoplasm (Fig. XVIII-5) or recognize it as potentially a bare nucleus (Fig. VII-1).

Malignant criteria are said to be lacking in small cell undifferentiated cancer. This is not true; but at times they are bland and their criteria must be carefully sought out. A degenerated cell must never serve as the basis for diagnosis.

Nuclear chromatin may be moderately clumped and angular, with somewhat irregular dispersion throughout abnormally cleared parachromatin, with a pattern typical of invasive cancer. While nucleoli are not prominent, they may be present.

In some cases the nuclear chromatin/parachromatin pattern may be very bland, finely granular and uniformly dispersed, even appearing 'ground-glass'. When this is encountered, one must carefully and meticulously search for significant nuclear membrane irregularities (Fig. XVIII-5), and carefully discern the intactness of the thin rim of cytoplasm to determine the usefulness of the N/C ratio apparent in that case.

3. Cytoplasm

Usually, the cytoplasm characteristically is extremely scanty. The cellular border may infrequently appear somewhat prominent (Fig. XVIII-5); but, most often, it is wispy and indistinct (Figs. XVIII-3, XVIII-4) or completely stripped off of the nucleus.

Such latter truly bare nuclei, even though occurring frequently with the delicate cytoplasm of small cell carcinoma, should *never* be the basis for a diagnosis either of *cancer* or of *cell type*. The high nucleo-cytoplasmic ratio must be found only in cells having well-preserved cytoplasm with intact cell borders (Fig. XVIII-5) for it to constitute a valuable criterion of cancer. Any cells which have lost their cytoplasm are able to be degenerated enough to squash together, mold their nuclei and appear, to the *unwary,* as a tissue fragment of small cell cancer (Table VI-1).

4. Tissue Fragments

When tissue fragments are encountered, they accentuate the cellular monotony, high nucleo-cytoplasmic ratio, and extreme nuclear molding. As the cells break away at the periphery of such fragments, they tend to retain their marked nuclear molding, rather than round up. In spite of a markedly irregular and molded nucleus, a uniformly thin rim of cytoplasm is frequently maintained between strikingly parallel nuclear and cell membranes.

References

1   ABU-L-FIDA, I. A.: Takwim al-buldan (descriptive geography of the Orient), 1321 A.D.; from Geographie d'Aboulfeda, traduite de l'arabe en français par M. Reinaud, Tome II, premiere partie (Imprimerie Nationale, Paris 1848); see also: F. M. Sandwith, The medical diseases of Egypt, title page (H. Klimpton, London 1905).

2   AKHTAR, M., ALI, M. A., SABBAH, R., BAKRY, M. and NASH, J. E.: Fine-needle aspiration biopsy diagnosis of round cell malignant tumors of childhood. Cancer 55: 1805–1817 (1985).

3   ALBERTS, B., BRAY, D., LEWIS, J., ROFF, M., ROBERTS, K. and WATSON, J.D.: Molecular biology of the cell (Garland, New York 1983).

4   ALBRIGHT, C. D., FROST, J. K. and PRESSMAN, N. J.: Cytologic preparations and objective morphologic analysis of cells from developing hamster squamous cell carcinomas. Analyt. Quant. Cytol. 4: 141 (1982).

5   ALLEN, R. D.: Amoeboid movement, in J. BRACHET and A. E. MIRSKY, The cell, vol. 2, chap. 3, pp. 135–216 (Academic Press, New York 1961).

6   ALLEN, R. D., METUZALS, J., TASAKI, I., BRADY, S. T. and GILBERT, S. P.: Fast axonal transport in squid giant axon. Science 218: 1127–1129 (1982).

7   American Society of Cytology, Educational Film Committee: The Papanicolaou stain: principles (film and text), National Committee for Careers in the Medical Laboratory, and The American Cancer Society (Wexler Films, Los Angeles 1974).

8   American Society of Cytology, Educational Film Committee: The Papanicolaou stain: methods (film and text), National Committee for Careers in the Medical Laboratory, and The American Cancer Society (Wexler Films, Los Angeles 1975).

9   American Society of Cytology, Educational Film Committee: Cytopreparation with microslides (film and text), National Committee for Careers in the Medical Laboratory, and The American Cancer Society (Wexler Films, Los Angeles 1975).

10   American Society of Cytology, Educational Film Committee: Cytopreparation with membrane filters (film and text), National Committee for Careers in the Medical Laboratory, and The American Cancer Society (Wexler Films, Los Angeles 1975).

11   American Society of Cytology, Educational Film Committee: Collection and preparation of sputum specimens (film and text), Health and Education Resources, and The American Cancer Society (Wexler Films, Los Angeles 1978).

12   American Society of Cytology, Educational Film Committee: Collection and preparation of bronchoscopy specimens (film and text), Health and Education Resources, and The American Cancer Society (Wexler Films, Los Angeles 1978).

13   American Society of Cytology, Educational Film Committee: Fixation in diagnostic cytology (film and text), Health and Education Resources, and The American Cancer Society (Wexler Films, Los Angeles 1979).

14 American Society of Cytology, Educational Film Committee: Coverslipping in diagnostic cytology (film and text), Health and Education Resources, and The American Cancer Society (Wexler Films, Los Angeles 1979).

15 ANASTASSOVA-KRISTEVA, M.: The nucleolar cycle in man. J. Cell Sci. *25:* 103–110 (1977).

16 ANDREW, W.: Cellular changes with age (Thomas, Springfield, Ill. 1952).

17 AUERBACH, O. and STOUT, A. P.: Histopathological aspects of occult cancer of the lung. Ann. N. Y. Acad. Sci. *114:* 803–810 (1964).

18 BAHR, G. F.: Chromosomes and chromatin structure, in J. J. YUNIS (ed.), Chromosomes in biology and medicine (Academic Press, New York 1977).

19 BAHR, G. F., BARTELS, P. A., BIBBO, M., DE NICOLAS, M. and WIED, G. L.: Evaluation of the Papanicolaou stain for computer assisted cellular pattern recognition. Acta Cytol. *17:* 106–112 (1973).

20 BAHR, G. F., BIBBO, M., OEHME, M., PULS, J. H., REALE, F. R. and WIED, G. L.: An automated device for the production of cell preparations suitable for automatic assessment. Acta Cytol. *22:* 243–249 (1978).

21 BAHR, G., MIKEL, D. and ENGELS, W. F.: Correlates of chromosome banding at the level of ultrastructure, in Nobel symposium 23 chromosome identification, pp. 280–289 (Academic Press, New York 1973).

22 BAHR, G. F., TESCHE, B. and ZEITLER, E.: Observations on the structure of mechanically stretched chromatin fibrils. Virchows Arch. (Cell Pathol.) *36:* 103–121 (1981).

23 BAKER, R. R. and WOOD, S.: Biologic assay of tumor cells in the blood. Surg. Gynec. Obstet. *124:* 742–746 (1967).

24 BALL, P. J., van der VALK, P., KURVER, P. H. J., LINDEMAN, J. and MEIJER, C. J. L. M.: Large cell lymphoma. II. Differential diagnosis of centroblastic and B-immunoblastic subtypes by morphometry on cytologic preparations. Cancer *55:* 486–492 (1985).

25 BALLENGER, J. J.: Effect of tobacco smoke on the human respiratory cilia (motion picture) (Northwestern University, Chicago 1960).

26 BANG, F. B.: Effects of invading organism on cells and tissues in culture, in E. N. WILLMER, Cells and tissues in culture, pp. 151–262 (Academic Press, New York 1966).

27 BARKER, B. E. and SANFORD, K. K.: Cytologic manifestations of neoplastic transformation *in vitro.* JNCI *44:* 39–63 (1970).

28 BARR, J. L.: Cellular changes induced by 3-methylcholanthrene on the vaginal epithelium of the Wistar rat (Thesis, Johns Hopkins School of Cytotechnology, Baltimore 1961).

29 BARR, M. L.: The sex chromosomes of man. Amer. J. Obstet. Gynec. *93:* 608–616 (1965).

30 BARR, M. L. and BERTRAM, E. G.: A morphological distinction between neurones of the male and female, and the behavior of the nucleolar satellite during accelerated nucleoprotein synthesis. Nature (London) *163:* 676–677 (1949).

31 BARTELS, P. H., BAHR, G. F., BIBBO, M., RICHARDS, D. L., SONEK, M. G. and WIED, G. L.: Analysis of variance of the Papanicolaou staining reaction. Acta Cytol. *18:* 522–531 (1974).

32 BARTELS, P. H., LAYTON, J. and SHOEMAKER, R. L.: Digital microscopy, in

S. D. GREENBERG (ed.), Computer-assisted image analysis cytology, pp. 28–61 (S. Karger, Basel 1984).

33 BEAVER, D. L.: The ultrastructure of the kidney in lead intoxication with particular reference to intracellular inclusions. Amer. J. Path. *39:* 195–208 (1961).

34 BEERMANN, W.: Chromosomal differentiation in insects, in D. RUDNICK, Developmental cytology, pp. 83–103 (Ronald Press, New York 1959).

35 BERENBLUM, I.: Cocarcinogenesis. Brit. Med. Bull. *4:* 343–344 (1947).

36 BEREZNEY, R.: The nuclear matrix: a structural milieu for the intranuclear attachment and replication of DNA, in H. G. SCHWEIGER (ed.), International cell biology 1980–1981, pp. 214–224 (Springer-Verlag, Berlin/New York 1981).

37 BEREZNEY, R. and COFFEY, D. S.: Identification of a nuclear protein matrix. Biochem. Biophys. Res. Commun. *60:* 1410–1417 (1974).

38 BERG, J. W. and DURFEE, G. R.: Cytological presentation of endometrial carcinoma. Cancer *11:* 158–172 (1958).

39 BERNABEU, C. and LAKE, J. A.: Nascent polypeptide chains emerge from the exit domain of the large ribosomal subunit: Immune mapping of the nascent chain. Proc. Natl. Acad. Sci. USA *79:* 3111–3115 (1982).

40 BESSIS, M.: The death of a cell (motion picture) (Squibb, Rariton, N. J. 1959).

41 BIBBO, M., BARTELS, P. H., DYTCH, H. E. and WIED, G. L.: Computed cell image information, in S. D. GREENBERG (ed.), Computer-assisted image analysis cytology, Monographs in clinical cytology, vol. 9, pp. 62–100 (Karger, Basel 1984).

42 BLOOM, W. and FAWCETT, D. W.: A textbook of histology, 10th ed. (Saunders, Philadelphia 1975).

43 BOONE, C. W., SANFORD, K. K., FROST, J. K., MANTEL, N., GILL, G. W. and JONES, G. M.: Cytomorphologic evaluation of the neoplastic potential of 28 cell culture lines by a panel of diagnostic cytopathologists. Cancer Res. (submitted for publication, 1986).

44 BOULIKAS, T., WISEMAN, J. M. and GARRARD, W. T.: Points of contact between histone H1 and the histone octamer. Proc. Natl. Acad. Sci. USA *77:* 127–131 (1980).

45 BOURGEOIS, C. A., HEMON, D. and BOUTEILLE, M.: Structural relationship between the nucleolus and the nuclear envelope. J. Ultrastruct. Res. *68:* 328–340 (1979).

46 BOURNE, G. H.: Cytology and cell physiology, 3rd ed. (Academic Press, New York 1964).

47 BRACHET, J. and MIRSKY, A. E. (eds.): The cell (Academic Press, New York 1959–1964).

48 BRODERS, A. C.: Carcinoma, grading and practical application. Arch. Path. *2:* 376–381 (1926).

49 BRODY, I.: An ultrastructural study on the role of the keratohyalin granules in the keratinization process. J. Ultrastruct. Res. *3:* 84–104 (1959).

50 BRYAN, W. T. K. and BRYAN, M. P.: Cytologic diagnosis in otolaryngology. Trans. Am. Acad. Ophthal. Otolaryngol. *63:* 597–615 (1959).

51 BURGESS, D. R.: Reactivation of intestinal epithelial cell brush border motility: ATP dependent contraction via a terminal web contractile ring. J. Cell Biol. *95:* 853–863 (1982).

52 CAMERON, R.: Pathological changes in cells, in G. H. BOURNE, Cytology and cell physiology (Academic Press, New York 1964).

53 CASPERSSON, T. O.: Cell growth and cell function: A cytochemical study (Norton, New York 1971).

54 CHATTERJEE, S., KWITEROVICH, P. O., Jr., GUPTA, P. K., EROZAN, Y. S., ALVING, C. R. and RICHARDS, R. L.: Localization of urinary lactosylceramide in cytoplasmic vesicles of renal tubular cells in homozygous familial hypercholesterol-emia. Proc. Natl. Acad. Sci. USA *80:* 1313–1317 (1983).

55 CHRISTOPHERSON, W. M. and BROGHAMER, W., Jr.: Progression of experimental cervical dysplasia in the mouse. Cancer *14:* 201–204 (1961).

56 COMINGS, D. E.: The rationale for an ordered arrangement of chromatin in the interphase nucleus. Am. J. Hum. Genet. *20:* 440–460 (1968).

57 COMINGS, D. E. and WALLACK, A. S.: DNA-binding properties of nuclear matrix proteins. J. Cell Sci. *34:* 233–246 (1978).

58 COOPER, D., SCHERMER, A. and SUN, T.-T.: Classification of human epithelia and their neoplasms using monoclonal antibodies to keratins: strategies, applications and limitations. Lab. Invest. *52:* 243–256 (1985).

59 COTTON, R. E., HARWOOD, T. R. and WARTMAN, W. B.: Regeneration of aortic endothelium. J. Path. Bact. *81:* 171–180 (1961).

60 COWDRY, E. V.: Cancer cells (Saunders, Philadelphia 1955).

61 COWDRY, E. V. (ed.): Special cytology, 2nd ed. (Hoeber, New York 1963).

62 CRABBE, J. G.: Cytology of voided urine with special reference to 'benign' papilloma and some of the problems encountered in the preparation of the smears. Acta Cytol. *5:* 233–240 (1961).

63 CURSCHMANN, H.: Beiträge zur Physiologie der Kleinhirnschenkel. Geissen Bruhleche Buchdruckerei 2.1–8°: 36 (1868).

64 DARZYNKIEWICZ, Z.: Acridine orange as a molecular probe in studies of nucleic acids *in situ*, in M. R. MELAMED, P. F. MULLANEY and M. L. MENDELSOHN, Flow cytometry and sorting, pp. 283–316 (John Wiley, New York 1979).

65 EARNSHAW, W. C., HALLIGAN, B., COOKE, C. A., HECK, M. M. S. and LIU, L. F.: Topoisomerase II is a structural component of mitotic chromosome scaffolds. J. Cell Biol. *100*: 1706–1715 (1985).

66 EARNSHAW, W. C. and HECK, M. M. S.: Localization of topoisomerase II in mitotic chromosomes. J. Cell Biol. *100*: 1716–1725 (1985).

67 EBNER, H. and SCHNEIDER, W.: Zur Zytologie eines menschlichen Aszitestumor. Zbl. Gynaek. *78:* 1486–1494 (1956).

68 EROZAN, Y. S. and FROST, J. K.: Cytopathologic diagnosis of lung cancer. Semin. Oncol. *1:* 191–198 (1974).

69 EROZAN, Y. S. and GUPTA, P. K.: Cytology in surgical pathology, Chapter 26 in Z. A. KARCIOGLU and A. SOMEREN (eds.), Practical surgical pathology, pp. 1094–1201 (Collamore Press, Lexington 1985).

70 EROZAN, Y. S., PRESSMAN, N. J., DONOVAN, P. A., GUPTA, P. K. and FROST, J. K.: A comparative cytopathologic study of noninvasive and invasive squamous cell carcinoma of the lung. Analyt. Quant. Cytol. *1:* 50–56 (1979).

71 EVANS, H. J., BUCKLAND, R. A. and PARDUE, M. L.: Location of the genes coding for 18S and 28S ribosomal RNA in the human genome. Chromosoma *48:* 405–426 (1974).

72 EWING, J.: Precancerous diseases and precancerous lesions, especially in the breast. Med. Rec. *86:* 951–958 (1914).

73   FAKAN, S. and PUVION, E.: The ultrastructural visualization of nucleolar and extranucleolar RNA synthesis and distribution. Int. Rev. Cytol. *65:* 255–299 (1980).
74   FARQUHAR, M. G. and PALADE, G. E.: Junctional complexes in various epithelia. J. Cell Biol. *17:* 375–412 (1963).
75   FARQUHAR, M. G. and PALADE, G. E.: The Golgi apparatus (complex) – (1954–1981) – from artifact to center stage. J. Cell Biol. *91:* 77s–103s (1981).
76   FAWCETT, D. W.: The cell (Saunders, Philadelphia/London 1966).
77   FELL, H. B. and MELLANBY, E.: Metaplasia produced in cultures of chick ectoderm by high Vitamin A. J. Physiol. *119:* 470–488 (1953).
78   FINCH, J. T. and KLUG, A.: Solenoidal model for superstructure in chromatin. Proc. Natl. Acad. Sci. USA *73:* 1897–1901 (1976).
79   FINCH, J. T., LUTTER, L. C., RHODES, D., BROWN, R. S., RUSHTON, B., LEVITT, M. and KLUG, A.: Structure of nucleosome core particles of chromatin. Nature *269:* 29–36 (1977).
80   FINCH, J. T., NOLL, M. and KORNBERG, R. D.: Electron microscopy of defined length of chromatin. Proc. Natl. Acad. Sci. USA *72:* 3320–3322 (1975).
81   FLESCH, P.: Biochemical data on physiological and pathological epidermal keratinization. J. Soc. Cosmet. Chem. *7:* 521–530 (1956).
82   FLUHMANN, C. G.: The cervix uteri and its diseases (Saunders, Philadelphia 1961).
83   FORNI, M., HOFMAN, F. M., PARKER, J. W., LUKES, R. J. and TAYLOR, C. R.: B- and T-lymphocytes in Hodgkin's disease. Cancer *55:* 728–737 (1985).
84   FOX, C. H., CASPERSSON, T., KUDYNOWSKI, J., SANFORD, K. K. and TARONA, R. E.: Morphometric analysis of neoplastic transformation in rodent fibroblast cell lines. Cancer Res. *37:* 892–897 (1977).
85   FRANKE, W. W.: Structure, biochemistry, and functions of the nuclear envelope. Int. Rev. Cytol. *4* (Supp.): 71–236 (1974).
86   FRANKE, W. W., KLEINSCHMIDT, J. A., SPRING, H., KROHNE, G., GRUND, C., TRENDELENBURG, M. F., STOEHR, M. and SCHEER, U.: A nucleolar skeleton of protein filaments demonstrated in amplified nucleoli of *Xenopus laevis.* J. Cell Biol. *90:* 289–299 (1981).
87   FRANKE, W. W., SCHEER, U., KROHNE, G. and JARASCH, E.-D.: The nuclear envelope and the architecture of the nuclear periphery. J. Cell Biol. *91:* 39s–50s (1981).
88   FRANKE, W. W., SCHMID, E., OSBORN, M., and WEBER, K.: Different intermediate-sized filaments distinguished by immunofluorescence microscopy. Proc. Natl. Acad Sci. USA *75:* 5034–5038 (1978).
89   FRENSTER, J. H., ALLFREY, V. G. and MIRSKY, A. E.: Repressed and active chromatin isolated from interphase lymphocytes. Proc. Nat. Acad. Sci. (Washington) *50:* 1026–1032 (1963).
90   FRIEDELL, G. H.: Terminology for epithelial abnormalities of the uterine cervix. Am. J. Clin. Pathol. *44:* 280–282 (1965).
91   FROST, J. K.: Cancer cells in circulating blood: A comparison with cytologic criteria for other sites. Acta Cytol. *9:* 83–93 (1965).
92   FROST, J. K.: The cancer cell: criteria of malignancy, Chapter 34 in A. M. LILIENFELD and A. J. GIFFORD, Chronic diseases and public health, pp. 360–374 (Johns Hopkins Press, Baltimore 1966).

93 FROST, J. K.: Presidential address: Cytology's challenge today – for a better tomorrow. Acta Cytol. *10:* 311–315 (1966).
94 FROST, J. K.: Concepts basic to general cytopathology, 4th ed. (Johns Hopkins Press, Baltimore 1972).
95 FROST, J. K.: Spinal fluid cytology, in L. J. RUBINSTEIN, Tumors of the central nervous system, Fascicle 6, Atlas of tumor pathology, second series, pp. 370–378 (Armed Forces Institute of Pathology, Washington, D. C. 1972).
96 FROST, J. K.: Cytologic evaluation of endocrine status and somatic sex, Chapter 36, in E. R. NOVAK, G. S. JONES and H. W. JONES Jr., Novak's textbook of gynecology, 9th ed., pp. 756–781 (Williams and Wilkins, Baltimore 1975).
97 FROST, J. K.: Gynecologic clinical cytopathology, Chapter 37, in E. R. NOVAK, G. S. JONES and H. W. JONES Jr., Novak's textbook of gynecology, 9th ed., pp. 782–812 (Williams and Wilkins, Baltimore 1975).
98 FROST, J. K.: Television microscopy for education and consultation in pathology via coaxial cable, laser beam, and COMSAT satellite. Pathologist *33:* 605–609 (1979).
99 FROST, J. K.: Gynecologic and obstetric cytopathology, in J. D. WOODRUFF, Novak's gynecologic and obstetric pathology, 9th ed. (Saunders, Philadelphia 1986).
100 FROST, J. K.: Manual for the twenty-seventh postgraduate institute for pathologists in clinical cytopathology (The Johns Hopkins University School of Medicine and The Johns Hopkins Hospital, Baltimore 1986).
101 FROST, J. K., BALL, W. C., Jr., LEVIN, M. L., TOCKMAN, M. S., BAKER, R. R., CARTER, D., EGGLESTON, J. C., EROZAN, Y. S., GUPTA, P. K., KHOURI, N. F., MARSH, B. R. and STITIK, F. P.: Early lung cancer detection: results of the initial (prevalence) radiologic and cytologic screening in the Johns Hopkins study. Am. Rev. Resp. Dis. *130:* 549–554 (1984).
102 FROST, J. K., BALL, W. C., Jr., LEVIN, M. L., TOCKMAN, M. S., EROZAN, Y. S., GUPTA, P. K., EGGLESTON, J. C., PRESSMAN, N. J., DONITHAN, M. P. and KIMBALL, A. W., Jr.: Sputum cytopathology: use and potential in monitoring the workplace environment by screening for biological effects of exposure. JOM *28* (1986).
103 FROST, J. K., EROZAN, Y. S., and GUPTA, P. K.: Cytopathology, in Atlas of Early Lung Cancer, National Cancer Institute cooperative early lung cancer group (Igaku-Shoin, New York/Tokyo 1983).
104 FROST, J. K., GUPTA, P. K., EROZAN, Y. S., CARTER, D., HOLLANDER, D. H., LEVIN, M. L. and BALL, W. C., Jr.: Pulmonary cytologic alterations in toxic environmental inhalation. Hum. Pathol. *4:* 521–536 (1973).
105 FROST, J. K., PETRAKIS, N. L., WOLDE, C. E., SANFORD, K. K., MacCARDLE, R. C., STEWART, W. E. and GEY, G. O.: Cell modulations, maturation and neoplastic transformation. Acta Cytol. *6:* 399–402 (1962).
106 FROST, J. K., PRESSMAN, N. J., ALBRIGHT, C. D., GILL, G. W. and VANSICKEL, M. H.: Centrifugal separation of cells in sputum. J. Histochem. Cytochem. *27:* 7–13 (1979).
107 FROST, J. K, PRESSMAN, N. J., GILL, G. W., SHOWERS, R. L. and ALBRIGHT, C. D.: Computerized image processing of clinical cytopathology specimens using scanning transmission optical microscopy. Analyt. Quant. Cytol. *2:* 295 (1980).

108 FROST, J. K. and STREET, C. M.: Cytology. Part I (non-malignant) (motion picture) (Wexler, Los Angeles 1962).

109 FROST, J. K. and STREET, C. M.: Cytology. Part II (malignant) (motion picture) (Wexler, Los Angeles 1962).

110 FROST, J. K., TYRER, H. W., PRESSMAN, N. J., ADAMS, L. A., VANSICKEL, M. H., ALBRIGHT, C. D., GILL, G. W. and TIFFANY, S. M.: Automatic cell identification and enrichment in lung cancer. III. Light scatter (size) and two fluorescent parameters (DNA, RNA) (in squamous cell carcinoma). J. Histochem. Cytochem. *27:* 557–559 (1979).

111 FU, Y. S., REAGAN, J. W., RICHART, R. M. and TOWNSEND, E. E.: Nuclear DNA and histological studies of genital lesions in diethylstilbesterol-exposed progeny. II. Intraepithelial glandular abnormalities. Am. J. Clin. Pathol. *72:* 515–520 (1979).

112 GALLAGHER, C. H., JUDAH, D. J. and REES, K. R.: Enzyme changes during liver autolysis. J. Path. Bact. *72:* 247–256 (1956).

113 GALVIN, G. A., JONES, H. W. and TeLINDE, R. W.: The significance of basal-cell hyperactivity in cervical biopsies. Amer. J. Obstet. Gynec. *70:* 808–821 (1955).

114 GEIRSSON, G., WOODWORTH, F. E., PATTEN, S. F. and BONFIGLIO, T. A.: Epithelial repair and regeneration in the uterine cervix. I. An analysis of the cells. Acta Cytol. *21:* 371–378 (1977).

115 GEORGIEV, G. P. and CHENTSOV, Yu. S.: The structure of the nucleus. Proc. Acad. Sci., USSR *132:* 199–202 (1960) (Transl.: DOLKAIDY, Biol. Sci. Sec. *132:* 357–360 (1960)).

116 GEORGIEV, G. P., NEDOSPASOV, S. A. and BAKAYEV, V. V.: Supranucleosomal levels of chromatin organization, in H. BUSCH (ed.), The cell nucleus (Academic Press, New York 1978).

117 GERACE, L. and BLOBEL, G.: The nuclear envelope lamina is reversibly depolymerized during mitosis. Cell *19:* 277–287 (1980).

118 GERACE, L., BLUM, A. and BLOBEL, G.: Immunocytochemical localization of the major polypeptides of the nuclear pore complex-lamina fraction. Interphase and mitotic distribution. J. Cell Biol. *79:* 546–566 (1978).

119 GEY, G. O.: Normal and malignant cells in tissue culture. Ann. N. Y. Acad. Sci. *76:* 547–549 (1958).

120 GHADIALLY, F. N.: Ultrastructural pathology of the cell and matrix, 2nd. ed. (Butterworths, London/Boston 1982).

121 GILBERT, W. and MULLER-HILL, B.: Isolation of the lac repressor. Proc. Nat. Acad. Sci. (Washington) *56:* 1891–1898 (1966).

122 GILL, G. W. and FROST, J. K.: In defense of cover glasses. Cytotechn. Bull. *20:* 51–54 (1983).

123 GILL, G. W., FROST, J. K. and MILLER, K. A.: A new formula for a half-oxidized hematoxylin solution that neither overstains nor requires differentiation. Acta Cytol. *18:* 300–311 (1974).

124 GILL, G. W. and PLOWDEN, K. M: Laboratory cytopathology techniques for specimen preparation, 7th ed., in J. K. FROST, Manual for the twenty-seventh postgraduate institute for pathologists in clinical cytopathology (The Johns Hopkins University School of Medicine and The Johns Hopkins Hospital, Baltimore 1986).

125 GOLDBERG, M. F., EROZAN, Y. S., DUKE, J. R. and FROST, J. K.: Cytopatho-

logic and histopathologic aspects of Fuchs' heterochromic iridocyclitis. Arch. Opthal. *74:* 604–609 (1965).

126 GRAHAM, R. M.: Cytologic diagnosis of cancer, 3rd ed. (Saunders, Philadelphia 1972).

127 GRAY, L. A.: Dysplasia, carcinoma in situ and micro-invasive carcinoma of the cervix uteri (Thomas, Springfield, Ill. 1964).

128 GREENBERG, S. D. (ed.): Computer-assisted image analysis cytology, Monographs in clinical cytology, vol. 9 (Karger, Basel 1984).

129 GRUNZE, H.: The comparative diagnostic accuracy, efficiency and specificity of cytologic techniques used in the diagnosis of malignant neoplasm in serous effusions of the pleural and pericardial cavities. Acta Cytol. *8:* 150–163 (1964).

130 GUPTA, P. K., ALLBRITTON, N., EROZAN, Y. S. and FROST, J. K.: Occurrence of cilia in exfoliated ovarian adenocarcinoma cells. Diag. Cytopathol. *1:* 223–225 (1985).

131 GUPTA, P. K. and FROST, J. K.: Cytologic changes associated with asbestos exposure. Semin. Oncol. *8:* 283–289 (1981).

132 GUPTA, P. K. MYERS, J. D., BAYLIN, S. B., MULSHINE, J. L., CUTTITTA, F. and GAZDAR, A. F.: Improved antigen detection in ethanol-fixed cytologic specimens: a modified avidin-biotin-peroxidase complex (ABC) method. Diag. Cytopathol. *1:* 133–136 (1985).

133 HAAM, E. von: Some observations in the field of exfoliative cytology. Am. J. Clin. Pathol. *24:* 652–662 (1954).

134 HAAM, E. von and SCARPELLI, D. G.: Experimental carcinoma of cervix; comparative cytologic and histologic study. Cancer Res. *15:* 449–455 (1955).

135 HAM, A. W.: Histology, rev. 8th ed. (Lippincott, Philadelphia 1979).

136 HAMKALO, B. A., MILLER, O. L., Jr. and BAKKEN, A. H.: Ultrastructure of active eukaryotic genomes. Cold Spring Harbor Symp. Quant. Biol. *38:* 915–919 (1973).

137 HANCOCK, R. and BOULIKAS, T.: Functional organization in the nucleus. Int. Rev. Cytol. *79:* 165–214 (1982).

138 HANCOCK, R. and HUGHES, M. E.: Organisation of DNA in the interphase nucleus. Biol. Cell *44:* 201–212 (1982).

139 HANDLEMAN, S. L., SANFORD, K. K., TARONE, R. E. and PARSHAD, R.: The cytology of spontaneous neoplastic transformation in culture. In Vitro *13:* 526–536 (1977).

140 HAYFLICK, L.: The cell biology of human aging. New Eng. J. Med. *295:* 1302–1308 (1976).

141 HERLAN, G., GIESE, G. and WUNDERLICH, F.: Influence of nuclear membrane lipid fluidity on nuclear RNA release. Exp. Cell Res. *118:* 305–309 (1979).

142 HERLAN, G., GIESE, G. and WUNDERLICH, F.: In vitro ribosomal ribonucleoprotein transport upon nuclear expansion. Biochemistry *19:* 3960–3966 (1980).

143 HEWISH, D. R. and BURGOYNE, L. A.: Chromatin sub-structure. The digestion of chromatin DNA at regularly spaced sites by a nuclear deoxyribonuclease. Biochem. Res. Comm. *52:* 504–510 (1973).

144 HOLLANDER, D. H. and GUPTA, P. K.: Detached ciliated tufts in cervico-vaginal smears. Acta Cytol. *18:* 367–369 (1974).

145 HOWDON, W. M., HOWDON, A. K., FROST, J. K. and WOODRUFF, J. D.: Cyto- and histopathologic correlation in mixed mesenchymal tumors of the uterus. Amer. J. Obstet. Gynec. *89:* 670–679 (1964).

146 HUMBOLDT, A. von: Cosmos (Gottascher 1845) (Transl.: E. C. OTTE (Harper and Bros., New York 1850)).

147 IDE, G., SUNTZEFF, V. and COWDRY, E. V.: A comparison of the histopathology of tracheal and bronchial epithelium of smokers and nonsmokers. Cancer *12:* 473–484 (1959).

148 KAMINETZKY, H. A. and McGREW, E. A.: The effect of podophyllin on the human endocervix. Obstet. Gynec. *18:* 255–258 (1961).

149 KAPLAN, L., MASIN, F., MASIN, M., CARLETON, R. and BERTALANFFY, L. von: Acridine orange fluorochrome in the study of normal and malignant epithelium of the uterine cervix. Amer. J. Obstet. Gynec. *80:* 1063–1073 (1960).

150 KATO, H., KONAKA, C., ONO, J., TAKAHASHI, M. and HAYATA, Y: Cytology of the lung (Igaku-Shoin, Tokyo/New York 1983).

151 KAUFMAN, S. H., COFFEY, D. S. and SHAPER, J. H.: Considerations in the isolation of rat liver nuclear matrix, nuclear envelope, and pore complex lamina. Exp. Cell Res. *132:* 105–123 (1981).

152 KEEBLER, C. M. and REAGAN, J. W. (eds.): A manual of cytotechnology (American Society of Clinical Pathologists Press, Chicago 1983).

153 KOEHLER, A.: Ein neues Beleuchtungsverfahren für mikrophotographische Zwecke. A. wiss. Mikr. *10:* 433–440 (1893).

154 KOPROWSKA, I., AN, S. H., CORSEY, D., DRACOPOULOS, I. and VASKELIS, P. S.: Cytologic patterns of developing bronchogenic carcinoma. Acta Cytol. *9:* 424–430 (1965).

155 KOSS, L. G.: Diagnostic cytology and its histologic bases, 3rd ed. (Lippincott, Philadelphia 1979).

156 KOSS, L. G. and SHERMAN, A. B.: Image analysis of cells in the sediment of voided urine, in S. D. GREENBERG (ed.), Computer-assisted image analysis cytology, Monographs in clinical cytology, vol. 9, pp. 148–162 (Karger, Basel 1984).

157 KOSS, L. G., WOYKE, S. and OLSZEWSKI, W.: Aspiration biopsy, cytologic interpretation and histologic bases (Igaku-Shoin, New York/Tokyo 1984).

158 KOTIN, P.: Carcinogenesis of the lung: Environmental and host factors, in A. LIEBOW and D. SMITH, The lung (Williams and Wilkins, Baltimore 1968).

159 KOTIN, P. and WISELEY, D. V.: Production of lung cancer in mice by inhalation exposure to influenza virus and aerosols of hydrocarbons. Progr. Exp. Tumor Res. *3:* 186–215 (1963).

160 LAEMMLI, U. K., CHENG, S. M., ADOLPH, K. W., PAULSON, J. R., BROWN, J. A. and BAUMBACK, W. R.: Metaphase chromosome structure: the role of non-histone proteins. Cold Spring Harbor Quant. Biol. *42:* 351–360 (1977).

161 LAKE, J. A.: Evolving ribosome structure: domains in archaebacteria, eubacteria, and eucaryotes. Cell *33:* 318–319 (1983).

162 LAMPE, V. A.: A cyto-histopathological correlation of dysplasia, carcinoma in situ, and invasive epidermoid carcinoma of the uterine cervix (Thesis, Johns Hopkins School of Cytotechnology, Baltimore 1967).

163 LEHNINGER, A. L.: The mitochondrion. Molecular basis of structure and function (W. A. Benjamin, New York/Amsterdam 1965).

164 LEHNINGER, A. L.: Biochemistry; the molecular basis of cell structure and function, 2nd ed. (Worth Publishers, Inc., New York 1975).

165 LEWIN, B.: Genes (John Wiley and Sons, New York 1983).

166 LIAO, L. W. and COLE, R. D.: Condensation of dinucleosomes by individual subfractions of H-1 histone. J. Biol. Chem. *256:* 10124–10128 (1981).

167 LILLIE, R. D.: H. J. Conn's biological stains, 9th ed. (Williams and Wilkins, Baltimore 1977).

168 LILLIE, R. D. and FULLMER, H. M.: Histopathologic technic and practical histochemistry, 4th ed. (McGraw-Hill, New York 1976).

169 LINMAN, J. W.: Principles of hematology, p. 4 (Macmillan, New York 1966).

170 LIU, L.: DNA topoisomerases – enzymes that catalyze the breaking and rejoining of DNA. CRC Crit. Rev. Biochem. *15:* 1–24 (1983).

171 LIU, L. F., ROWE, T. C., YANG, L., TEWEY, K. M. and CHEN, G. L.: DNA cleavage by mammalian DNA topoisomerase II. J. Biol. Chem. *258:* 15365–15370 (1983).

172 LJUNG, B.-M.: Fine needle aspiration of breast, in G. L. WIED, L. G. KOSS and J. W. REAGAN (eds.), Compendium on diagnostic cytology, 5th ed. (Tutorials of Cytology, Chicago 1983).

173 LOPES CARDOZO, P.: Advantages and disadvantages of the May-Gruenwald Giemsa procedure in exfoliative cytology. Acta Cytol. *2:* 284–288 (1958).

174 LOPES CARDOZO, P.: Atlas of clinical cytology (Targa b.v. 's-Hertogenbosch, Leiden, The Netherlands 1975).

175 LOWHAGEN, T. and WILLEMS, J.-S.: Aspiration biopsy cytology in diseases of the thyroid, in L. G. KOSS and D. COLEMAN (eds.), Advances in clinical cytology (Butterworths, London 1981).

176 LUKES, R. J. and COLLINS, R. D.: Immunological characterization of human malignant lymphomas. Cancer *34:* 1488–1503 (1974).

177 LUTTER, L. C.: Precise location of DNase I cutting sites in the nucleosome core determined by high resolution gel electrophoresis. Nucleic Acids Res. *6:* 41–56 (1979).

178 MANN, R. B., JAFFE, E. D. and BERARD, C. W.: Malignant lymphomas: a conceptual understanding of morphologic diversity. Am. J. Pathol. *94:* 105–192 (1979).

179 MARSDEN, M. P. F. and LAEMMLI, U. K.: Metaphase chromosome structure: Evidence for a radial loop model. Cell *17:* 849–858 (1979).

180 MASIN, F. and MASIN, M.: Gradients of maturation and basic dye desorption in normal and malignant cervical cells. Cancer *16:* 1408–1424 (1963).

181 MATHOG, D., HOCHSTRASSER, M., GRUENBAUM, Y., SAUMWEBER, H. and SEDAT, J.: Characteristic folding pattern of polytene chromosomes in *Drosophila* salivary gland nuclei. Nature *308*: 414–421 (1984).

182 MATOLTSY, A. G. and MATOLTSY, M. N.: Cytoplasmic droplets of pathologic horny cells. J. Invest. Derm. *38:* 323–325 (1962).

183 MATOLTSY, A. G. and MATOLTSY, M. N.: A study of morphological and chemical properties of keratohyalin granules. J. Invest. Derm. *38:* 237–247 (1962).

184 MATOLTSY, A. G. and MATOLTSY, M. N.: The chemical nature of keratohyalin granules of the epidermis. J. Cell Biol. *47:* 593–603 (1970).

185 McDOWELL, E. M. and TRUMP, B. F.: Histogenesis of preneoplastic and neoplas-

tic lesions in tracheobronchial epithelium. Surv. Synth. Path. Res. *2:* 235–279 (1983).

186 McKAY, D. G., TERJANIAN, B., POSCHYACHINDA, D., YOUNGE, P. A. and HERTIG, A. T.: Clinical and pathologic significance of anaplasia (atypical hyperplasia) of the cervix uteri. Obstet. Gynec. *13:* 2–22 (1959).

187 McMANUS, J. F. A. and MOWRY, R. W.: Staining methods; histologic and histochemical (Hoeber, New York 1960).

188 MELAMED, M. R., VOUTSA, N. G. and GRABSTOLD, H.: Natural history and clinical behavior of in situ carcinoma of the human bladder. Cancer *17:* 1533–1545 (1964).

189 MELLORS, R. S., KEANE, J. F., Jr. and PAPANICOLAOU, G. N.: Nucleic acid content of the squamous cancer cell. Science *116:* 265–269 (1952).

190 MITTWOCH, U.: Sex chromosomes (Academic Press, New York 1967).

191 MUKAWA, A., KAMITSUMA, Y., JISAKI, F., TANAKA, N., IKEDA, H., UENO, T. and TSUNEKAWA, S.: Progress report on experimental use of CYBEST Model 2 for practical gynecologic mass screening. Analyt. Quant. Cytol. *5:* 31–34 (1983).

192 MURCIA, G. de and KOLLER, Th.: The electron microscopic appearance of soluble rat liver chromatin mounted on different supports. Biol. Cell *40:* 165–174 (1981).

193 NAIB, Z. M.: Exfoliative cytopathology, 3rd ed. (Little, Brown and Company, Boston 1985).

194 NASIELL, M., KATO, H., AUER, G., ZETTERBERG, A., ROGER, V. and KARLEN, L.: Cytomorphological grading and Feulgen DNA-analysis of metaplastic and neoplastic bronchial cells. Cancer *41:* 1511–1521 (1978).

195 NAYLOR, B.; RAILEY, C.: A pitfall in the cytodiagnosis of sputum of asthmatics. J. clin. Path. *17:* 84–89 (1964).

196 National Cancer Institute cooperative early lung cancer group: manual of procedures, 2nd ed. (NIH Publication No. 79–1972, Washington, D.C. 1979).

197 National Cancer Institute sponsored study of classifications of non-Hodgkin's lymphomas. Summary and description of a working formulation for clinical usage. Cancer *49:* 2112–2135 (1982).

198 NG, A. B. P. and REAGAN, J. W.: Pathology and cytopathology of microinvasive squamous cell carcinoma of the uterine cervix, in G. L. WIED, L. G. KOSS and J. W. REAGAN (eds.), Compendium of diagnostic cytology (Tutorials in Cytology, Chicago 1983).

199 NUNEZ, C. and REAGAN, J. W.: Fine needle aspiration biopsy of lymph nodes, in G. L. WIED, L. G. KOSS and J. W. REAGAN (eds.), Compendium of diagnostic cytology (Tutorials in Cytology, Chicago 1983).

200 OLINS, A. L. and OLINS, D. E.: Spheroid chromatin units (*v* bodies). Science *183:* 330–332 (1974).

201 PAPANICOLAOU, G. N.: A survey of the actualities and potentialities of exfoliative cytology in cancer diagnosis. Ann. Intern. Med. *31:* 661–674 (1949).

202 PAPANICOLAOU, G. N.: Atlas of exfoliative cytology; suppl. 1956, 1960 (Harvard Univ. Press, Cambridge, Mass. 1954).

203 PAPANICOLAOU, G. N.: Degenerative changes in ciliated cells exfoliating from bronchial epithelium as cytologic criterion in diagnosis of diseases of lung ciliocytophthoria. N. Y. State J. Med. *56:* 2647–2650 (1956).

204 PAPANICOLAOU, G. N. and TRAUT, H. F.: Diagnosis of uterine cancer by the vaginal smear (Commonwealth Fund, New York 1949)

205 PAPANICOLAOU, G. N., TRAUT, H. F. and MARCHETTI, A. A.: The epithelia of woman's reproductive organs (Commonwealth Fund, New York 1948).

206 PATTEN, S. F., Jr.: Diagnostic cytopathology of the uterine cervix, 2nd ed. (S. Karger, Basel 1978).

207 PATTEN, S. F., Jr.: Benign proliferative reactions of the uterine cervix, in G. L. WIED, L. G. KOSS and J. W. REAGAN (eds.), Compendium on diagnostic cytology, 5th ed. (Tutorials of Cytology, Chicago 1983).

208 PAUL, J.: Metabolic processes in normal and cancer cells, in E. J. AMBROSE and F. J. C. ROE, The biology of cancer (van Nostrand, London 1966).

209 PAUL, J.: Template activity of DNA is restricted in chromatin. J. Molec. Biol. *16:* 242–244 (1966).

210 PAUL, J. and GILMOUR, R. S.: Restriction of deoxyribonucleic acid template activity in chromatin is organ-specific. Nature (London) *210:* 992–993 (1966).

211 PEARSE, A. G. E.: Histochemistry, theoretical and applied, 4th ed. vol. 2 (Churchill-Livingston, Edinburgh/New York 1985).

212 PEDERSON, T.: Chromatin structure and gene transcription: Nucleosomes permit a new synthesis. Int. Rev. Cytol. *55:* 1–22 (1978).

213 PEHRSON, J. and COLE, R. D.: Histone H1 accumulates in growth-inhibited cultured cells. Nature (London) *285:* 43–44 (1980).

214 POLLARD, T. D. and CRAIG, S. W.: Mechanism of actin polymerization. Trends Biochem. Sci. *7:* 55–58 (1982).

215 PRESSMAN, N. J.: Implementation of an automated high-resolution audio-visual teleconferencing system for medical image communications, processing and analysis. Comput. Biol. Med. *16* (1986).

216 PRESSMAN, N. J., ALBRIGHT, C. D. and FROST, J. K.: Centrifugal separation of cells in sputum specimens from patients with adenocarcinoma. Cytometry *1:* 260–264 (1981).

217 PRESSMAN, N. J. and FROST, J. K.: BIDAS: a microscope imaging communications processing and analysis system for cytopathology, in Proceedings IEEE international symposium on medical images and icons, pp. 372–379 (1984).

218 PRESSMAN, N. J. and FROST, J. K.: Design of a high-resolution microcomputer-controlled audio-video teleconferencing system for medical education, research, and patient care. Analyt. Quant. Cytol. *8* (1986).

219 PRESSMAN, N. J., FROST, J. K., ALBRIGHT, C. D., MAKOWSKA, W. K., SHOWERS, R. L., BITMAN, W. R., JEFFRIES, L. H., FROST, J. K., Jr. and GILL, G. W.: Quantitative nuclear analysis of developing squamous cell carcinoma of the hamster cheek pouch. Analyt. Quant. Cytol. *5:* 216 (1983).

220 PRESSMAN, N. J., FROST, J. K., SHOWERS, R. L., GILL, G. W., FROST, J. K., Jr. and ALBRIGHT, C. D.: Scanning transmission optical microscopy for automated cytopathology. Cytometry *7* (1986).

221 PRESSMAN, N. J. and WIED, G. L. (eds.): The automation of cancer cytology and cell image analysis (Tutorials of Cytology, Chicago 1979).

222 PUVION-DUTILLEUL, F. and PUVION, E.: New aspects of intracellular structures following partial decondensation of chromatin: a cytochemical and high-resolution autoradiographical study. J. Cell Sci. *42:* 305–321 (1981).

223 QUASTEL, J. H.: Symposium on transport reactions at the cell membrane: Introductory survey. Canad. J. Biochem. *42:* 907–916 (1964).

224 REAGAN, J. W.: Carcinoma in situ and the pathologist. Amer. J. Clin. Path. *25:* 791–794 (1955).

225 REAGAN, J. W.: Presidential address. Acta Cytol. *9:* 265–267 (1965).

226 REAGAN, J. W., HARMONIC, J. M. and WENTZ, W. B.: Analytical study of the cells in cervical squamous cell cancer. Lab. Invest. *6:* 241–250 (1957).

227 REAGAN, J. W., HICKS, D. J. and SCOTT, R. B.: Atypical hyperplasia of the uterine cervix. Cancer *8:* 42–52 (1955).

228 REAGAN, J. W. and NG, A. B. P.: The cells of uterine adenocarcinoma, 2nd, revised ed. (Karger/Williams and Wilkins, Basel/Baltimore 1973).

229 REAGAN, J. W. and PATTON, S. F., Jr.: Dysplasia: A basic reaction to injury in the uterine cervix. Ann. N. Y. Acad. Sci. *97:* 662–682 (1962).

230 REAGAN, J. W. and WENTZ, W. B.: Changes in desquamated cells in carcinogenesis. Arch. Path. *67:* 287–292 (1959).

231 REAGAN, J. W. and WENTZ, W. B.: Changes in mouse cervix antedating induced cancer. Cancer *12:* 389–395 (1959).

232 REAGAN, J. W., WENTZ, W. B. and MACHICAO, N.: Induced cancer of the cervix uteri in the mouse. Arch. Path. *60:* 451–456 (1955).

233 REBUCK, J. W.: Technic for the study of leukocytic functions in man. Meth. Med. Res. *7:* 161–164 (1958).

234 RENZ, M., NEHLS, P. and HOZIER, J.: Involvement of histone H1 in the organization of the chromosome fiber. Proc. Natl. Acad. Sci. USA *74:* 1879–1883 (1977).

235 RHUMBLER, L.: Physikalische Analyse von Lebenserscheinungen der Zelle. 1. Arch. Entwickl. Mech. Org. *7:* 103–198 (1898).

236 RICHARDSON, J. C. W. and MADDY, A. H.: The polypeptides of rat liver nuclear envelope. I. Examination of nuclear pore complex polypeptides by solid-state lactoperoxidase labelling. J. Cell Sci. *43:* 253–267 (1980).

237 RICHART, R. M.: Natural history of cervical intraepithelial neoplasia. Clin. Obstet. Gynecol. *10:* 748–784 (1967).

238 RICHART, R. M., LERCH, V. and BARRON, B. A.: A time-lapse cinematographic study *in vitro* in mitosis of normal human cervical epithelium, dysplasia, and carcinoma *in situ.* JNCI *39:* 571–577 (1967).

239 RIEDER, C. L.: The formation, structure and composition of the mammalian kinetochore and kinetochore fiber. Int. Rev. Cytol. *79:* 1–58 (1982).

240 RITOSSA, F. M. and SPIEGELMAN, S.: Localization of DNA complementary to ribosomal RNA in the nucleolus organizer region of Drosophilia melanogaster. Proc. Natl. Acad. Sci. USA *53:* 737–745 (1965).

241 ROBERTIS, E. D. P. de, NOWINSKI, W. W. and SAEZ, F. A.: General cytology, 3rd ed. (Saunders, Philadelphia 1960).

242 ROBERTIS, E. M. De: Nucleocytoplasmic segregation of proteins and RNA. Cell *32:* 1021–1025 (1983).

243 ROSENTHAL, D. L. and SUFFIN, S. C.: Predictive value of digitized cell images for the prognosis of cervical neoplasia, in S. D. GREENBERG (ed.), Computer-assisted image analysis cytology, Monographs in clinical cytology, vol. 9, pp. 163–180 (Karger, Basel 1984).

244 ROUS, P. and KIDD, J. G.: Conditional neoplasms and subthreshold neoplastic states. J. Exp. Med. *73:* 365–390 (1941).
245 RUSCH, H. P.: Extrinsic factors that influence carcinogenesis. Physiol. Rev. *24:* 177–204 (1944).
246 SACCOMANNO, G., ARCHER, V. E., AUERBACH, O., SAUNDERS, R. P. and BRENNEN, L. M.: Development of carcinoma of the lung as reflected in exfoliated cells. Cancer *33:* 256–270 (1974).
247 SACCOMANNO, G., SAUNDERS, R. P., ARCHER, V. E., AUERBACH, O., KUSCHNER, M. and BECKLER, P. A.: Cancer of the lung: the cytology of sputum prior to the development of carcinoma. Acta Cytol. *9:* 413–423 (1965).
248 SALAMAN, M. H.: Cocarcinogenesis. Brit. Med. Bull. *14:* 116–120 (1958).
249 SANFORD, K. K., BOONE, C. W., MERWIN, R. M., JONES, G. M. and GARRISON, C. U.: The plate-implant as a bioassay for the neoplastic potential of cultured cells. Int. J. Cancer *25:* 509–516 (1980).
250 SANFORD, K. K., PARSHAD, R., STANBRIDGE, E. J., FROST, J. K., JONES, G.M., WILKINSON, J. E. and TARONE, R. E.: Analysis of chromosomal radiosensitivity during the G_2 cell cycle period and cytopathology of human normal X tumor cell hybrids. Cancer Res. (submitted for publication, 1986).
251 SAPHIR, O., LEVENTHAL, M. L. and KLINE, T. S.: Podophyllin-induced dysplasia of the cervix uteri; its histological resemblance to carcinoma in situ. Amer. J. Clin. Path. *32:* 446–456 (1959).
252 SCHAEFFER, A. A.: Amoeboid movement (Princeton Univ. Press, Princeton, N. J. 1921).
253 SCHMID, F. and STEIN, J. (eds.): Cell research and cellular therapy, 2nd ed. (Ott, Thoune, Switzerland 1967).
254 SCHWARTZ, L. M. and AZAR, M. M. (eds.): Advanced cell biology (van Nostrand Reinhold, New York 1981).
255 SEALE, R. L. and ARONSON, A. I.: Chromatin-associated proteins of the developing sea urchin embryo. II. Acid-soluble proteins. J. Mol. Biol. *75:* 647–658 (1973).
256 SEYEDIN, S. M. and COLE, R. D.: H1 histones of trout. J. Biol. Chem. *256:* 442–444 (1981).
257 SHELTON, K. R., EGLE, P. M. and COCHRAN, D. L.: The nuclear lamins: their properties and functions, in G. G. MAUL (ed.), The nuclear envelope and the nuclear matrix, Wistar Symposium Series, vol. 2, pp. 157–168 (Liss, New York 1982).
258 SIEGLER, E. E.: Microdiagnosis of carcinoma in situ of the uterine cervix. A comparative study of pathologists' diagnoses. Cancer *9:* 463–469 (1956).
259 SIMPSON, R. T.: Structure of the chromatosome, a chromatin particle containing 160 base pairs of DNA and all the histones. Biochemistry *17:* 5524–5531 (1978).
260 SINDEN, R. R. and PETTIJOHN, D. E.: Chromosomes in living *E. coli* cells are segregated into domains of supercoiling. Proc. Natl. Acad. Sci. USA *78:* 224–228 (1981).
261 SMETANA, K., STEELE, W. J. and BUSCH, H.: A nuclear ribonucleoprotein network. Exp. Cell Res. *31:* 198–201 (1963).
262 SMITH, G. R.: DNA supercoiling: another level for regulating gene expression. Cell *24:* 599–600 (1981).
263 SPRIGGS, A. I.: The cytology of effusions in the pleural, pericardial and peritoneal cavities, 2nd ed., reissued with revisions (William Heinemann, London 1976).

264 STANTON, M. F. and GLACKWELL, R.: Induction of epidermoid carcinoma in lungs of rats: A new method based upon deposition of methylcholanthrene in areas of pulmonary infarction. J. Nat. Cancer Inst. *27:* 375–407 (1961).

265 STOWELL, R. E., CHANG, J. P. and BERENBOM, M.: Histochemical studies of necrosis of mouse liver tissue in the peritoneal cavity. Lab. Invest. *10:* 111–128 (1961).

266 STREHLER, B. L.: Time cells and aging (Academic Press, New York 1962).

267 SUN, T.-T., SHIH, C., and GREEN, H.: Keratin cytoskeletons in epithelial cells of internal organs. Proc. Natl. Acad. Sci. USA *76:* 2813–2817 (1979).

268 SUNDIN, O. and VARSHAVSKY, A.: Arrest of segregation leads to accumulation of highly intertwined catenated dimers: dissection of the final stages of SV40 DNA replication. Cell *25:* 659–669 (1981).

269 SWANSON, C. P.: Cytology and cytogenetics (Prentice-Hall, Englewood Cliffs, N. J. 1957).

270 SWANSON, C. P. and WEBSTER, P. L.: The cell, 5th ed. (Prentice-Hall, Englewood Cliffs, N. J. 1985).

271 SWIFE, H.: Report of the nucleolus nomenclature committee. Nat. Cancer Inst. Monogr. *23:* 573–574 (1966).

272 TAHMISIAN, T. N.: Cytological effects, in M. ERRERO and A. FORSSBERG, Mechanisms in radiology, vol. I, ch. 4 (Academic Press, New York 1961).

273 TAKAHASHI, M.: Color atlas of cancer cytology, 2nd ed. (Igaku-Shoin, Tokyo/New York 1981).

274 TAKAHASHI, K. and TASHIIO, Y.: Binding of antibodies against Histone H1 to unfolded and folded nucleofilaments. Eur. J. Biochem. *97:* 353–360 (1979).

275 TANAKA, N., IKEDA, H., UENO, T., MUKAWA, A. and KAMITSUMA, K.: Field test and experimental use of CYBEST Model 2 for practical gynecologic mass screening. Analyt. Quant. Cytol. *1:* 122–126 (1979).

276 TASSONI, E. M.: Pools of lymphocytes: Significance in pulmonary secretions. Acta Cytol. *7:* 168–173 (1963).

277 THOMA, F. and KOLLER, Th.: Influence of histone H1 on chromatin structure. Cell *12:* 101–107 (1977).

278 THOMA, F. and KOLLER, Th.: Unravelled nucleosomes, nucleosome beads and higher order structures of chromatin: Influence of non-histone components and Histone H1. J. Mol. Biol. *149:* 709–733 (1981).

279 THOMA, F., KOLLER, Th. and KLUG, A.: Involvement of histone H1 in the organization of the nucleosome and of the salt-dependent superstructures of chromatin. J. Cell Biol. *83:* 403–427 (1979).

280 TRUMP, B. F. and ARSTILA, A. U.: Cellular reaction to injury, in M. LaVIA and R. HILL (eds.), Principles of pathobiology (Oxford University Press, New York 1975).

281 TRUMP, B. F., GOLDBLATT, P. J. and STOWELL, R. E.: Nuclear and cytoplasmic changes during necrosis *in vitro* (autolysis); an electron microscopic study. Am. J. Pathol. *43:* 23a (1963).

282 TYRER, H. W., FROST, J. K., PRESSMAN, N. J., ADAMS, L. A., ALBRIGHT, C. D., VANSICKEL, M. H. and TIFFANY, S. M.: Automatic cell identification and enrichment in lung cancer: IV. Small cell carcinoma analysis by light scatter and two fluorescence parameters. Flow Cytometry *IV:* 462–472 (Universitetsforlaget 1980).

283 TYRER, H. W., GOLDEN, J. F., VANSICKEL, M. H., ECHOLS, C. K., FROST,

J. K., WEST, S. S., PRESSMAN, N. J., ALBRIGHT, C. D., ADAMS, L. A. and GILL, G. W.: Automatic cell identification and enrichment in lung cancer. II. Acridine orange for cell sorting of sputum. J. Histochem. Cytochem. *27:* 552–556 (1979).

284 TYRER, H. W., PRESSMAN, N. J., ALBRIGHT, C. D. and FROST, J. K.: Automatic cell identification and enrichment in lung cancer. V. Adenocarcinoma and large cell undifferentiated carcinoma. Cytometry *6:* 37–46 (1985).

285 ULTMAN, J. E., KOPROWSKA, I. and ENGLE, R. L., Jr.: A cytologic study of lymph node imprints. Cancer *11:* 507–524 (1958).

286 VIGORITA, V. J., GUPTA, P. K., BARGERON, D. B. and FROST, J. K.: Occurrence and identification of intracellular calcium crystals in pulmonary specimens. Acta Cytol. *23:* 49–52 (1979).

287 VILLARREAL, L.: A paranuclear extract contains a unique set of viral transcripts late in SV40 infection. Virology *113:* 663–671 (1981).

288 VINCENT, W. S. and MILLER, O. L., Jr.: International symposium on the nucleolus, its structure and function. Nat. Cancer Inst. Monogr. *23* (1966).

289 VIRCHOW, R.: Die Cellularpathologie (Berlin 1858).

290 VORBRODT, A. and MAUL, G. G.: Cytochemical studies on the relation of nucleoside triphosphatase activity to ribonucleoproteins in isolated rat liver nuclei. J. Histochem. Cytochem. *28:* 27–35 (1980).

291 WADDINGTON, C. H.: The nucleolus – retrospect and prospect. Nat. Cancer Inst. Monogr. *23:* 563–572 (1966).

292 WALLER, C. E.: Cellular changes induced by podophyllin on the vaginal epithelium of the Wistar rat (Thesis, Johns Hopkins School of Cytotechnology, Baltimore 1961).

293 WANG, J. C.: DNA topoisomerases. Ann. Rev. Biochem. *54:* 665–697 (1985).

294 WATSON, J. D.: Molecular biology of the gene, 3rd ed. (W. A. Benjamin, Menlo Park, Calif. 1976).

295 WEISS, P.: The dynamics of the membrane-bound incompressible body: A mechanism of cellular and subcellular motility. Proc. Nat. Acad. Sci. (Washington) *52:* 1024–1029 (1964).

296 WEISS, P.: Deterioration in cells, in W. L. MARXER and G. R. COWGILL, The art of predictive medicine, pp. 125–132 (Thomas, Springfield, Ill. 1967).

297 WENTZ, W. B. and REAGAN, J. W.: Survival in cervical cancer with respect to cell type. Cancer *12:* 384–388 (1959).

298 WHEELESS, L. L., PATTEN, S. F., BERKAN, T. K., BROOKS, C. L., GORMAN, K. M., LESH, S. R., LOPEZ, P. A., and WOOD, J. C. S.: Multidimensional slit-scan prescreening system: preliminary results of a single blind clinical study. Cytometry *5:* 1–8 (1984).

299 WIED, G. L.: Suggested standard for karyopyknosis. Fertil. Steril. *6:* 61–65 (1955).

300 WIED, G. L. (ed.): Definition, morphology, cytochemistry and diagnostic importance of dyskaryotic cells (Symposium). Acta Cytol. *1:* 23–61 (1957).

301 WIED, G. L.: Opinion poll on cytological definitions. Acta Cytol. *2:* 26–62 (1958).

302 WIED, G. L.: Opinion poll on cytological terminology. Acta Cytol. *2:* 63–139 (1958).

303 WIED, G. L. (ed.): An international agreement on histological terminology for lesions of the uterine cervix. Acta Cytol. *6:* 235–236 (1962).

304 WIED, G. L.: Introduction to quantitative cytochemistry II (Academic Press, New York 1970).

305 WIED, G. L., BARTELS, P. H., DYTCH, H. E. and BIBBO, M.: Rapid DNA evaluation in clinical diagnosis. Acta Cytol. *27:* 33–37 (1983).

306 WOOD, S., BAKER, R. R. and MAZZOCCHI, B.: Locomotion of cancer cells in vivo compared with normal cells (motion picture) (Johns Hopkins University, Baltimore 1966).

307 YANNARELL, A., SCHUMM, D. and WEBB, T. E.: Nature of the facilitated messenger ribonucleic acid transport from isolated nuclei. Biochem. J. *154:* 379–385 (1976).

308 ZAJICEK, J.: Aspiration biopsy cytology. Part I. Cytology of supradiaphragmatic organs, in G. L. WIED (ed.), Monographs in clinical cytology (S. Karger, Basel 1974).

309 ZOLLINGER, H. U. von: Beitrag zur Pathogenese der Einschlusskörper (lead). Schweiz. Z. allg. Path. Bakt. *14:* 446–455 (1951).

Subject Index